Homogenization Methods for
Multiscale Mechanics

Homogenization Methods for Multiscale Mechanics

Chiang C Mei
Massachusetts Institute of Technology, USA

Bogdan Vernescu
Worcester Polytechnic Institute, USA

NEW JERSEY • LONDON • SINGAPORE • BEIJING • SHANGHAI • HONG KONG • TAIPEI • CHENNAI

Published by

World Scientific Publishing Co. Pte. Ltd.

5 Toh Tuck Link, Singapore 596224

USA office: 27 Warren Street, Suite 401-402, Hackensack, NJ 07601

UK office: 57 Shelton Street, Covent Garden, London WC2H 9HE

British Library Cataloguing-in-Publication Data
A catalogue record for this book is available from the British Library.

HOMOGENIZATION METHODS FOR MULTISCALE MECHANICS

ISBN-13 978-981-4282-44-4
ISBN-10 981-4282-44-8

Typeset by Stallion Press
Email: enquiries@stallionpress.com

Printed in Singapore by World Scientific Printers.

Dedicated to

Caroline and Deborah Mei, and my parents (CCM)

Cornelia and Mihail Vernescu, my parents (BV)

Acknowledgments

I wish to thank US National Science Foundation, Office of Naval Research, Air Force Office of Scientific Research and MIT for financial support over many years on various research topics. My foray into several aspects of homogenization theory beyond wave dynamics began by collaborations with Professor Jean-Louis Auriault, Université Joseph Fourier, Grenoble, France. Materials in parts of this book have been extensively drawn from work done by Professor Auriault and his associates. I am also indebted to all former students, now colleagues, for the wonderful journey of learning with them and from them. Finally it is a pleasure to acknowledge the fortunes received in abundance from MIT for challenges and opportunities, and from Caroline and Deborah for sharing the same voyage of life.

CCM

I would like to thank the National Science Foundation, the Alfred P. Sloan Foundation and WPI for their support over the years. I am grateful to my mentor Horia I. Ene, Institute of Mathematics, Romanian Academy, who first introduced me to the topic of homogenization and multiscale materials, and to my collaborators and students with whom, over the years, I have shared ideas, and long hours of work. First and foremost I would like to thank my parents and friends for their support and encouragement.

BV

Acknowledgments

[illegible]

[illegible]

Preface

In applied mechanics, and in other physical and engineering sciences, the first task for examining a problem theoretically is to formulate the governing equations accurate enough to describe the dominant physical processes, and the second is to find effective means of solution. For both tasks, understanding the scales of physical relevance and utilizing them for expediting the mathematics are essential. In many physical problems several scales can be present either in space or in time, caused by either inhomogeneity of the medium, or complexity of the mechanical process. Depending on the objectives, one may focus on the process in a particular range of scales. For example, in the study of gas dynamics, if we are primarily concerned with the physics over a length scale much greater than the distances separating the particles or the molecules, detailed interactions of the particles are often overlooked and the medium is regarded as a continuum. Conservation laws of mass, energy, and momentum are derived by extending the basic laws of particle mechanics. Constitutive relations among macroscale variables are added on the basis of some basic experiments. Examples of these relations are those between heat flux rate and the temperature gradient, between stresses and strains in a solid, and between stresses and strain rates in a fluid, etc. However, a more fundamental approach is to first construct microscale models, and then to deduce the macroscale laws and the constitutive relations by properly averaging over the microscale. Thus in the kinetic theory of gases, models of colliding molecules are constructed to yield the hydrodynamical equations for the continuum as well as theoretical formulas of viscosity and thermal diffusivity, etc.

For many multiphase media such as fluid-saturated porous solids (soil, bones, or tissues), or of laminated or fiber-reinforced elastic solids, the primary interest is often restricted to the large-scale behavior of the composite. Theoretical derivation of the macroscale equations based on microscale considerations and prediction of the constitutive coefficients have been a challenge in both computational mathematics and theoretical mechanics. Under certain conditions, analytical theories have been advanced for physical problems governed by differential equations. These latter theories have the advantage of enhancing physical understanding and reducing the computational labor. In particular, for materials or processes with a periodic microstructure, the perturbation method of multiple scales can be used to derive averaged equations for a much larger scale from considerations of the small scales. In the mechanics of multiscaled media, the analytical scheme of upscaling is known as the *Theory of Homogenization.* Closely parallel methods have also been developed in the vast literature of wave propagation, where averages over the nearly periodic short waves are taken and the main attention is shifted to the long-scale evolution of wave envelopes. In wave theories, the technique is variably known as the envelope theory, or coupled-mode theory, etc.

Started by mathematicians in the late 1970s, the literature of homogenization theory for inhomogeneous media has mushroomed in recent decades. Attempts to derive or rederive the macroscale equations have been applied to numerous fields of applications. Several monographs have also appeared. While the key tool of the theory is the perturbation method of multiple scales now widely used in a variety of problems in applied analysis, existing treatises are often presented in abstract mathematical language which is not easily accessible to many students and researchers.

The present authors share the view that the general methods of homogenization deserve to be more widely understood and practiced by applied scientists and engineers. Hence this book is aimed at providing a less abstract treatment of the basic theory and its applications to a diverse range of problems involving inhomogeneity originated from either material structures, motion, or boundary geometry. Each chapter deals with a different class of physical problems. We hope the content can be useful not only to newcomers wishing to learn the essence of the method, but also to specialists of one field who may wish to extend their expertise to other fields. To tackle a new problem, we adopt the approach of first discussing the physically relevant scales, then identifying the small parameters and their roles in the normalized governing equations. The details of asymptotic analysis are explained only afterward. Whenever possible we shall include known quantitative results of the constitutive coefficients, which can be obtained analytically only in a few cases, and must in general be obtained by solving numerically the so-called cell problems. Since these numerical tasks are often not trivial, we discuss in Chapter 2 the mathematical alternative of variational bounds for the relatively simple problem of heat conduction, and illustrate how to estimate the range of possible values of the constitutive coefficients. Applications are then extended to seepage in porous media in Chapter 3, where the empirical law of Darcy is derived theoretically. Extensions to three-scale seepage flow are discussed and the nonlinear effects of convection are sketched. Relevant to environmental applications is the topic of shear-enhanced diffusion, i.e., dispersion. Two examples are analyzed in Chapter 4: one for a spatially periodic porous medium and the other for wave motion periodic in both space and time. Multiscale problems have a long history in elastic composites which are of great importance to construction and manufacturing industries. Comprehensive treatises describing various approximations and computational schemes are already available. In Chapter 5 we discuss the use of homogenization theory to two examples in elastic composites. A fairly detailed account is again given to the prediction of variational bounds of the constitutive coefficients defined by homogenization theory. The mechanics of poro-elasticity is of basic interest in soil mechanics, geophysics, and biomechanics. The coupling between pore fluid and the solid matrix is often treated as two interacting continua. We discuss in Chapter 6 the derivation of Terzaghi–Biot theory from micromechanical considerations. In the final chapter we illustrate how the method of homogenization can be employed in wave dynamics in an inhomogeneous environment. In particular, we shall also show that the asymptotic theory can be straightforwardly extended to random media where the random fluctuations are weaker than the stochastic average. A few examples involving linear or weakly nonlinear properties will be treated to predict the mean field. On the whole, our primary emphasis is on analytical and approximate derivations of macroscale equations. Less effort is devoted to their solutions and the implied physical significance. Details of numerical computations are left to the cited references.

In most of this book the mathematical prerequisite is kept at the level comfortable for graduate students in theoretical engineering sciences. Some prior exposure to perturbation methods is helpful but not necessary, since the details are explained from ground zero. The parts on variational bounds in Chapters 2 and 5 are likely to be more demanding to many engineering students. To introduce to them one of the theoretical topics which has been enriched by both engineers and mathematicians alike, derivations have been described in considerable detail in order to make the material more accessible. We hope that the scope of coverage here is sufficiently broad to appeal to theoretical researchers who are interested in either the analytical tools or the variety of applications.

Contents

Introductory Examples of Homogenization Method

In this introductory chapter we shall use several one-dimensional examples to demonstrate the homogenization theory for deriving effective equations. To illustrate the importance of scales we show in Secs. 1.1 and 1.2 that, for waves in a space-wise periodic medium, different relations among length scales lead to different effective equations, representing different physics. In the last section, an example of multiple time scales is demonstrated through the celebrated theory of Taylor (1953, 1954) on shear-enhanced diffusion in a pipe flow.

1.1. Long Waves in a Layered Elastic Medium

Let us start with an example where the material inhomogeneity and the physical process are characterized by two very different length scales. Consider the longitudinal vibrations or wave propagation in an elastic rod, governed by the following partial differential equation,

$$\frac{\partial}{\partial x}\left(E\frac{\partial v}{\partial x}\right) = \rho\frac{\partial^2 v}{\partial t^2}, \tag{1.1.1}$$

where v, E, and ρ denote, respectively, the longitudinal displacement, Young's modulus of elasticity, and the mass per unit length. We assume that the material properties E and ρ are spatially periodic with the period ℓ. As a short-hand expression, we shall say that E and ρ are ℓ-periodic in x.

If ω is the characteristic frequency, then the characteristic length of the elastic waves can be estimated by

$$\frac{2\pi}{k} = \sqrt{\frac{E_o}{\rho_o}}\frac{1}{\omega}, \tag{1.1.2}$$

where k is the wavenumber and E_o and ρ_o are the characteristic values of $E(x)$ and $\rho(x)$, respectively. There are at least two possible scenarios: (i) $k\ell \ll 1$, and (ii) $k\ell = \mathcal{O}(1)$.

We consider here the first case of long waves and designate the wavelength as the macroscale $\ell' = 2\pi/k$ so that

$$\epsilon \equiv \frac{\ell}{\ell'} \ll 1\,. \tag{1.1.3}$$

If one is primarily interested in the averaged variation over a scale of a wavelength, can the details on the microscale be bypassed and an equation obtained for the global behavior on the scale of ℓ'?

Let us introduce the following normalization,

$$v = v_o v^\dagger\,, \quad x = \ell x^\dagger\,, \quad t = \frac{t^\dagger}{\omega}\,, \quad E = E_o E^\dagger\,, \quad \rho = \rho_o \rho^\dagger\,, \tag{1.1.4}$$

where dimensionless quantities are distinguished by $\{\cdot\}^\dagger$ and u_0, E_0, and ρ_0 are the characteristic scales of u, E, and ρ, respectively. Note that x is normalized by the microlength ℓ. In dimensionless variables, Eq. (1.1.1) becomes

$$\frac{\partial}{\partial x^\dagger}\left(E^\dagger \frac{\partial v^\dagger}{\partial x^\dagger}\right) = \left(\frac{\ell^2}{E_o/\rho_o\omega^2}\right)\rho^\dagger \frac{\partial^2 v^\dagger}{\partial t^{\dagger 2}} = \epsilon^2 \rho^\dagger \frac{\partial^2 v^\dagger}{\partial t^{\dagger 2}}\,. \tag{1.1.5}$$

Clearly inertia is much smaller than the stress force on the microscale. For exhibiting the physical origin of each term, it is often convenient to keep the equations in dimensional form, but to indicate the relative weight of each term by the ordering parameter ϵ, with the normalizing scales defined. From here on we shall omit the daggers and simply write Eq. (1.1.5) as

$$\frac{\partial}{\partial x}\left(E \frac{\partial v}{\partial x}\right) = \epsilon^2 \rho \frac{\partial^2 v}{\partial t^2}\,. \tag{1.1.6}$$

Using the fact that there exist two characteristic lengths, let us employ the perturbation method of multiple scales by introducing the fast and slow variables x and $x' = \epsilon x$. The unknown displacement is now expanded in the form of a power series in ϵ:

$$v = v_0 + \epsilon v_1 + \epsilon^2 v_2 + \cdots\,, \tag{1.1.7}$$

where $v_i, i = 0, 1, 2, \ldots$ are functions of both x and x'. The original derivative then becomes, according to the chain rule,

$$\frac{\partial}{\partial x} \to \frac{\partial}{\partial x} + \epsilon \frac{\partial}{\partial x'}\,.$$

It follows from Eq. (1.1.6) that

$$\left(\frac{\partial}{\partial x}+\epsilon\frac{\partial}{\partial x'}\right)\left[E\left(\frac{\partial}{\partial x}+\epsilon\frac{\partial}{\partial x'}\right)(v_0+\epsilon v_1+\epsilon^2 v_2+\cdots)\right]$$
$$=\epsilon^2\rho\frac{\partial^2}{\partial t^2}(v_0+\epsilon v_1+\epsilon^2 v_2+\cdots).$$

Equating the coefficient of each power of ϵ to zero leads to a sequence of perturbation equations. At the order $\mathcal{O}(\epsilon^0)$, we get,

$$\frac{\partial}{\partial x}\left(E\frac{\partial v_0}{\partial x}\right)=0\,, \tag{1.1.8}$$

which governs the microscale variation of v_0. Because the microstructure is assumed to be periodic on the ℓ-scale, v_0 should be likewise periodic. The general solution to the homogeneous equation (1.1.8) is

$$v_0=A_1(x')\int_{x_0}^{x}\frac{dx}{E(x,x')}+A_2(x')\,,$$

where x_0 is an arbitrary starting point and $A_1(x')$ and $A_2(x')$ are integration constants. To ensure periodicity over the distance ℓ (i.e., $v_0(x_0)=v_0(x_0+\ell)$), $A_1(x')$ must vanish, implying that the leading-order displacement depends only on the macroscale, i.e.,

$$v_0=v_0(x')=A_2(x')\,. \tag{1.1.9}$$

Therefore, v_0 must represent the ℓ-average of v with an error of $\mathcal{O}(\epsilon)$. At the leading order, variations within a few spatial periods ($\mathcal{O}(\ell)$) are insignificant.

At the next order $\mathcal{O}(\epsilon)$, the perturbation equation governing v_1 is

$$\frac{\partial}{\partial x}\left[E\left(\frac{\partial v_1}{\partial x}+\frac{\partial v_0}{\partial x'}\right)\right]=0\,. \tag{1.1.10}$$

As a boundary condition, v_1 must also be ℓ-periodic. Equation (1.1.10) is a linear inhomogeneous equation for v_1, which may be represented formally by

$$v_1=S(x,x')\frac{\partial v_0}{\partial x'}\,, \tag{1.1.11}$$

so that

$$\frac{\partial}{\partial x}\left[E\left(1+\frac{\partial S}{\partial x}\right)\right]=0\,, \tag{1.1.12}$$

where S is ℓ-periodic in x. The solution of the above problem is determined up to an additive constant. To ensure that v_0 represents the wavelength-average with an error of $\mathcal{O}(\epsilon^2)$, we add the condition that

$$\langle S \rangle = \frac{1}{\ell} \int_{x_0}^{x_0+\ell} S \, dx = 0 \,, \tag{1.1.13}$$

where angle brackets signify the wavelength-average. Successive integrations of Eq. (1.1.12) yield

$$\frac{\partial S}{\partial x} = \frac{\mathcal{E}}{E} - 1 \,, \tag{1.1.14}$$

and

$$S = S_0 - \int_{x_0}^{x} \frac{E - \mathcal{E}}{E} dx \,, \tag{1.1.15}$$

where $\mathcal{E}(x')$ and $S_0(x')$ are constants of integration. Since v_1 must be ℓ-periodic, it is necessary that $S(x_0) = S(x_0 + \ell)$, i.e.,

$$S_0 = S_0 - \ell + \mathcal{E} \int_{x_0}^{x_0+\ell} \frac{dx}{E} \,,$$

therefore

$$\mathcal{E} = \left(\frac{1}{\ell} \int_{x_0}^{x_0+\ell} \frac{dx}{E} \right)^{-1} , \quad \text{or} \quad \frac{1}{\mathcal{E}} = \frac{1}{\ell} \int_{x_0}^{x_0+\ell} \frac{dx}{E} \,. \tag{1.1.16}$$

Thus $\mathcal{E}$ is the harmonic mean of E over the period ℓ. $S(x_0)$ can be fixed by applying the auxiliary condition (Eq. (1.1.13)),

$$S(x_0) = \left\langle \int_{x_0}^{x} \frac{E - \mathcal{E}}{E} dx \right\rangle = \frac{1}{\ell} \int_{x_0}^{x_0+\ell} dx \int_{x_0}^{x} \frac{E - \mathcal{E}}{E} d\bar{x} \,. \tag{1.1.17}$$

Next we proceed to the perturbation equation at the order $\mathcal{O}(\epsilon^2)$:

$$\frac{\partial}{\partial x}\left(E \frac{\partial v_2}{\partial x} \right) + \frac{\partial}{\partial x'}\left(E \frac{\partial v_0}{\partial x'} \right) + \frac{\partial}{\partial x}\left(E \frac{\partial v_1}{\partial x'} \right) + \frac{\partial}{\partial x'}\left(E \frac{\partial v_1}{\partial x} \right) = \rho \frac{\partial^2 v_0}{\partial t^2} \,, \tag{1.1.18}$$

which is again an inhomogeneous equation for v_2, similar to Eq. (1.1.10) for v_1. Furthermore, v_2 is ℓ-periodic. Since from Eqs. (1.1.11) and (1.1.14),

$$\frac{\partial v_1}{\partial x} = -\frac{\partial v_0}{\partial x'} + \frac{\mathcal{E}}{E} \frac{\partial v_0}{\partial x'} \,, \tag{1.1.19}$$

it follows that

$$\frac{\partial}{\partial x'}\left(E\frac{\partial v_1}{\partial x}\right)+\frac{\partial}{\partial x'}\left(E\frac{\partial v_0}{\partial x'}\right)=\frac{\partial}{\partial x'}\left(\mathcal{E}\frac{\partial v_0}{\partial x'}\right).$$

With this result Eq. (1.1.18) may be rewritten as

$$\frac{\partial}{\partial x}\left(E\frac{\partial v_2}{\partial x}\right)+\frac{\partial}{\partial x'}\left(\mathcal{E}\frac{\partial v_0}{\partial x'}\right)+\frac{\partial}{\partial x}\left(E\frac{\partial v_1}{\partial x'}\right)=\rho\frac{\partial^2 v_0}{\partial t^2}.$$

By taking the wavelength-average of the preceding equation and invoking periodicity, we get

$$\frac{\partial}{\partial x'}\left(\mathcal{E}\frac{\partial v_0}{\partial x'}\right)=\langle\rho\rangle\frac{\partial^2 v_0}{\partial t^2}, \tag{1.1.20}$$

where

$$\langle\rho\rangle=\frac{1}{\ell}\int_{x_0}^{x_0+\ell}\rho\,dx. \tag{1.1.21}$$

Equation (1.1.20) governs the macroscale variation of the mean displacement. While the effective Young's modulus is the harmonic mean of E, the effective density is the arithmetic mean of ρ. Subject to further boundary conditions on the macroscale, v_0 can be solved.

Once v_0 is found, the fluctuation about the mean, v_1, can be found from Eq. (1.1.11).

The procedure described above is typical of the *method of homogenization*. A boundary-value problem is solved for each typical period (cell) at each order. At the leading order the cell problem is homogeneous and the solution is indeterminate. At higher orders the problems are inhomogeneous; their solvability imposes constraints on the lower order solution. It is one of the constraints (*solvability conditions*) that leads to the averaged equation governing the global behavior of the lowest order. As an important byproduct, the effective constitutive coefficient is also derived in terms of the known material property on the microscale.

So far the elastic coefficient E has been treated as a continuous function of x. For laminated materials E is discontinuous. Thus within each ℓ-period we must add the jump conditions representing, respectively, the continuity of displacement and stress:

$$[v]_\xi=0\,,\quad \left[E\frac{\partial v}{\partial x}\right]_\xi=0\,, \tag{1.1.22}$$

where $[F]_\xi=F|_{\xi+}-F|_{\xi-}$ denotes the discontinuity of F at $x=\xi$ within each ℓ-period. In the perturbation analysis, one must add the following jump conditions at successive orders:

$$
\begin{aligned}
&[v_0]_\xi = 0\,, \quad \left[E\frac{\partial v_0}{\partial x}\right]_\xi = 0\,, \\
&[v_1]_\xi = 0\,, \quad \left[E\left(\frac{\partial v_1}{\partial x} + \frac{\partial v_0}{\partial x'}\right)\right]_\xi = 0\,, \\
&[v_2]_\xi = 0\,, \quad \left[E\left(\frac{\partial v_2}{\partial x} + \frac{\partial v_1}{\partial x'}\right)\right]_\xi = 0\,, \\
&\quad\vdots
\end{aligned}
\tag{1.1.23}
$$

It can be shown that the results (Eq. (1.1.20)) with Eqs. (1.1.16) and (1.1.21) are the same as if E were continuous. This conclusion can be reached alternatively by employing the theory of generalized functions (see Bakhvalov and Panasenko, 1989).

As a special case, let each ℓ period be composed of two distinct layers, within which Young's moduli and thicknesses are, respectively, (E_1, E_2) and $(1-\theta\ell, \theta\ell)$. From Eq. (1.1.16)

$$
\mathcal{E} = \left\langle \frac{1}{E} \right\rangle^{-1} = \left(\frac{1-\theta}{E_1} + \frac{\theta}{E_2} \right)^{-1} = \frac{E_1 E_2}{(1-\theta)E_2 + \theta E_1}\,, \tag{1.1.24}
$$

i.e., $\mathcal{E}$ is the harmonic mean of E_1 and E_2.

In particular, if θ or $1-\theta$ is not too small compared to unity, then

$$
\mathcal{E} \sim \frac{E_2}{\theta} \quad \text{if} \quad E_1 \gg E_2; \qquad \mathcal{E} \sim \frac{E_1}{(1-\theta)} \quad \text{if} \quad E_1 \ll E_2\,.
$$

1.2. Short Waves in a Weakly Stratified Elastic Medium

We now consider the case where $k\ell = \mathcal{O}(1)$ for a scenario that gives rise to contrasting scales and calls for homogenization. In many dynamical problems, contrasting space and time scales arise as a result of resonance, even when the material inhomogeneity is weak. As an example let us consider the slow evolution of waves through a slightly periodic medium, by taking Eq. (1.1.1) with

$$
\rho = \text{constant}\,, \quad E = E_o(1 + \epsilon D \cos Kx)\,, \tag{1.2.1}
$$

where D is of order unity, i.e.,

$$
E_0 \frac{\partial}{\partial x}\left[(1 + \epsilon D \cos Kx)\frac{\partial v}{\partial x}\right] = \rho \frac{\partial^2 v}{\partial t^2}\,. \tag{1.2.2}
$$

We now assume that the spatial period of inhomogeneity $\ell \equiv 2\pi/K$ and the elastic wavelength $2\pi/k = \sqrt{E_o/\rho}/\omega$ are comparable. As a consequence, wave reflection can be significant.

Let us first try a naive expansion, $v = v_0 + \epsilon v_1 + \cdots$. The crudest solution is easily found to be

$$v_0 = \frac{A}{2} e^{ikx - i\omega t} + \text{c.c.}, \tag{1.2.3}$$

where c.c. signifies the complex conjugate of the preceding term, and

$$\frac{2\pi}{k} \equiv \sqrt{\frac{E_o}{\rho}\frac{1}{\omega}}. \tag{1.2.4}$$

At the next order the governing equation is

$$\begin{aligned} \frac{\partial}{\partial x}\left(E_o \frac{\partial v_1}{\partial x}\right) - \rho \frac{\partial^2 v_1}{\partial t^2} &\\ = -\frac{E_o D}{2}\frac{\partial}{\partial x}\left[\left(e^{iKx} + e^{-iKx}\right)\frac{\partial v_0}{\partial x}\right] &\\ = -\frac{E_o D}{2}\frac{\partial}{\partial x}\left[\left(e^{iKx} + e^{-iKx}\right)\left(\frac{ikA}{2} e^{ikx - i\omega t} - \frac{ikA^\dagger}{2} e^{-ikx + i\omega t}\right)\right], & \end{aligned} \tag{1.2.5}$$

where $A^\dagger$ denotes the complex conjugate of A. For general K, v_1 can be found in terms of the harmonics $\exp(\pm i(K \pm k) \pm \omega t)$. Higher order improvements can be proceeded straightforwardly. However, when

$$K = 2k + \delta, \quad \delta \ll k, \tag{1.2.6}$$

some of the forcing terms on the right-hand side will be close to a natural mode $\exp(\pm i(kx + \omega t))$. Resonance of the reflected waves must be expected. The relation $K = 2k$ (cf. Eq. (1.2.6)) is the well-known condition for Bragg resonance. Let us consider the response to forcing of reflected waves

$$E_o \frac{\partial^2 v_1}{\partial x^2} - \rho \frac{\partial^2 v_1}{\partial t^2} = R e^{i\phi_o} e^{i\delta x} + \text{c.c.}$$

with

$$R = -\frac{E_0 D k^2 A^\dagger}{4} \quad \text{and} \quad \phi_o = kx + \omega t,$$

where c.c. denotes the complex conjugate of the preceding term. Combining homogeneous and inhomogeneous solutions and requiring that $v_1(0, t) = 0$, we get

$$v_1 = \frac{R e^{i\phi_o}\left(1 - e^{i\delta x}\right)}{E_o((k+\delta)^2 - k^2)} + \text{c.c.}.$$

Clearly if $\delta = \mathcal{O}(\epsilon)$, $\epsilon v_1 \sim \mathcal{O}(\epsilon/\delta)$ and is not small compared to v_0 when $\epsilon x = \mathcal{O}(1)$. Furthermore as x increases, v_1 grows as ϵx. This implies that the reflected waves are resonated and are no longer much smaller than the incident waves in the distance $\epsilon x = \mathcal{O}(1)$. The naive expansion is no longer useful.

To get a solution uniformly valid for all x, we focus on the neighborhood of resonance. Since the spatial scale of resonance is characterized by $\epsilon x = \mathcal{O}(1)$ we introduce fast and slow variables in space

$$x \text{ and } x' = \epsilon x\,. \tag{1.2.7}$$

Let the detuning from exact resonance be small, i.e., the incident wave frequency is $\omega + \epsilon\omega'$, where $\epsilon\omega'$ is the frequency detuning. This amounts to a very slow variation in time. Therefore two time variables are needed,

$$t \text{ and } t' = \epsilon t\,. \tag{1.2.8}$$

The following multiple-scale expansion is then proposed,

$$v = v_0(x, x'; t, t') + \epsilon v_1(x, x'; t, t') + \cdots\,. \tag{1.2.9}$$

After making the changes

$$\frac{\partial}{\partial x} \to \frac{\partial}{\partial x} + \epsilon\frac{\partial}{\partial x'}\,, \quad \frac{\partial}{\partial t} \to \frac{\partial}{\partial t} + \epsilon\frac{\partial}{\partial t'}\,, \tag{1.2.10}$$

and substituting Eqs. (1.2.9) and (1.2.10) into Eq. (1.2.2), we get

$$\frac{\partial}{\partial x}\left(E_o\frac{\partial v_0}{\partial x}\right) - \rho\frac{\partial^2 v_0}{\partial t^2} = 0 \tag{1.2.11}$$

at $\mathcal{O}(\epsilon^0) = \mathcal{O}(1)$. Anticipating strong but finite reflection, we take the solution to be

$$v_0 = \frac{A}{2}e^{ikx-i\omega t} + \text{c.c.} + \frac{B}{2}e^{-ikx-i\omega t} + \text{c.c.}\,, \tag{1.2.12}$$

where $A(x', t')$ and $B(x', t')$ vary slowly in space and time. At the order $O(\epsilon)$ we have

$$\frac{\partial}{\partial x}\left(E_o\frac{\partial v_1}{\partial x}\right) - \rho\frac{\partial^2 v_1}{\partial t^2}$$
$$= -2E_o\frac{\partial^2 v_0}{\partial x\partial x'} + 2\rho\frac{\partial^2 v_0}{\partial t\partial t'} - \frac{E_o D}{2}\frac{\partial}{\partial x}\left[(e^{2ikx} + e^{-2ikx})\frac{\partial v_0}{\partial x}\right]$$

$$
\begin{aligned}
&= -E_o \left[\frac{\partial A}{\partial x'}(ik)e^{ikx-i\omega t} + \text{c.c.} + \frac{\partial B}{\partial x'}(-ik)e^{-ikx-i\omega t} + \text{c.c.}\right] \\
&\quad + \rho \left[\frac{\partial A}{\partial t'}(-i\omega)e^{ikx-i\omega t} + \text{c.c.} + \frac{\partial B}{\partial t'}(-i\omega)e^{-ikx-i\omega t} + \text{c.c.}\right] \\
&\quad - \frac{E_o D}{4}\frac{\partial}{\partial x}\left\{(e^{2ikx} + \text{c.c.})\frac{\partial}{\partial x}[Ae^{ikx-i\omega t} + \text{c.c.} + Be^{-ikx-i\omega t} + \text{c.c.}]\right\}.
\end{aligned}
\tag{1.2.13}
$$

The last line can be reduced to

$$
\begin{aligned}
-\frac{E_o D}{4}(&k^2 Be^{ikx-i\omega t} + \text{c.c.} + k^2 Ae^{-ikx-i\omega t} + \text{c.c.} \\
&- 3k^2 Ae^{3ikx-i\omega t} + \text{c.c.} - 3k^2 Be^{-3ikx-i\omega t} + \text{c.c.})\,.
\end{aligned}
$$

To avoid unbounded resonance of v_1, i.e., to ensure the solvability of v_1, we equate to zero the coefficients of terms $e^{\pm i(kx-\omega t)}$ and $e^{\pm i(kx+\omega t)}$ on the right-hand side of Eq. (1.2.13). The following equations are then obtained:

$$\frac{\partial A}{\partial t'} + C\frac{\partial A}{\partial x'} = \frac{ikCD}{4}B\,, \tag{1.2.14}$$

$$\frac{\partial B}{\partial t'} - C\frac{\partial B}{\partial x'} = \frac{ikCD}{4}A\,, \tag{1.2.15}$$

where $\sqrt{E_o/\rho} = C = \omega/k$ denotes the phase speed. These equations govern the macroscale variation of the envelopes of the incident and reflected waves, and can be combined to give the Klein–Gordon equation

$$\frac{\partial^2 A}{\partial t'^2} - C^2\frac{\partial^2 A}{\partial x'^2} + \Omega_0^2 A = 0\,, \quad \text{where } \Omega_0 = \frac{kCD}{4}\,. \tag{1.2.16}$$

If the domain is infinite, the following is a simple solution

$$A = A_0 e^{i(Kx'-\Omega t')}\,. \tag{1.2.17}$$

The corresponding right-going wave

$$v_+ = \frac{A_0}{2}\exp\left[i((k+\epsilon K)x - (\omega+\epsilon\Omega)t)\right] \tag{1.2.18}$$

is slightly detuned from (k, ω). Substituting Eq. (1.2.17) into Eq. (1.2.16) we get

$$K = \pm\frac{1}{C}\sqrt{\frac{\Omega^2}{\Omega_0^2} - 1}\,. \tag{1.2.19}$$

In the ranges $\Omega > \Omega_0$ and $\Omega < -\Omega_0$, K is real; the envelope is a propagating wave. The relations between K and Ω are two branches of hyperbola, and hence nonlinear. The envelope is a dispersive wave whose wave speed varies with the wavelength. In the range $-\Omega_0 < \Omega < \Omega_0$, K is pure imaginary. Propagation is forbidden; the envelope must decay exponentially in space. Hence the range $[-\Omega_0, \Omega_0]$ is called the *bandgap*.

With suitable boundary conditions on the macroscale, interesting physics can be deduced for the slow variation of wave envelopes, and hence the global behavior of wave motion.

The use of multiple scales to study the slow modulation of nearly sinusoidal waves is very common in the wave dynamics literature, but is usually known as the WKB or geometrical optics approximation instead of the homogenization theory (see e.g., Nayfeh, 1981; Mei, 1985, 1989; Mei *et al.*, 2005).

1.3. Dispersion of Passive Solute in Pipe Flow

How does one predict the transport of dissolved chemicals in a blood vessel, the spreading of pollutants in a pipe or a river, in shallow estuaries, or in the ever-moving atmosphere? For dilute concentration one can in principle find the fluid motion first and then the convective diffusion of the solute concentration next. In general the computational task can be quite demanding. In some cases, such as in a pipe or channel flow, the velocity of the two- or three-dimensional flow is essentially in one direction, with transverse variations. Since mixing tends to homogenize the solute concentration in the transverse direction, why do not we focus attention on the macroscale spreading in the longitudinal direction by averaging over the microscale in the transverse direction?

It was discovered half a century ago for pipe flows by Taylor (1953, 1954) that the cross-sectionally averaged concentration of a dye cloud is not simply convected by the mean velocity and diffused by molecular or eddy viscosity. Instead, the velocity shear across the pipe tends to augment the effective diffusion in the direction of flow; the enhanced diffusion is now known as *Taylor dispersion*. This theory and its extensions are now the scientific foundation of nearly all studies on the transport of contaminants in the environment, and on the exchange of oxygen or carbon dioxide between blood-carrying arteries or capillaries and surrounding tissues. For the one-dimensional pipe flow Taylor's original analysis was quite heuristic. A more formal procedure called the method of moments was introduced later by

Aris (1960) who also studied the case of pulsating flows in a pipe. Here we demonstrate the use of the homogenization theory by multiple scales.

1.3.1. *Scale Estimates*

Consider the laminar flow in a long pipe with radius a. Let x axis be the axis of the pipe, and r be the radial distance from the axis. Due to a spatially constant pressure gradient the profile of fluid velocity is given by

$$u = U_s(r) + \mathcal{R}e[U_w(r)e^{-i\omega t}], \tag{1.3.1}$$

where $\mathcal{R}e(F)$ denotes the real part of the complex quantity F. The concentration of the solute is governed by the convection–diffusion equation

$$\frac{\partial C}{\partial t} + \frac{\partial (uC)}{\partial x} = D\left[\frac{\partial^2 C}{\partial x^2} + \frac{1}{r}\frac{\partial}{\partial r}\left(r\frac{\partial C}{\partial r}\right)\right], \quad 0 < r < a, \tag{1.3.2}$$

where D is the molecular diffusivity. We shall be interested in the transport over a distance L (macroscale) much greater than the radius a (microscale). Let us first assume the pipe radius to be so small that lateral diffusion is completed within a few periods, i.e.,

$$\frac{2\pi}{\omega} \sim \frac{a^2}{D}, \tag{1.3.3}$$

and choose the following normalizations,

$$x = Lx^\dagger, \quad r = ar^\dagger, \quad u = U_o u^\dagger, \quad t = \frac{a^2}{D}t^\dagger, \tag{1.3.4}$$

where U_o is the scale of u, which can be the center line velocity in either the steady flow or the oscillatory flow. Equation (1.3.2) is then normalized to

$$\frac{\partial C}{\partial t^\dagger} + \frac{U_o a}{D}\frac{a}{L}\frac{\partial (u^\dagger C)}{\partial x^\dagger} = \frac{a^2}{L^2}\frac{\partial^2 C}{\partial {x^\dagger}^2} + \frac{1}{r^\dagger}\frac{\partial}{\partial r^\dagger}\left(r^\dagger\frac{\partial C}{\partial r^\dagger}\right). \tag{1.3.5}$$

Let

$$Pe = \frac{U_o a}{D}, \quad \epsilon = \frac{a}{L} \tag{1.3.6}$$

be defined as the Péclet number and the aspect ratio, respectively. We shall next assume for generality that $Pe = \mathcal{O}(1)$ but $\epsilon \ll 1$. Equation (1.3.5) becomes,

$$\frac{\partial C}{\partial t^\dagger} + \epsilon Pe\frac{\partial (u^\dagger C)}{\partial x^\dagger} = \epsilon^2\frac{\partial^2 C}{\partial {x^\dagger}^2} + \frac{1}{r^\dagger}\frac{\partial}{\partial r^\dagger}\left(r^\dagger\frac{\partial C}{\partial r^\dagger}\right). \tag{1.3.7}$$

Again let us return to physical variables with dimensions and insert the order symbol ϵ to indicate the relative magnitude of each term,

$$\frac{\partial C}{\partial t} + \epsilon \frac{\partial (uC)}{\partial x} = \epsilon^2 D \frac{\partial^2 C}{\partial x^2} + \frac{D}{r} \frac{\partial}{\partial r} \left(r \frac{\partial C}{\partial r} \right) . \tag{1.3.8}$$

1.3.2. *Multiple-Scale Analysis*

Associated with two sharply different length scales a and L, there are three sharply distinct time scales whose ratios are:

$$\left(\frac{1}{\omega} \sim \frac{a^2}{D} \right) : \frac{L}{U_o} : \frac{L^2}{D} = \frac{a^2}{D} \left(1 : \frac{1}{\epsilon} : \frac{1}{\epsilon^2} \right) . \tag{1.3.9}$$

Let us introduce the multiple time coordinates

$$t \, , \, t' = \epsilon t \, , \, t'' = \epsilon^2 t \, , \tag{1.3.10}$$

and assume

$$C = C_0 + \epsilon C_1 + \epsilon^2 C_2 + \cdots \, , \tag{1.3.11}$$

where $C_i = C_i(x, r, t, t', t'')$. The original time derivative becomes, according to the chain rule:

$$\frac{\partial}{\partial t} \to \frac{\partial}{\partial t} + \epsilon \frac{\partial}{\partial t'} + \epsilon^2 \frac{\partial}{\partial t''} . \tag{1.3.12}$$

A sequence of perturbation problems is obtained. At the leading order of $\mathcal{O}(\epsilon^0)$, C_0 is governed by

$$\frac{\partial C_0}{\partial t} = \frac{D}{r} \frac{\partial}{\partial r} \left(r \frac{\partial C_0}{\partial r} \right) \tag{1.3.13}$$

with the boundary conditions:

$$\frac{\partial C_0}{\partial r} = 0 \, , \quad r = 0, a \, . \tag{1.3.14}$$

Here x is just a parameter. For any given initial $C_0(x, r, 0, 0, 0)$ nonuniform in r, the general solution at $\mathcal{O}(1)$ is

$$C_0 = C_{00}(x, t', t'') + \sum_{n=1}^{\infty} C_{0n}(x) e^{-(k_n)^2 t} J_0(k_n r) \, ,$$

where k_n is the nth root of $J_0'(ka) = 0$, with J_0 being the Bessel function of first kind and order zero. The series terms die out exponentially fast with t and are insignificant for $t' \geq \mathcal{O}(1)$. Limiting ourselves to the behavior long after the time for transverse diffusion to complete or periodicity to be achieved, i.e., after $t \sim \mathcal{O}(1/\omega) \sim \mathcal{O}(a^2/D)$, we shall omit the series part and take the solution to be

$$C_0 = C_0(x, t', t''), \tag{1.3.15}$$

which amounts to replacing Eq. (1.3.13) by

$$0 = \frac{D}{r}\frac{\partial}{\partial r}\left(r\frac{\partial C_0}{\partial r}\right). \tag{1.3.16}$$

Thus C_0 is the nontrivial solution to the homogeneous boundary value problem governed by Eqs. (1.3.16) and (1.3.14).

At $\mathcal{O}(\epsilon)$, C_1 is governed by:

$$\frac{\partial C_0}{\partial t'} + \frac{\partial C_1}{\partial t} + \frac{\partial (uC_0)}{\partial x} = \frac{D}{r}\frac{\partial}{\partial r}\left(r\frac{\partial C_1}{\partial r}\right) \tag{1.3.17}$$

with the boundary conditions

$$\frac{\partial C_1}{\partial r} = 0, \quad r = 0, a. \tag{1.3.18}$$

At $\mathcal{O}(\epsilon^2)$, C_2 must satisfy

$$\frac{\partial C_0}{\partial t''} + \frac{\partial C_1}{\partial t'} + \frac{\partial C_2}{\partial t} + \frac{\partial (uC_1)}{\partial x} = D\frac{\partial^2 C_0}{\partial x^2} + \frac{D}{r}\frac{\partial}{\partial r}\left(r\frac{\partial C_2}{\partial r}\right) \tag{1.3.19}$$

with

$$\frac{\partial C_2}{\partial r} = 0, \quad r = 0, a. \tag{1.3.20}$$

Let the known velocity be the sum of the steady and oscillatory parts,

$$u = U_s(r) + \mathcal{R}e(U_w(r)\, e^{-i\omega t}), \tag{1.3.21}$$

then at $\mathcal{O}(\epsilon)$,

$$\frac{\partial C_0}{\partial t'} + \frac{\partial C_1}{\partial t} + \{U_s + \mathcal{R}e[U_w(r)e^{-i\omega t}]\}\frac{\partial C_0}{\partial x} = \frac{D}{r}\frac{\partial}{\partial r}\left(r\frac{\partial C_1}{\partial r}\right). \tag{1.3.22}$$

In both Eqs. (1.3.13) and (1.3.17), we shall be interested in the time-harmonic response after the initial transient relative to the shortest time scale. Denoting the time(period)-average by overline, i.e.,

$$\bar{f} = \frac{\omega}{2\pi}\int_t^{t+2\pi/\omega} f\, dt,$$

and taking the time-average of Eqs. (1.3.22) and (1.3.18), we get

$$\frac{\partial C_0}{\partial t'} + U_s \frac{\partial C_0}{\partial x} = \frac{D}{r}\frac{\partial}{\partial r}\left(r \frac{\partial \overline{C}_1}{\partial r}\right) \tag{1.3.23}$$

with

$$\frac{\partial \overline{C}_1}{\partial r} = 0\,, \quad r = 0, a\,. \tag{1.3.24}$$

Thus $\overline{C}_1$ is governed by an inhomogeneous but steady boundary value problem. Denoting the area-average of f by $\langle f \rangle$, i.e.,

$$\langle f \rangle = \frac{1}{\pi a^2} \int_0^a 2\pi r f \, dr\,,$$

we obtain from Eq. (1.3.23)

$$\frac{\partial C_0}{\partial t'} + \langle U_s \rangle \frac{\partial C_0}{\partial x} = 0\,. \tag{1.3.25}$$

Mathematically, Eq. (1.3.25) is the solvability condition for the inhomogeneous boundary value problem for $\overline{C}_1$. Physically, over the time scale $t' = \mathcal{O}(1)$, or $t = \mathcal{O}(1/\epsilon)$, the solute is simply convected by the mean flow.

Let us subtract Eq. (1.3.25) from Eq. (1.3.22) to get

$$\frac{\partial C_1}{\partial t} + \left\{\widetilde{U}_s + \mathcal{R}e\big[U_w e^{-i\omega t}\big]\right\} \frac{\partial C_0}{\partial x} = \frac{D}{r}\frac{\partial}{\partial r}\left(r \frac{\partial C_1}{\partial r}\right), \tag{1.3.26}$$

where

$$\widetilde{U}_s = U_s(r) - \langle U_s(r) \rangle \tag{1.3.27}$$

is the velocity deviation from its area-average. Thus the fluctuation C_1 satisfies an inhomogeneous diffusion equation. In view of linearity the solution can be formally expressed as

$$C_1 = \frac{\partial C_0}{\partial x}\{B_s(r) + \mathcal{R}e[B_w(r)e^{-i\omega t}]\}\,, \tag{1.3.28}$$

and substituted in Eqs. (1.3.26) and (1.3.18), leading to two cell problems for B_s and B_w. The cell problem for the steady part B_s is governed by:

$$\frac{D}{r}\frac{d}{dr}\left(r \frac{dB_s}{dr}\right) = \widetilde{U}_s(r)\,, \tag{1.3.29}$$

and the boundary conditions

$$\frac{dB_s}{dr} = 0\,, \quad r = 0, a\,. \tag{1.3.30}$$

For the oscillatory part B_w we have instead,

$$\frac{D}{r}\frac{d}{dr}\left(r\frac{dB_w}{dr}\right) + i\omega B_w = U_w(r) \tag{1.3.31}$$

with

$$\frac{dB_w}{dr} = 0\,, \quad r = 0, a\,. \tag{1.3.32}$$

These two cell problems are solved explicitly in later sections for the circular pipe. For any other cross-section, numerical solution is not difficult.

After solving for $B_s(r)$ and $B_w(r)$, we go to $\mathcal{O}(\epsilon^2)$, i.e., Eq. (1.3.19), and get

$$\begin{aligned}&\frac{\partial C_0}{\partial t''} + \frac{\partial C_1}{\partial t'} + \frac{\partial C_2}{\partial t}\\ &\quad + \{\langle U_s\rangle + \widetilde{U}_s + \mathcal{R}e[U_w e^{-i\omega t}]\}\{B_s + \mathcal{R}e[B_w(r)e^{-i\omega t}]\}\frac{\partial^2 C_0}{\partial x^2}\\ &\quad = D\frac{\partial^2 C_0}{\partial x^2} + \frac{D}{r}\frac{\partial}{\partial r}\left(r\frac{\partial C_2}{\partial r}\right).\end{aligned} \tag{1.3.33}$$

From Eqs. (1.3.28) and (1.3.25) we find

$$\frac{\partial C_1}{\partial t'} = -\frac{\partial^2 C_0}{\partial x^2}\langle U_s\rangle\{B_s(r) + \mathcal{R}e[B_w(r)e^{-i\omega t}]\}\,. \tag{1.3.34}$$

It follows from Eq. (1.3.33) that

$$\begin{aligned}&\frac{\partial C_0}{\partial t''} + \frac{\partial C_2}{\partial t} + \{\widetilde{U}_s + \mathcal{R}e[U_w e^{-i\Omega t}]\}\{B_s + \mathcal{R}e[B_w e^{-i\omega t}]\}\frac{\partial^2 C_0}{\partial x^2}\\ &\quad = D\frac{\partial^2 C_0}{\partial x^2} + \frac{D}{r}\frac{\partial}{\partial r}\left(r\frac{\partial C_2}{\partial r}\right).\end{aligned} \tag{1.3.35}$$

By taking the time-average over a period,[1] we get a differential equation for $\overline{C}_2$

$$\frac{\partial C_0}{\partial t''} + \left\{\widetilde{U}_s B_s + \frac{1}{2}\mathcal{R}e[U_w B_w^*]\right\}\frac{\partial^2 C_0}{\partial x^2} = D\frac{\partial^2 C_0}{\partial x^2} + \frac{D}{r}\frac{\partial}{\partial r}\left(r\frac{\partial \overline{C}_2}{\partial r}\right) \tag{1.3.36}$$

with B_w^* denoting the complex conjugate of B_w, and the boundary conditions

$$\frac{\partial \overline{C}_2}{\partial r} = 0\,, \quad r = 0, a\,. \tag{1.3.37}$$

[1]There is a handy formula for the period-average of a quadratic product of two simple harmonic functions. If $a = \mathcal{R}e[a_o e^{-i\omega t}]$ and $b = \mathcal{R}e[b_o e^{-i\omega t}]$, then $\overline{ab} = (1/2)\mathcal{R}e\{a_o b_o^*\} = (1/2)\mathcal{R}e\{a_o^* b_o\}$.

Note that $\overline{C}_2$ is governed by an inhomogeneous steady boundary-value problem. Finally the area-average of Eq. (1.3.36) across the pipe gives

$$\frac{\partial C_0}{\partial t''} = (D + \mathcal{D})\frac{\partial^2 C_0}{\partial x^2}, \tag{1.3.38}$$

where

$$\mathcal{D} = -\left\{\langle \tilde{U}_s B_s\rangle + \frac{1}{2}\mathcal{R}e\langle U_w B_w^*\rangle\right\}. \tag{1.3.39}$$

Mathematically Eq. (1.3.36) is the solvability condition for the inhomogeneous problem of $\overline{C}_2$ on the microscale. Physically, over the time scale $t'' = \mathcal{O}(1)$, or $t = \mathcal{O}(1/\epsilon^2)$, the solvent also undergoes longitudinal diffusion where the effective diffusivity is the sum of the molecular diffusivity D and the dispersivity $\mathcal{D}$ which owes its existence to transverse shear.

To combine the effects of convection and diffusion over the long time scale, we add Eq. (1.3.25) to $\epsilon \times$ Eq. (1.3.38) to get:

$$\left(\frac{\partial}{\partial t'} + \epsilon\frac{\partial}{\partial t''}\right) C_0 + \langle U_s\rangle\frac{\partial C_0}{\partial x} = \epsilon(D + \mathcal{D})\frac{\partial^2 C_0}{\partial x^2}. \tag{1.3.40}$$

Now the artifice of two times is no longer needed and can be removed so that

$$\frac{\partial C_0}{\partial t} + \langle U_s\rangle\frac{\partial C_0}{\partial x} = (D + \mathcal{D})\frac{\partial^2 C_0}{\partial x^2}, \tag{1.3.41}$$

which is a one-dimensional convective diffusion equation describing the averaged behavior of a two-dimensional phenomenon.

The expression of the dispersion coefficient will be worked out in the next subsections for steady and oscillatory flows separately by solving for B_s and B_w. Even without their explicit solutions, it can be shown that $\mathcal{D}$ must be positive. We demonstrate below that $\langle \widetilde{U}_s B_s\rangle < 0$. Note by definition that

$$\langle \widetilde{U}_s B_s\rangle = \frac{2\pi}{\pi a^2}\int_0^a r\widetilde{U}_s B_s\, dr\,.$$

Using Eq. (1.3.29) and omitting the factor D/a^2, the right-hand side may be written as

$$\begin{aligned}\int_0^a B_s\frac{d}{dr}\left(r\frac{dB_s}{dr}\right) dr &= \int_0^a \frac{d}{dr}\left(rB_s\frac{dB_s}{dr}\right) dr - \int_0^a r\left(\frac{dB_s}{dr}\right)^2 dr\\ &= \left[rB_s\frac{dB_s}{dr}\right]_0^a - \int_0^a r\left(\frac{dB_s}{dr}\right)^2 dr\\ &= -\int_0^a r\left(\frac{dB_s}{dr}\right)^2 dr < 0\,,\end{aligned}$$

by partial integration and by virtue of the boundary conditions (Eq. (1.3.30)). Hence

$$\langle \widetilde{U}_s B_s \rangle < 0 \tag{1.3.42}$$

which implies that the steady part of the dispersion coefficient is positive. By a similar reasoning it is easy to demonstrate that

$$\frac{1}{2}\mathcal{R}e\langle U_w B^*{}_w \rangle < 0\,, \tag{1.3.43}$$

and hence $\mathcal{D}$ is positive-definite.

We leave it as an exercise to show that the dispersion coefficient for a pipe of any cross-section is always positive.

1.3.3. *Dispersion Coefficient for Steady Flow*

Let us work out the details for the special case of a steady flow.

With $U_w = 0$, the steady velocity profile is parabolic

$$U_s(r) = \frac{2\langle U_s \rangle}{a^2}(a^2 - r^2)\,, \tag{1.3.44}$$

where $\langle U_s \rangle$ is the cross-sectional average, which is related to the steady part of the applied pressure gradient by

$$\langle U_s \rangle = -\frac{a^2}{8\rho\nu}\frac{\partial p_s}{\partial x}\,. \tag{1.3.45}$$

The equation for B_s is

$$\frac{D}{r}\frac{d}{dr}\left(r\frac{dB_s}{dr}\right) = \frac{2\langle U_s \rangle}{a^2}\left(\frac{a^2}{2} - r^2\right). \tag{1.3.46}$$

It follows by integration that

$$B_s = \frac{\langle U_s \rangle}{2a^2 D}\left(\frac{a^2 r^2}{2} - \frac{r^4}{4}\right) + B_o\,. \tag{1.3.47}$$

For uniqueness we can impose the condition that

$$\langle B_s \rangle = 2\int_0^a r B_s dr = 0\,, \tag{1.3.48}$$

yielding,

$$B_o = -\frac{\langle U_s \rangle a^2}{12D}\,, \tag{1.3.49}$$

so that

$$B_s = \frac{\langle U_s \rangle}{2a^2 D}\left(\frac{a^2 r^2}{2} - \frac{r^4}{4} - \frac{a^4}{6}\right). \tag{1.3.50}$$

Further integration gives

$$\mathcal{D}_s = -\langle \widetilde{U}_s B_s \rangle = \frac{\langle U_s \rangle^2 a^2}{48D} = \frac{Pe^2 D}{48}, \tag{1.3.51}$$

which was first obtained by Taylor (1953) using a different reasoning. The greater the Péclet number, $Pe = \langle U_s a \rangle / D$, the more dominant is dispersion over molecular diffusion.

1.3.4. *Dispersion Coefficient for Oscillatory Flow*

Let the oscillatory part of the applied pressure gradient be

$$-\frac{1}{\rho}\frac{\partial p_w}{\partial x} = Q\,\mathcal{R}e[e^{-i\omega t}], \tag{1.3.52}$$

where Q is the amplitude of the pressure gradient. It is straightforward to show that the velocity profile is

$$U_w(r) = \frac{iQ}{\omega}\left[1 - \frac{J_0(\alpha r)}{J_0(\alpha a)}\right], \tag{1.3.53}$$

where

$$\alpha = \sqrt{\frac{i\omega}{\nu}} = \frac{1+i}{\delta}, \quad \text{with } \delta = \sqrt{\frac{2\omega}{\nu}} \tag{1.3.54}$$

being the Stokes boundary layer thickness of momentum. Bessel functions with a complex argument can be expressed in terms of Kelvin functions.

The cell problem for B_w is then governed by

$$\frac{D}{r}\frac{d}{dr}\left(r\frac{dB_w}{dr}\right) + i\omega B_w = U_w(r) = \frac{iQ}{\omega}\left[1 - \frac{J_0(\alpha r)}{J_0(\alpha a)}\right]. \tag{1.3.55}$$

Let us write

$$B_w = \frac{Q}{\omega^2} + B'_w, \tag{1.3.56}$$

so that

$$\frac{D}{r}\frac{d}{dr}\left(r\frac{dB'_w}{dr}\right) + i\omega B'_w = -\frac{iQ}{\omega}\frac{J_0(\alpha r)}{J_0(\alpha a)}. \tag{1.3.57}$$

B'_w can be decomposed into a homogeneous and an inhomogeneous part. The homogeneous part is

$$AJ_o(\beta r), \quad \text{where } \beta = \sqrt{\frac{i\omega}{D}},$$

and A is a constant yet to be determined. The inhomogeneous part can be readily shown to be

$$-\frac{iQ}{\omega D}\frac{1}{\beta^2-\alpha^2}\frac{J_0(\alpha r)}{J_0(\alpha a)}.$$

Hence

$$B_w = AJ_0(\beta r) + \frac{Q}{\omega^2}\left[1 - \frac{\beta^2}{\beta^2-\alpha^2}\frac{J_0(\alpha r)}{J_0(\alpha a)}\right]. \tag{1.3.58}$$

Applying the no-flux condition on the pipe wall $r = a$, we easily find A,

$$A = \frac{\beta}{\alpha}\frac{J_1(\alpha a)}{J_1(\beta a)} \tag{1.3.59}$$

and obtain

$$B_w = \frac{Q}{\omega^2}\left\{1 + \frac{\beta^2}{\beta^2-\alpha^2}\frac{(\alpha/\beta)J_1(\alpha a)J_0(\beta r) - J_0(\alpha r)J_1(\beta a)}{J_0(\alpha a)J_1(\beta a)}\right\}. \tag{1.3.60}$$

We leave it as an exercise to derive the dispersion coefficient. Indeed, one can show that the dispersion coefficient in a pure oscillatory flow is

$$\mathcal{D}_w = \frac{Q^2}{\omega^3}\mathcal{R}e\left\{\frac{A^*[\alpha J_0(\beta^* a)J_1(\alpha a) - \beta^* J_0(\alpha a)J_1(\beta^* a)]}{(\omega a/\nu)(\nu^2/D^2)|J_0(\alpha a)|^2}\right\}. \tag{1.3.61}$$

This result was first found by Aris (1960),[2] and is the form given by Ng (2006) who also studied the effects of chemical reactions of the pipe wall (Ng, 2000, 2004). The theoretical result has been confirmed in laboratory experiments by Joshi *et al.* (1983).

Further extensions have been made by Hydon and Pedley (1993) to dispersion in a tube with elastic wall which is of interest to transport in blood flow.

1.4. Typical Procedure of Homogenization Analysis

The elementary examples in this chapter demonstrate the basic ideas of the homogenization theory which can be extended to many problems with a sharp contrast between micro- and macroscales. Developed

[2] See also Watson (1983).

mostly for periodic microstructures, the typical steps can be summarized as follows

(i) Identify the micro- and macroscales.
(ii) Introduce multiple-scale variables and expansions and deduce boundary-value problems for a typical period at successive orders. The leading-order ($\mathcal{O}(\epsilon^0)$) problem is homogeneous; either the solution itself or the coefficient of the homogeneous solution is indeterminate and independent of the microscale coordinates.
(iii) At the next order $\mathcal{O}(\epsilon)$, the inhomogeneous microscale problem is forced by the leading-order solution. Solve a canonical microscale problem for unit forcing.
(iv) Taking the average of the inhomogeneous microscale problem at the order $\mathcal{O}(\epsilon^2)$, one gets the equation governing the macroscale behavior of the leading order unknown. The constitutive coefficients in the macroscale equation are obtained from the solution of the canonical cell problem.

We now turn to other extensions.

References

Aris, A. (1960). On the dispersion of a solute in pulsating flow through a tube. *Proc. R. Soc. Lond. A* **259**: 370–376.

Bakhvalov, N. and G. Panasenko (1989). *Homogenization: Averaging Processes in Periodic Media*, Kluwer, Dordrecht.

Hydon, P. E. and T. J. Pedley (1993). Axial dispersion in a channel with oscillating walls. *J. Fluid Mech.* **249**: 535–555.

Joshi, C. H., R. D. Kamm, J. M. Drazen and A. S. Slutsky (1983). An experimental study of gas flow through a tube. *J. Fluid Mech.* **133**: 245–254.

Mei, C. C. (1985). Resonant scattering of water waves by periodic bars. *J. Fluid Mech.* **152**: 315–335.

Mei, C. C. (1989). *Applied Dynamics of Ocean Surface Waves,* World Scientific, 700 pp.

Mei, C. C., M. Stiassnie and D. K.-P. Yue (2005). *Theory and Applications of Ocean Surface Waves, Vols. I and II*, World Scientific, Singapore.

Nayfeh, A. (1981). *Introduction to Perturbation Methods*, Wiley-Interscience, 519 pp.

Ng, C. O. (2000). A note on Aris dispersion in a tube with phase exchange and reaction. *Int. J. Eng. Sci.* **38**: 1639–1649.

Ng, C. O. (2004). A time-varying diffusivity model for shear dispersion in oscillatory channel flow. *Fluid Dyn. Res.* **34**: 335–355.

Ng, C. O. (2006). Dispersion in steady and oscillatory flows through a tube with reversible and irreversible wall reactions. *Proc. R. Soc. Lond. A* **462**: 481–515.

Taylor, G. I. (1953). Dispersion of soluble matter in solvent flowing slowly through a tube. *Proc. R. Soc. Lond. A* **219**: 186–203.

Taylor, G. I. (1954). The dispersion of matter in turbulent flow through a pipe. *Proc. R. Soc. Lond. A* **223**: 446–468.

Watson, E. J. (1983). Diffusion in oscillatory pipe flow. *J. Fluid Mech.* **133**: 233–244.

Diffusion in a Composite

2

In this chapter we shall consider the diffusion of heat or mass in a composite medium. Typical applications are heat conduction in steel-reinforced concrete or fiber-reinforced plastic, and diffusion of a solvent in an aggregated soil. The ideal case of a composite formed by only two component materials in perfect contact will be studied in Secs. 2.1–2.8, where the material properties are different but comparable in magnitude. Effective equations for the composite will be derived. The task of finding the constitutive coefficients in the effective equation is reduced to the solution of certain boundary-value problem in a unit cell. For laminated materials the cell problem is one dimensional and governed by an ordinary differential equation for which explicit solution can be found. For two- or three-dimensional cell problems, numerical methods are in general necessary. A more mathematical section on variational bounds is included to give the theoretical range of the effective coefficients. In Secs. 2.9–2.11 the theory is then extended to composites with imperfect contact at the interfaces, resulting in thermal resistance. Our last example described in Sec. 2.12 is concerned with a soil remediation technology called *Soil–Vapor Extraction* (SOE) where the constituent phases have sharply different diffusivities. Modification of the homogenization procedure and physical implications of the resulting equations are discussed.

2.1. Basic Equations for Two Components in Perfect Contact

Let T_α, ρ_α, c_α, and $(K_\alpha)_{ij}$ denote, respectively, the temperature, density, specific heat, and heat conductivity tensor in the material α, where

$\alpha = 1, 2$. For generality we assume that the component materials are nonhomogeneous and anisotropic. The heat conductivity $(K_\alpha)_{ij}$, $\alpha = 1, 2$, for each component is a symmetric and positive-definite tensor that depends on the position vector $\boldsymbol{x}$. In three dimensions it has the form:

$$K_\alpha(\boldsymbol{x}) = \begin{pmatrix} (K_\alpha)_{11} & (K_\alpha)_{12} & (K_\alpha)_{13} \\ (K_\alpha)_{12} & (K_\alpha)_{22} & (K_\alpha)_{23} \\ (K_\alpha)_{13} & (K_\alpha)_{23} & (K_\alpha)_{33} \end{pmatrix}, \tag{2.1.1}$$

where all components $(K_\alpha)_{ij}$ are functions of $\boldsymbol{x}$.

In each component the temperature changes according to the diffusion equation

$$\rho_\alpha c_\alpha \frac{\partial T_\alpha}{\partial t} = \frac{\partial}{\partial x_i}\left((K_\alpha)_{ij}\frac{\partial T_\alpha}{\partial x_j}\right), \quad \alpha = 1, 2. \tag{2.1.2}$$

On the interface Γ between two components, let us assume perfect contact so that the temperature and heat flux must be continuous,

$$T_1 = T_2, \quad \text{on } \Gamma; \tag{2.1.3}$$

$$(K_1)_{ij}\frac{\partial T_1}{\partial x_i}n_j = (K_2)_{ij}\frac{\partial T_2}{\partial x_i}n_j, \quad \text{on } \Gamma, \tag{2.1.4}$$

where $\boldsymbol{n} = (n_i)$ denotes the unit vector normal to the interface pointing from component 1 to component 2. Imperfect contact will be dealt with later in this chapter. Here and throughout the book we employ Einstein's convention of summation over repeated indices for space variables only. The convention does not apply to the index α which distinguishes the materials. In Eqs. (2.1.1)–(2.1.4) $i, j = 1, 2$ for two-dimensional problems and $i, j = 1, 2, 3$ for three-dimensional problems.

2.2. Effective Equation on the Macroscale

Let the microstructure be spatially periodic, so that the composite can be divided into periodic cells of typical dimension ℓ. We let Ω denote a typical cell, Ω_α the part of Ω occupied by material α, $\alpha = 1, 2$ so that $\Omega_1 \cup \Omega_2 = \Omega$. Let the characteristic dimension of the bulk composite be ℓ'.

From the characteristic values K, ρ, and c for the heat conductivity, density, and specific heat, respectively, two vastly different time scales can be identified:

$$\tau = \frac{\ell^2}{K/\rho c} \quad \text{and} \quad \tau' = \frac{\ell'^2}{K/\rho c}, \tag{2.2.1}$$

which characterize the diffusion times over the microscale ℓ and macroscale ℓ', respectively. We are interested only in the macroscale behavior due to external or internal constraints and forcing. Let us normalize all variables as well as the material properties as follows:

$$x^\dagger = \frac{x}{\ell}, \quad t^\dagger = \frac{t}{\tau'}, \quad T_\alpha^\dagger = \frac{T_\alpha}{\Delta T}, \quad \rho_\alpha^\dagger = \frac{\rho_\alpha}{\rho},$$
$$c_\alpha^\dagger = \frac{c_\alpha}{c}, \quad K_\alpha^\dagger = \frac{K_\alpha}{K}, \tag{2.2.2}$$

with ΔT representing the scale of temperature variation. Again the coordinates are normalized by the microscale length. The diffusion equation (2.1.2) becomes

$$\left(\frac{\ell}{\ell'}\right)^2 \rho_\alpha^\dagger c_\alpha^\dagger \frac{\partial T_\alpha^\dagger}{\partial t^\dagger} = \frac{\partial}{\partial x_i^\dagger}\left((K_\alpha^\dagger)_{ij} \frac{\partial T_\alpha^\dagger}{\partial x_j^\dagger}\right), \tag{2.2.3}$$

and the boundary conditions (Eqs. (2.1.3) and (2.1.4)) preserve their forms under the normalization. We assume that the microscopic scale ℓ is much smaller than the macroscopic scale ℓ', i.e.,

$$\frac{\ell}{\ell'} = \epsilon \ll 1, \tag{2.2.4}$$

then the left-hand side of Eq. (2.2.3) is multiplied by a factor $\mathcal{O}(\epsilon^2)$. From here on we return to physical variables but retain the order symbols, for the sake of brevity, so that

$$\epsilon^2 \rho_\alpha c_\alpha \frac{\partial T_\alpha}{\partial t} = \frac{\partial}{\partial x_i}\left((K_\alpha)_{ij} \frac{\partial T_\alpha}{\partial x_j}\right), \tag{2.2.5}$$

$$T_1 = T_2, \quad \boldsymbol{x} \in \Gamma; \tag{2.2.6}$$

and

$$(K_1)_{ij} \frac{\partial T_1}{\partial x_i} n_j = (K_2)_{ij} \frac{\partial T_1}{\partial x_i} n_j, \quad \boldsymbol{x} \in \Gamma. \tag{2.2.7}$$

Let us introduce the fast and slow coordinates x_i and $x_i' = \epsilon x_i$ and the multiple-scale expansion for T_α:

$$T_\alpha = T_\alpha^{(0)}(t, x_i, x_i') + \epsilon T_\alpha^{(1)}(t, x_i, x_i') + \epsilon^2 T_\alpha^{(2)}(t, x_i, x_i') + \cdots \tag{2.2.8}$$

with $T^{(0)}$, $T^{(1)}$, $T^{(2)}, \ldots$ being Ω-periodic functions in the variable $\boldsymbol{x}$.

Upon substituting the expansion in Eq. (2.2.5), we have, at the order $\mathcal{O}(\epsilon^0)$:

$$\frac{\partial}{\partial x_i}\left((K_\alpha)_{ij}\frac{\partial T_\alpha^{(0)}}{\partial x_j}\right) = 0\,, \quad \boldsymbol{x} \in \Omega_\alpha\,, \;\; \alpha = 1, 2\,. \tag{2.2.9}$$

On the interface Γ between the two different materials we require continuity of temperature and heat flux:

$$T_1^{(0)} = T_2^{(0)}\,, \quad \boldsymbol{x} \in \Gamma\,, \tag{2.2.10}$$

$$(K_1)_{ij}\frac{\partial T_1^{(0)}}{\partial x_j}n_i = (K_2)_{ij}\frac{\partial T_2^{(0)}}{\partial x_j}n_i\,, \quad \boldsymbol{x} \in \Gamma\,. \tag{2.2.11}$$

Equations (2.2.9)–(2.2.11), together with the periodicity condition for $T^{(0)}$ on the boundary of the Ω-cell, define a homogeneous boundary-value problem coupling $T_1^{(0)}$ and $T_2^{(0)}$ in the two materials. It is evident that the solution is a constant with respect to $\boldsymbol{x}$, i.e.,

$$T_1^{(0)} = T_2^{(0)} = T^{(0)}(t, \boldsymbol{x}')\,. \tag{2.2.12}$$

At order $\mathcal{O}(\epsilon)$ an inhomogeneous boundary-value problem is obtained for $T_\alpha^{(1)}$, $\alpha = 1, 2$:

$$\frac{\partial}{\partial x_i}\left((K_\alpha)_{ij}\left(\frac{\partial T^{(0)}}{\partial x_j'} + \frac{\partial T_\alpha^{(1)}}{\partial x_j}\right)\right) = 0\,, \quad x_i \in \Omega_\alpha\,, \;\; x_i' \in \Omega\,, \tag{2.2.13}$$

$$T_1^{(1)} = T_2^{(1)}, \quad x_i \in \Gamma\,, \tag{2.2.14}$$

$$(K_1)_{ij}\left(\frac{\partial T^{(0)}}{\partial x_j'} + \frac{\partial T_1^{(1)}}{\partial x_j}\right)n_i = (K_2)_{ij}\left(\frac{\partial T^{(0)}}{\partial x_j'} + \frac{\partial T_2^{(1)}}{\partial x_j}\right)n_i\,, \quad x_i \in \Gamma\,, \tag{2.2.15}$$

$$T_\alpha^{(1)} \text{ is } \Omega\text{-periodic}\,. \tag{2.2.16}$$

Use has been made of Eq. (2.2.12). Since this problem for $T^{(1)}$ is linear and is forced by terms proportional to the macroscale gradient of $T^{(0)}$, we represent the solution as follows:

$$T_\alpha^{(1)} = w_{\alpha l}\frac{\partial T^{(0)}}{\partial x_l'}\,. \tag{2.2.17}$$

Thus the coefficient vector $\boldsymbol{w}_\alpha = (w_{\alpha_j}(\boldsymbol{x}))$ is governed by the following boundary-value problem:

$$\frac{\partial}{\partial x_i}\left((K_\alpha)_{ij}\left(\delta_{jl}+\frac{\partial w_{\alpha_l}}{\partial x_j}\right)\right)=0\,,\quad x_i\in\Omega_\alpha\,,\ \alpha=1,2\,, \tag{2.2.18}$$

$$w_{1_l}=w_{2_l}\,,\quad x_i\in\Gamma\,, \tag{2.2.19}$$

$$(K_1)_{ij}\left(\delta_{jl}+\frac{\partial w_{1_l}}{\partial x_j}\right)n_i=(K_2)_{ij}\left(\delta_{jl}+\frac{\partial w_{2_l}}{\partial x_j}\right)n_i\,,\quad x_i\in\Gamma\,, \tag{2.2.20}$$

$$\boldsymbol{w}_\alpha \text{ is } \Omega\text{-periodic}\,. \tag{2.2.21}$$

To render the solution for $\boldsymbol{w}_\alpha$ unique, we impose further a normalization condition:

$$\langle \boldsymbol{w}_\alpha\rangle = 0\,, \tag{2.2.22}$$

where the angle brackets denote the volume average over the cell Ω:

$$\langle \boldsymbol{w}_\alpha\rangle=\frac{1}{\Omega}\left(\iiint_{\Omega_1}\boldsymbol{w}_1 d\Omega+\iiint_{\Omega_2}\boldsymbol{w}_2 d\Omega\right)\,. \tag{2.2.23}$$

Equations (2.2.18)–(2.2.22) define the *cell problem* for $\boldsymbol{w}_\alpha$ in Ω. The physical meaning is especially simple for constant conductivities. The vector function $\boldsymbol{w}_\alpha$ is the solution to a static heat conduction problem in a periodic cell, forced by a vector flux discontinuity at the interface equal to the difference in conductivities. For a general geometry the cell problem must be solved numerically. A numerical technique for a similar cell problem will be explained in Sec. 4.2, Chapter 4.

At the next order $\mathcal{O}(\epsilon^2)$, we have an inhomogeneous boundary-value problem for $T_\alpha^{(2)}$:

$$\begin{aligned}\rho_\alpha c_\alpha\frac{\partial T_\alpha^{(0)}}{\partial t}&=\frac{\partial}{\partial x_i'}\left((K_\alpha)_{ij}\frac{\partial T_\alpha^{(0)}}{\partial x_j'}\right)+\frac{\partial}{\partial x_i'}\left((K_\alpha)_{ij}\frac{\partial T_\alpha^{(1)}}{\partial x_j}\right)\\&\quad+\frac{\partial}{\partial x_i}\left((K_\alpha)_{ij}\left(\frac{\partial T_\alpha^{(1)}}{\partial x_j'}+\frac{\partial T_\alpha^{(2)}}{\partial x_j}\right)\right)\,,\quad x_i\in\Omega_\alpha\,,\end{aligned} \tag{2.2.24}$$

$$T_1^{(2)}=T_2^{(2)}\,,\quad x_i\in\Gamma\,, \tag{2.2.25}$$

$$(K_1)_{ij}\left(\frac{\partial T_1^{(1)}}{\partial x'_j}+\frac{\partial T_1^{(2)}}{\partial x_j}\right)n_i=(K_2)_{ij}\left(\frac{\partial T_2^{(1)}}{\partial x'_j}+\frac{\partial T_2^{(2)}}{\partial x_j}\right)n_i\,,\quad x_i\in\Gamma\,, \tag{2.2.26}$$

$$T_\alpha^{(2)} \text{ is } \Omega\text{-periodic}\,. \tag{2.2.27}$$

Next we integrate Eq. (2.2.24) over the partial volumes Ω_α, $\alpha=1,2$ and add the two volume integrals,

$$\begin{aligned}
&\frac{1}{\Omega}\sum_\alpha\iiint_{\Omega_\alpha}\rho_\alpha c_\alpha\frac{\partial T^{(0)}}{\partial t}d\Omega\\
&\quad=\frac{1}{\Omega}\sum_\alpha\frac{\partial}{\partial x'_i}\left\{\frac{\partial T^{(0)}}{\partial x'_j}\iiint_{\Omega_\alpha}(K_\alpha)_{ij}d\Omega\right\}\\
&\quad+\frac{1}{\Omega}\sum_\alpha\frac{\partial}{\partial x'_i}\left\{\frac{\partial T^{(0)}}{\partial x'_\ell}\iiint_{\Omega_\alpha}(K_\alpha)_{ij}\frac{\partial(w_\alpha)_\ell}{\partial x_j}d\Omega\right\}\\
&\quad+\frac{1}{\Omega}\sum_\alpha\iint_{\partial\Omega_\alpha}(K_\alpha)_{ij}\left(\frac{\partial T_\alpha^{(1)}}{\partial x'_j}+\frac{\partial T_\alpha^{(2)}}{\partial x_j}\right)(n_\alpha)_i dS\,,
\end{aligned} \tag{2.2.28}$$

where $\partial\Omega_\alpha$ represents the bounding surface of Ω_α. The left-hand side of the preceding equation is

$$\sum_\alpha\left\{\iiint_{\Omega_\alpha}\rho_\alpha c_\alpha d\Omega\right\}\frac{\partial T^{(0)}}{\partial t}=\langle\rho c\rangle\frac{\partial T^{(0)}}{\partial t}\,,$$

where $\langle\cdot\rangle$ denotes the volume average as defined in Eq. (2.2.23). The first and second terms on the right-hand side can be combined as

$$\sum_\alpha\frac{\partial}{\partial x'_i}\left\{\frac{\partial T^{(0)}}{\partial x'_\ell}\iiint_{\Omega_\alpha}(K_\alpha)_{ij}\left(\delta_{j\ell}+\frac{\partial(w_\alpha)_\ell}{\partial x_j}\right)d\Omega\right\}.$$

The last term on the right-hand side of Eq. (2.2.28) is the sum of two surface integrals. The sum over the granular boundary Γ_α vanishes because of the boundary condition (Eq. (2.2.11)) and $(n_1)_j=n_j=-(n_2)_j$. The other sum over the cell boundary $\partial\Omega$ vanishes by periodicity.

Let us define the effective conductivity tensor by,

$$\begin{aligned}
\mathcal{K}_{ij}&=\left\langle K_{il}\left(\delta_{lj}+\frac{\partial w_j}{\partial x_l}\right)\right\rangle=\frac{1}{\Omega}\left\{\iiint_{\Omega_1}\left[(K_1)_{il}\left(\delta_{lj}+\frac{\partial(w_1)_j}{\partial x_l}\right)\right]d\Omega\right.\\
&\left.\quad+\iiint_{\Omega_2}\left[(K_2)_{il}\left(\delta_{lj}+\frac{\partial(w_2)_j}{\partial x_l}\right)\right]d\Omega\right\},
\end{aligned} \tag{2.2.29}$$

and

$$\langle \rho c \rangle = \frac{1}{\Omega} \left\{ \iiint_{\Omega_1} \rho_1 c_1 d\Omega + \iiint_{\Omega_2} \rho_2 c_2 d\Omega \right\} \tag{2.2.30}$$

with Ω being the volume of the Ω-cell. We have finally the following effective diffusion equation on the macroscale for the composite:

$$\langle \rho c \rangle \frac{\partial T^{(0)}}{\partial t} = \frac{\partial}{\partial x'_i} \left(\mathcal{K}_{ij} \frac{\partial T^{(0)}}{\partial x'_j} \right) . \tag{2.2.31}$$

Note that the effective conductivity is not simply the volume average of the thermal conductivity of the two materials, i.e., $\langle K_{ij} \rangle \neq \mathcal{K}_{ij}$.

2.3. Effective Boundary Condition

The analysis has so far been given for typical cells in the interior of the composite medium, Ω', i.e, far away from the bounding surface $\partial\Omega'$. In particular the governing equation for the average temperature $T^{(0)}$ and the effective diffusivity are all defined for the interior of Ω'. How can we express the boundary condition on the surface $\partial\Omega'$ in terms of $T^{(0)}$? This question was studied by Prat (1989, 1992) for heat conduction.

If the ambient temperature is prescribed on $\partial\Omega'$, the natural Dirichlet condition is simply $T^{(0)}(\boldsymbol{x}, t) = F(\boldsymbol{x})$, for all $\boldsymbol{x} \in \partial\Omega'$. If instead the rate of heat flux is prescribed, we expect the effective Neumann condition to be

$$\mathcal{K}_{ij} \frac{\partial T^{(0)}}{\partial x'_j} n_i = \langle q \rangle = q \, , \quad \boldsymbol{x} \in \partial\Omega' \, . \tag{2.3.1}$$

In physical variables, let the microscale boundary condition on the surface of each species be of Neumann type

$$(K_\alpha)_{ij} \frac{\partial T_\alpha}{\partial x_j} n_i = q(\boldsymbol{x}') \, , \quad \boldsymbol{x} \in \partial\Omega' \, , \;\; \alpha = 1, 2 \text{ (no summation over } \alpha) \, , \tag{2.3.2}$$

where the flux rate q is imposed externally as a prescribed function of the macroscale $\boldsymbol{x}'$. Since the analysis in the following does not involve time derivatives, the results apply to both steady-state and transient diffusion; the variable t will be omitted for brevity.

The geometry of the boundary surface $\partial\Omega$ is assumed to depend on $\boldsymbol{x}'$ only. Since the scale of the flux rate must be comparable to the macroscale

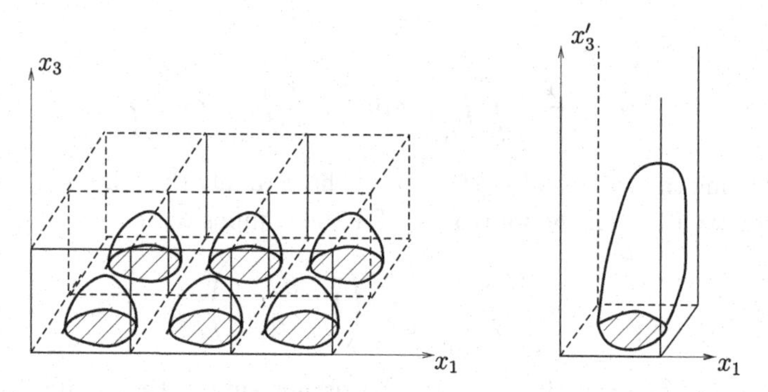

Fig. 2.1. The boundary layer $\hat{\Omega}$.

rate of conduction $K\Delta T/\ell'$, the normalized form (with daggers omitted) of Eq. (2.3.2) is

$$(K_\alpha)_{ij}\frac{\partial T_\alpha}{\partial x_j}n_i = \epsilon q_\alpha\,, \quad \boldsymbol{x} \in \partial\Omega'\,, \;\; \alpha = 1,2 \text{ (no summation over } \alpha). \tag{2.3.3}$$

Along the surface $\partial\Omega'$ let us define a thin layer of periodic cells of height $\delta = \mathcal{O}(\ell)$ and base area $S = \mathcal{O}(\ell^2)$, as sketched in Fig. 2.1. We assume that the boundary of Ω is in the plane $x'_3 = 0$ (otherwise one can make a change of variables).

Away from the thin layer the solution derived Sec. 2.2 for the interior applies,

$$T_\alpha = T^{(0)}(t, x'_i) + \epsilon(w_\alpha)_i\frac{\partial T_\alpha^{(0)}}{\partial x'_i} + \cdots. \tag{2.3.4}$$

At the leading order the temperature is the same in both phases. Inside the thin layer we assume the approximation to be of the form

$$T_\alpha = \widehat{T}_\alpha^{(0)}(t, x_1, x_2, x_3, x'_1, x'_2) + \epsilon\widehat{T}_\alpha^{(1)}(t, x_1, x_2, x_3, x'_1, x'_2) + \cdots, \tag{2.3.5}$$

where $x_3 = x'_3/\epsilon$ and $\widehat{T}_\alpha^{(i)}$ are periodic only in (x_1, x_2). The dependence of $\widehat{T}_\alpha^{(i)}$ on x'_1, x'_2 reflects the influence of the macroscopic data and the dependence on x_1, x_2 and x_3 reflects the influence of the local microstructure.

We now consider a unit cell $\widehat{\Omega}$ at the boundary, which is a box of height $\delta = \mathcal{O}(\ell)$ in the x_3 direction. From the above, the leading-order temperature $\widehat{T}_\alpha^{(0)}$ in each phase satisfies

$$\frac{\partial}{\partial x_i}\left((K_\alpha)_{ij}\frac{\partial \widehat{T}_\alpha^{(0)}}{\partial x_j}\right) = 0 \quad \text{in } \widehat{\Omega}\,. \tag{2.3.6}$$

The boundary conditions are:

$$\left[\widehat{T}_\alpha^{(0)}\right]_-^+ = 0\,, \quad \left[(K_\alpha)_{ij}\frac{\partial \widehat{T}_\alpha^{(0)}}{\partial x_j}n_i\right]_-^+ = 0 \quad \text{on } \Gamma \tag{2.3.7}$$

on fluid–grain interfaces,

$$\widehat{T}_\alpha^{(0)}(t, x_1, x_2, \delta, x_1', x_2') = T_\alpha^{(0)}(t, x_1', x_2', 0)\,, \quad \text{on } I \tag{2.3.8}$$

on the border between the layer and the outer region, and

$$(K_\alpha)_{3j}\frac{\partial \widehat{T}_\alpha^{(0)}}{\partial x_j}(t, x_1, x_2, 0, x_1', x_2') = 0\,, \quad x_3 = 0 \tag{2.3.9}$$

on the physical boundary. In addition, we require that

$$\widehat{T}_\alpha^{(0)} \text{ is } (x_1, x_2) \text{ periodic}\,. \tag{2.3.10}$$

The square brackets symbolize the jump across the interface between two phases, $[F]_-^+ \equiv F|_{\Gamma_+} - F|_{\Gamma_-}$. The above problem has a unique solution that is constant with respect to the fast variables x_1, x_2, x_3,

$$\widehat{T}_\alpha^{(0)} = T_\alpha^{(0)}(t, x_1', x_2', 0) = T^{(0)}(t, x_1', x_2', 0)\,, \quad \alpha = 1, 2 \tag{2.3.11}$$

which satisfies conditions (Eqs. (2.3.6)–(2.3.10)) with $T_1^{(0)} = T_2^{(0)} = T^{(0)}$, again being the same in both phases. Note that the normal derivative $n_i(\partial T^{(0)}/\partial x_i)$ is also automatically continuous across the interface.

At the next order the boundary-layer correction $\widehat{T}_\alpha^{(1)}$ is governed by the following inhomogeneous boundary-value problem

$$\frac{\partial}{\partial x_i}\left((K_\alpha)_{ij}\left(\frac{\partial \widehat{T}_\alpha^{(0)}}{\partial x_j'} + \frac{\partial \widehat{T}_\alpha^{(1)}}{\partial x_j}\right)\right) = 0\,, \quad \text{in } \widehat{\Omega}\,, \tag{2.3.12}$$

$$[\widehat{T}_\alpha^{(0)}]_-^+ = 0\,, \quad \left[(K_\alpha)_{ij}\left(\frac{\partial \widehat{T}_\alpha^{(0)}}{\partial x_j'} + \frac{\partial \widehat{T}_\alpha^{(1)}}{\partial x_j}\right)n_i\right]_-^+ = 0\,, \quad \text{on } \Gamma\,. \tag{2.3.13}$$

At the edge of the boundary layer at $x_3 = \delta$ we require smooth matching so that the temperature and its normal gradient are continuous, i.e.,

$$\widehat{T}_\alpha^{(1)}(t, x_1, x_2, \delta, x_1', x_2') = T_\alpha^{(1)}(t, x_1', x_2', 0) \quad \text{on } A\,, \tag{2.3.14}$$

and

$$\frac{\partial T^{(0)}}{\partial x'_j}(t, x_1, x_2, \delta, x'_1, x'_2) + \frac{\partial \widehat{T}_\alpha^{(1)}}{\partial x_j}(t, x_1, x_2, \delta, x'_1, x'_2)$$

$$= \left(\frac{\partial (w_\alpha)_i}{\partial x_j} + \delta_{ij}\right) \frac{\partial T^{(0)}}{\partial x'_i}(t, x'_1, x'_2, 0)\,, \quad \text{on } A\,. \tag{2.3.15}$$

At the physical boundary we require

$$(K_\alpha)_{3j}\left(\frac{\partial \widehat{T}_\alpha^{(0)}}{\partial x'_j} + \frac{\partial \widehat{T}_\alpha^{(1)}}{\partial x_j}\right) = q(x'_1, x'_2) \quad \text{on } x_3 = 0\,. \tag{2.3.16}$$

In addition,

$$\widehat{T}_\alpha^{(1)} \text{ is periodic in } (x_1, x_2)\,. \tag{2.3.17}$$

Applying Gauss' theorem to Eq. (2.3.12) and making use of the constant flux condition (Eq. (2.3.13)), we get, after cancellation on opposite faces

$$\iint_{x_3=0} K_{3j}\left(\frac{\partial \hat{T}^{(0)}}{\partial x'_j} + \frac{\partial \hat{T}^{(1)}}{\partial x_j}\right) dS = \iint_{x_3=\infty} K_{3j}\left(\frac{\partial \hat{T}^{(0)}}{\partial x'_j} + \frac{\partial \hat{T}^{(1)}}{\partial x_j}\right) dS\,. \tag{2.3.18}$$

By using Eq. (2.3.15) on the left-hand side of the above equation and the matching condition (Eq. (2.3.16)) on the right-hand side we get

$$q = \frac{1}{S}\iint_{x_3=0} K_{3j}\left(\frac{\partial \hat{T}^{(0)}}{\partial x'_j} + \frac{\partial \hat{T}^{(1)}}{\partial x_j}\right) dS$$

$$= \frac{1}{S}\iint_{x'_3=0} K_{3j}\left(\frac{\partial (w_\alpha)_i}{\partial x_j} + \delta_{ij}\right) \frac{\partial T^{(0)}}{\partial x'_i} dS\,. \tag{2.3.19}$$

Next we follow Prat (1989) and show that the area integral on the right-hand side can be converted to a volume integral

$$\frac{1}{S}\iint_{x'_3=0} K_{3j}\left(\frac{\partial (w_\alpha)_i}{\partial x_j} + \delta_{ij}\right) \frac{\partial T^{(0)}}{\partial x'_i} dS$$

$$= \frac{1}{\Omega}\iiint_{\Omega} K_{3j}\left(\frac{\partial (w_\alpha)_i}{\partial x_j} + \delta_{ij}\right) \frac{\partial T^{(0)}}{\partial x'_i} d\Omega\,, \tag{2.3.20}$$

where Ω is a periodic cell next to and on the interior side of A; its opposite faces A and A' are orthogonal to the x_3 axis, then

$$\frac{dS}{A} = \frac{dS}{\Omega}(x_3(I) - x_3(I')) = \frac{d\Omega}{\Omega}\,,$$

where A also represents of the area of cell face A. Upon introducing the notation $W_i = w_i + x_i$, the left-hand side in Eq. (2.3.20) becomes

$$\begin{aligned}
&\frac{1}{S}\iint_A K_{3j}\frac{\partial (W_\alpha)_i}{\partial x_j}\frac{\partial T^{(0)}}{\partial x'_i}\,dS \\
&\quad = \frac{1}{\Omega}\left(\iint_A K_{3j}\frac{\partial (W_\alpha)_i}{\partial x_j}\frac{\partial T^{(0)}}{\partial x'_i}x_3\,dS - \iint_{A'} K_{3j}\frac{\partial (W_\alpha)_i}{\partial x_j}\frac{\partial T^{(0)}}{\partial x'_i}x_3\,dS\right) \\
&\quad = \frac{1}{\Omega}\iint_{\partial\Omega} K_{\ell j}\frac{\partial (W_\alpha)_i}{\partial x_j}\frac{\partial T^{(0)}}{\partial x'_i}x_3 n_\ell\,dS\,,
\end{aligned}$$

where the integrals on opposite faces of the cell Ω, except for A and A', cancel by periodicity. The last surface integral can be written by Gauss' theorem

$$\begin{aligned}
&\frac{\partial T^{(0)}}{\partial x'_i}\iint_{\partial\Omega} K_{\ell j}\frac{\partial (W_\alpha)_i}{\partial x_j}x_3 n_\ell\,dS \\
&\quad = \frac{\partial T^{(0)}}{\partial x'_i}\iiint_\Omega \frac{\partial}{\partial x_\ell}\left(K_{\ell j}\frac{\partial (W_\alpha)_i}{\partial x_j}x_3\right)d\Omega \\
&\quad = \frac{\partial T^{(0)}}{\partial x'_i}\iiint_\Omega x_3\frac{\partial}{\partial x_\ell}\left(K_{\ell j}\frac{\partial (W_\alpha)_i}{\partial x_j}\right)d\Omega \\
&\qquad + \frac{\partial T^{(0)}}{\partial x'_i}\iiint_\Omega K_{3j}\frac{\partial (W_\alpha)_i}{\partial x_j}d\Omega\,.
\end{aligned} \tag{2.3.21}$$

Now the first integral on the right-hand side vanishes because of Eq. (2.2.18). Since the volume integral covers both phases, it follows from Eq. (2.2.29) defining the effective conductivity that

$$\mathcal{K}_{3j}\frac{\partial T^{(0)}}{\partial x'_j} = q\,, \tag{2.3.22}$$

which implies in general

$$\mathcal{K}_{ij}\frac{\partial T^{(0)}}{\partial x'_j}n_i = q\,. \tag{2.3.23}$$

This conclusion confirms the intuitive expectation.

2.4. Symmetry and Positiveness of Effective Conductivity

Aside from finding a way for predicting the effective coefficients, we can further deduce two important properties of the effective conductivity tensor ($\mathcal{K}_{ij}$): symmetry and positiveness, before actual computations are carried out.

In the definition of $\mathcal{K}_{ij}$ (Eq. (2.2.29)), the first part representing the volume average is clearly symmetric because K_{ij} is symmetric for each constituent.

As for the second part let us introduce the short-hand notation $\boldsymbol{w}$ and K without the subscript α to mean

$$\boldsymbol{w} = \begin{cases} \boldsymbol{w}_1 = (w_{1_j}) & \text{if } \boldsymbol{x} \in \Omega_1 , \\ \boldsymbol{w}_2 = (w_{2_j}) & \text{if } \boldsymbol{x} \in \Omega_2 , \end{cases} \qquad K_{ij} = \begin{cases} (K_1)_{ij} & \text{if } \boldsymbol{x} \in \Omega_1 , \\ (K_2)_{ij} & \text{if } \boldsymbol{x} \in \Omega_2 . \end{cases} \tag{2.4.1}$$

Similarly we omit the subscript α in $W_{\alpha_i} = w_{\alpha_i} + x_i$ and define W_l to be $(W_1)_l$ in phase Ω_1 and $(W_2)_l$ in phase Ω_2 (as in Eq. (2.4.1)), Eq. (2.2.18) becomes

$$\frac{\partial}{\partial x_i}\left(K_{ij}\frac{\partial W_l}{\partial x_j}\right) = 0 \tag{2.4.2}$$

in each phase Ω_α. Let Φ be any Ω-periodic function continuous across Γ. Then Eq. (2.4.2) implies that:

$$\frac{\partial}{\partial x_i}\left(K_{ij}\Phi\frac{\partial W_l}{\partial x_j}\right) = K_{ij}\frac{\partial W_l}{\partial x_j}\frac{\partial \Phi}{\partial x_i} . \tag{2.4.3}$$

By integrating Eq. (2.4.3) over each phase Ω_α and applying the divergence theorem, we get:

$$\iint_{\partial\Omega} K_{ij}\Phi\frac{\partial W_l}{\partial x_j} n_i dS - \iiint_{\Omega} K_{ij}\frac{\partial W_l}{\partial x_j}\frac{\partial \Phi}{\partial x_i} d\Omega = 0 . \tag{2.4.4}$$

In the surface integral above the integrand vanishes on Γ due to the condition (Eq. (2.2.20)) for flux continuity. Because of periodicity and opposite signs of the unit normal, the integrals on opposite sides of the Ω-cell also cancel. Thus we have

$$\left\langle K_{ij}\frac{\partial W_l}{\partial x_j}\frac{\partial \Phi}{\partial x_i}\right\rangle = 0 . \tag{2.4.5}$$

The effective conductivity is from Eq. (2.2.29)

$$\mathcal{K}_{ij} = \left\langle K_{il}\frac{\partial W_j}{\partial x_l}\right\rangle = \left\langle K_{kl}\frac{\partial W_j}{\partial x_l}\frac{\partial x_i}{\partial x_k}\right\rangle = \left\langle K_{kl}\frac{\partial W_j}{\partial x_l}\frac{\partial (W_i - w_i)}{\partial x_k}\right\rangle . \tag{2.4.6}$$

In the last expression we have

$$\left\langle K_{kl}\frac{\partial W_j}{\partial x_l}\frac{\partial w_i}{\partial x_k}\right\rangle = 0 \tag{2.4.7}$$

in view of Eq. (2.4.5), hence

$$\mathcal{K}_{ij} = \left\langle K_{kl} \frac{\partial W_j}{\partial x_l} \frac{\partial W_i}{\partial x_k} \right\rangle . \tag{2.4.8}$$

Assume symmetry of (K_{kl}) in each phase, the right-hand side is clearly symmetric in i and j. Therefore the effective conductivity tensor must be symmetric.

To prove that $(\mathcal{K}_{ij})$ is positive let us consider a constant vector $\boldsymbol{\xi} = (\xi_i)$ (independent of the variable $\boldsymbol{x}$). It follows from Eq. (2.4.8) that

$$\mathcal{K}_{ij}\xi_i\xi_j = \left\langle K_{kl} \frac{\partial (W_j\xi_j)}{\partial x_l} \frac{\partial (W_i\xi_i)}{\partial x_k} \right\rangle . \tag{2.4.9}$$

Since ξ_i and ξ_j are arbitrary scalars, the positiveness of (K_{kl}) guarantees the same for $(\mathcal{K}_{ij})$.

To prove that $(\mathcal{K}_{ij})$ is positive definite we need to show that $\mathcal{K}_{ij}\xi_i\xi_j = 0$ only if $\xi_i = 0$ identically. We assume that (K_{kl}) is positive definite. Then the condition for $\mathcal{K}_{ij}\xi_i\xi_j$ to vanish requires that

$$\frac{\partial (W_i\xi_i)}{\partial x_k} = \xi_i \left(\frac{\partial w_i}{\partial x_k} + \delta_{ik} \right) = 0 .$$

By integrating this equation over the Ω-cell and taking into account the periodicity of w_j, we find

$$\xi_i\delta_{ik} = \xi_k = 0$$

identically. Thus $\mathcal{K}_{ij}$ must be positive definite.

2.5. Laminated Composites

In general the cell problem for $\boldsymbol{w}_l$ hence the effective conductivity must be solved numerically. For a periodic composite formed by two materials with a laminated microgeometry, the task can be explicitly carried out.

Let us consider a laminate composed of two homogeneous materials of constant conductivities $((K_1)_{ij})$ and $((K_2)_{ij})$ and volume fractions θ_1 and θ_2 $(\theta_1 + \theta_2 = 1)$. Let x_1 coordinate be normal to the laminate. The cell geometry is one-dimensional; the governing equations reduce to ordinary differential equations with constant coefficients, and can be readily solved.

Equations (2.2.18)–(2.2.21) describe two vectors $\boldsymbol{w}_1$ and $\boldsymbol{w}_2$, and can be solved for one scalar component at a time. Equivalently we can solve for their scalar products with any constant vector $\boldsymbol{\xi}$, for which the cell problem can be obtained from Eqs. (2.2.18)–(2.2.21) as:

$$(K_1)_{ij}\frac{\partial}{\partial x_i}\left(\frac{\partial(\boldsymbol{w_1}\cdot\boldsymbol{\xi})}{\partial x_j}+\xi_j\right)=0 \quad \text{in } \Omega_1\,, \tag{2.5.1}$$

$$(K_2)_{ij}\frac{\partial}{\partial x_i}\left(\frac{\partial(\boldsymbol{w_2}\cdot\boldsymbol{\xi})}{\partial x_j}+\xi_j\right)=0 \quad \text{in } \Omega_2\,, \tag{2.5.2}$$

$$\boldsymbol{w_1}\cdot\boldsymbol{\xi}=\boldsymbol{w_2}\cdot\boldsymbol{\xi} \quad \text{on } \Gamma\,, \tag{2.5.3}$$

$$(K_1)_{i1}\left(\frac{\partial(\boldsymbol{w_1}\cdot\boldsymbol{\xi})}{\partial x_i}+\xi_i\right)=(K_2)_{i1}\left(\frac{\partial(\boldsymbol{w_2}\cdot\boldsymbol{\xi})}{\partial x_i}+\xi_i\right) \quad \text{on } \Gamma\,, \tag{2.5.4}$$

$$\boldsymbol{w}\cdot\boldsymbol{\xi} \text{ is } \Omega\text{-periodic}\,. \tag{2.5.5}$$

Equation (2.2.29) for the effective conductivity can be rewritten as

$$\mathcal{K}_{ij}\xi_j=\left\langle K_{ij}\left(\frac{\partial(\boldsymbol{w}\cdot\boldsymbol{\xi})}{\partial x_j}+\xi_j\right)\right\rangle. \tag{2.5.6}$$

By assigning $\xi_1=1, \xi_2=\xi_3=0$, i.e., $\boldsymbol{\xi}=(1,0,0)$, one gets $\mathcal{K}_{i1}$. Similarly, by assigning $\boldsymbol{\xi}=(0,1,0)$ and $\boldsymbol{\xi}=(0,0,1)$ in turn, one gets $\mathcal{K}_{i2}$ and $\mathcal{K}_{i3}$, respectively.

Since the geometry depends only on x_1 and the coefficients are constants, the solutions to Eqs. (2.5.1) and (2.5.2) are linear functions in x_1 in each domain:

$$\boldsymbol{w_1}\cdot\boldsymbol{\xi}=Ax_1+a \quad \text{in } \Omega_1\,, \tag{2.5.7}$$

$$\boldsymbol{w_2}\cdot\boldsymbol{\xi}=Bx_1+b \quad \text{in } \Omega_2\,. \tag{2.5.8}$$

Since the solution of the cell problem is determined uniquely up to a constant, we can choose $b=0$. Let us define Ω_1 to be $0<x_1<\theta_1\ell$ and Ω_2 to be $\theta_1\ell<x_1<\ell$ so that θ_1 represents the volume fraction of material 1, as shown in Fig. 2.2.

It follows by imposing the continuity conditions (Eqs. (2.5.3) and (2.5.4)) on Γ

$$A\theta_1\ell+a=B\theta_1\ell\,, \tag{2.5.9}$$

and

$$(K_1)_{11}A+(K_1)_{i1}\xi_i=(K_2)_{11}B+(K_2)_{i1}\xi_i\,. \tag{2.5.10}$$

The periodicity condition (Eq. (2.5.5)) on the cell sides yields

$$a=B(\theta_1+\theta_2)\ell\,. \tag{2.5.11}$$

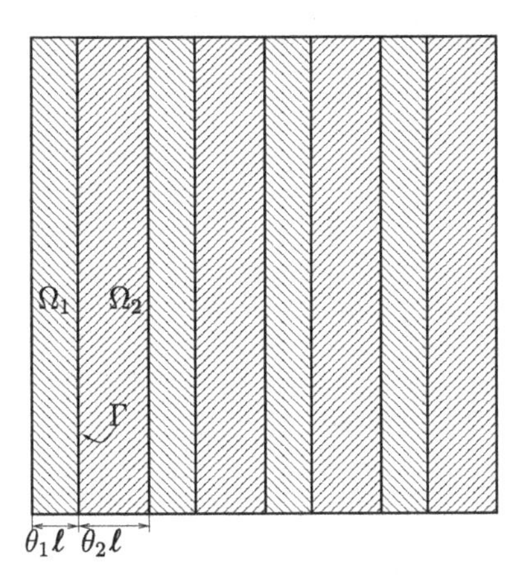

Fig. 2.2. A two-phase laminate composite.

The system (Eqs. (2.5.9)–(2.5.11)) can be solved to yield

$$A = \frac{[(K_2)_{i1} - (K_1)_{i1}]\theta_2\xi_i}{(K_1)_{11}\theta_2 + (K_2)_{11}\theta_1}, \tag{2.5.12}$$

$$B = -\frac{[(K_2)_{i1} - (K_1)_{i1}]\theta_1\xi_i}{(K_1)_{11}\theta_2 + (K_2)_{11}\theta_1}, \tag{2.5.13}$$

$$a = -\frac{[(K_2)_{i1} - (K_1)_{i1}]\theta_1\xi_i\ell}{(K_1)_{11}\theta_2 + (K_2)_{11}\theta_1}. \tag{2.5.14}$$

Equation (2.5.6) becomes

$$\mathcal{K}_{ij}\xi_j = \theta_1(K_1)_{i1}A + \theta_2(K_2)_{i1}B + \langle K_{ij}\rangle\xi_j\,. \tag{2.5.15}$$

An important limit is the case when the conductivity tensors $(K_1)_{ij}$ and $(K_2)_{ij}$ are diagonal, then the coefficients A and B simplify further to:

$$A = \frac{((K_2)_{11} - (K_1)_{11})\theta_2\xi_1}{(K_1)_{11}\theta_2 + (K_2)_{11}\theta_1}, \quad B = \frac{((K_1)_{11} - (K_2)_{11})\theta_1\xi_1}{(K_1)_{11}\theta_2 + (K_2)_{11}\theta_1}. \tag{2.5.16}$$

By using $\theta_1 + \theta_2 = 1$ and defining the volume average as $\langle K_{ij}\rangle = (K_1)_{ij}\theta_1 + (K_2)_{ij}\theta_2$, it is straightforward to show that the effective conductivity tensor has the form

$$\mathcal{K} = \begin{pmatrix} \left(\dfrac{\theta_1}{(K_1)_{11}} + \dfrac{\theta_2}{(K_2)_{11}}\right)^{-1} & 0 & 0 \\ 0 & \langle K_{22}\rangle & 0 \\ 0 & 0 & \langle K_{33}\rangle \end{pmatrix}. \tag{2.5.17}$$

In particular, for isotropic phases, $(K_1)_{ij} = K_1\delta_{ij}$ and $(K_2)_{ij} = K_2\delta_{ij}$, the effective conductivity becomes

$$\mathcal{K} = \begin{pmatrix} \left(\dfrac{\theta_1}{K_1} + \dfrac{\theta_2}{K_2}\right)^{-1} & 0 & 0 \\ 0 & \theta_1 K_1 + \theta_2 K_2 & 0 \\ 0 & 0 & \theta_1 K_1 + \theta_2 K_2 \end{pmatrix}. \tag{2.5.18}$$

Thus the normal component $\mathcal{K}_{11}$ is equal to the harmonic mean, while the transverse components $\mathcal{K}_{22} = \mathcal{K}_{33}$ are equal to the arithmetic mean of the conductivities of the two materials.

2.6. Bounds for Effective Conductivity

In Sec. 2.2 we have obtained a general formula (Eq. (2.2.29)) for the effective conductivity of a composite in terms of the conductivities of the component materials, the corresponding volume fractions and the solutions of the cell problems that depend on the microscale geometry. Except for laminated materials the cell problems call for nontrivial numerical computations in general. Crude estimates of bounds can therefore be of practical value for providing ranges of the effective diffusivity and also checks for computations or experiments. In designing new materials it is of interest to know beforehand the range of effective conductivities achievable by different arrangements of given materials.

We demonstrate below the simplest bounds with the help of two variational principles.

2.6.1. *First Variational Principle and the Upper Bound*

For any Ω-periodic function φ, let us define the following functional $\mathcal{F}$:

$$\mathcal{F}(\varphi) = \frac{1}{\Omega} \iiint_\Omega K_{ij}(x) \left(\frac{\partial \varphi}{\partial x_i} + \xi_i\right) \left(\frac{\partial \varphi}{\partial x_j} + \xi_j\right) d\Omega\,. \tag{2.6.1}$$

Then the effective conductivity satisfies the following variational principle: among all admissible φ's, the one that minimizes $\mathcal{F}(\varphi)$ gives the bilinear form $\mathcal{K}_{ij}\xi_i\xi_j$

$$\boxed{\mathcal{K}_{ij}\xi_i\xi_j = \min_\varphi \mathcal{F}(\varphi)}\,, \tag{2.6.2}$$

for any constant vector $\boldsymbol{\xi} = (\xi_i)$, where $\mathcal{K}_{ij}$ is the effective conductivity. The symbol $\min_\varphi$ denotes the minimum (greatest lower bound) of $\mathcal{F}$ for all functions φ which are Ω-periodic. Thus $\mathcal{F}(\varphi)$ with any other φ is greater than, hence an upper bound of, $\mathcal{K}_{ij}\xi_i\xi_j$.

The proof goes as follows: recall from Eq. (2.4.9) that

$$\mathcal{K}_{ij}\xi_i\xi_j = \left\langle K_{kl}\left(\frac{\partial(w_i\xi_i)}{\partial x_l} + \xi_l\right)\left(\frac{\partial(w_i\xi_i)}{\partial x_k} + \xi_k\right)\right\rangle . \tag{2.6.3}$$

In view of Eq. (2.6.1), this expression is just $\mathcal{F}(w_i\xi_i)$ and thus it is no less than the minimum, i.e.,

$$\mathcal{K}_{ij}\xi_i\xi_j = \mathcal{F}(w_i\xi_i) \geq \min_{\varphi} \mathcal{F}(\varphi) . \tag{2.6.4}$$

Now let

$$\delta = w_i\xi_i - \varphi = \boldsymbol{w}\cdot\boldsymbol{\xi} - \varphi , \tag{2.6.5}$$

which is also Ω-periodic. Substituting $\varphi = w_i\xi_i - \delta$ in Eq. (2.6.1) we have

$$\begin{aligned}
&\left\langle K_{ij}\left(\frac{\partial\varphi}{\partial x_j} + \xi_j\right)\left(\frac{\partial\varphi}{\partial x_i} + \xi_i\right)\right\rangle \\
&= \left\langle K_{ij}\left(\frac{\partial(\boldsymbol{w}\cdot\boldsymbol{\xi} - \delta)}{\partial x_j} + \xi_j\right)\left(\frac{\partial(\boldsymbol{w}\cdot\boldsymbol{\xi} - \delta)}{\partial x_i} + \xi_i\right)\right\rangle \\
&= \left\langle K_{ij}\left(\frac{\partial(\boldsymbol{w}\cdot\boldsymbol{\xi})}{\partial x_j} + \xi_j\right)\left(\frac{\partial(\boldsymbol{w}\cdot\boldsymbol{\xi})}{\partial x_i} + \xi_i\right)\right\rangle + \left\langle K_{ij}\frac{\partial\delta}{\partial x_j}\frac{\partial\delta}{\partial x_i}\right\rangle \\
&\quad - 2\left\langle K_{ij}\frac{\partial\delta}{\partial x_i}\left(\frac{\partial(\boldsymbol{w}\cdot\boldsymbol{\xi})}{\partial x_j} + \xi_j\right)\right\rangle .
\end{aligned} \tag{2.6.6}$$

By rearranging terms we obtain

$$\begin{aligned}
\mathcal{K}_{ij}\xi_i\xi_j = &\left\langle K_{ij}\left(\frac{\partial\varphi}{\partial x_j} + \xi_j\right)\left(\frac{\partial\varphi}{\partial x_i} + \xi_i\right)\right\rangle \\
&- \left\langle K_{ij}\frac{\partial\delta}{\partial x_j}\frac{\partial\delta}{\partial x_i}\right\rangle + 2\left\langle K_{ij}\frac{\partial\delta}{\partial x_i}\left(\frac{\partial(\boldsymbol{w}\cdot\boldsymbol{\xi})}{\partial x_j} + \xi_j\right)\right\rangle .
\end{aligned} \tag{2.6.7}$$

By partial integration and using Gauss' formula we can write the last term above as:

$$\begin{aligned}
&2\left\langle\frac{\partial}{\partial x_i}\left[\delta K_{ij}\left(\frac{\partial(\boldsymbol{w}\cdot\boldsymbol{\xi})}{\partial x_j} + \xi_j\right)\right]\right\rangle - 2\left\langle\delta\frac{\partial}{\partial x_i}\left[K_{ij}\left(\frac{\partial(\boldsymbol{w}\cdot\boldsymbol{\xi})}{\partial x_j} + \xi_j\right)\right]\right\rangle \\
&= \frac{2}{\Omega}\iint_{\partial\Omega} K_{ij}\delta\left(\frac{\partial(\boldsymbol{w}\cdot\boldsymbol{\xi})}{\partial x_j} + \xi_j\right) n_i dS \\
&\quad - 2\xi_\ell\left\langle\delta\frac{\partial}{\partial x_i}\left(K_{ij}\frac{\partial w_\ell}{\partial x_j} + \delta_{j\ell}\right)\right\rangle .
\end{aligned}$$

The second term on the right-hand side of the preceding equation is zero because w_ℓ satisfies Eq. (2.2.18) while the first term cancels on opposite faces of Ω because of periodicity. From the positivity of (K_{ij}) the second term on the right-hand side of Eq. (2.6.7) is negative. It follows that

$$\mathcal{K}_{ij}\xi_i\xi_j \leq \left\langle K_{ij}\left(\frac{\partial\varphi}{\partial x_j}+\xi_j\right)\left(\frac{\partial\varphi}{\partial x_i}+\xi_i\right)\right\rangle, \tag{2.6.8}$$

for all admissible φ, i.e.,

$$\mathcal{K}_{ij}\xi_i\xi_j \leq \min \mathcal{F}(\varphi). \tag{2.6.9}$$

The two inequalities (Eqs. (2.6.4) and (2.6.9)) cannot both be true except for strict equality

$$\mathcal{K}_{ij}\xi_i\xi_j = \min_{\varphi} \mathcal{F}(\varphi), \tag{2.6.10}$$

hence the variational principle is proven.

Let us see how to obtain upper bounds for the effective conductivity $(\mathcal{K}_{ij})$ from this variational principle, by making particular choices for the Ω-periodic function φ. The simplest possible choice for the Ω-periodic function φ in Eq. (2.6.2) is $\varphi =$ constant, then

$$\mathcal{K}_{ij}\xi_i\xi_j \leq \frac{1}{\Omega}\left(\iiint_\Omega K_{ij}(\mathbf{x})d\Omega\right)\xi_i\xi_j = \langle K_{ij}\rangle\xi_i\xi_j, \tag{2.6.11}$$

where $\langle K_{ij}\rangle$ is the volume average of the conductivity tensor.

Consider the special case where both component materials are isotropic, i.e., $K_{ij} = K_1\delta_{ij}$ in material 1 and $K_{ij} = K_2\delta_{ij}$ in material 2. Let us denote the volume fractions of the two materials by θ_1 and θ_2. Then the volume average also has a diagonal form

$$\langle K_{ij}\rangle = (\theta_1 K_1 + \theta_2 K_2)\delta_{ij}. \tag{2.6.12}$$

Since $(\mathcal{K}_{ij})$ is symmetric one can find its eigenvalues $\lambda_{(n)}, n = 1, 2, 3$ from

$$(\mathcal{K}_{ij} - \lambda\delta_{ij})u_i = 0 \tag{2.6.13}$$

and the eigenvectors $\boldsymbol{u}$ so as to diagonalize $\mathcal{K}_{ij}$. In general $\lambda_{(1)} \neq \lambda_{(2)} \neq \lambda_{(3)}$ so that the diagonal elements are not the same, i.e., the effective conductivity tensor is anisotropic. Now from Eq. (2.6.12)

$$\lambda_{(n)} \leq \theta_1 K_1 + \theta_2 K_2, \quad n = 1, 2, 3. \tag{2.6.14}$$

Thus the volume average of the component conductivities is an upper bound for the largest principal conductivity on the macroscale.

The upper bound may be improved by using a φ with several coefficients which are selected to minimize the functional $\mathcal{F}(\varphi)$. In particular for homogeneous and isotropic component materials the largest of the eigenvalues $\lambda_{(n)}$ can be calculated numerically by using finite-element approximation for $\varphi(\boldsymbol{x})$ in the unit cell Ω and minimizing $\mathcal{F}$ to get the nodal coefficients.

2.6.2. *Dual Variational Principle and the Lower Bound*

Let us use the short-hand notation of Eq. (2.4.1) to denote a vector function $\boldsymbol{\sigma} = (\sigma_i)$:

$$\boldsymbol{\sigma} = \begin{cases} \boldsymbol{\sigma_1} = (\sigma_{1_j}) & \text{if } \boldsymbol{x} \in \Omega_1 \,, \\ \boldsymbol{\sigma_2} = (\sigma_{1_j}) & \text{if } \boldsymbol{x} \in \Omega_2 \,. \end{cases} \tag{2.6.15}$$

We shall call a vector $\boldsymbol{\sigma}$ admissible if it is solenoidal:

$$\nabla \cdot \boldsymbol{\sigma} = 0 \,, \quad \boldsymbol{x} \in \Omega \,, \tag{2.6.16}$$

and has a given average $\langle \sigma_i \rangle$.

Defining the following new functional $\mathcal{G}(\boldsymbol{\sigma})$:

$$\mathcal{G}(\boldsymbol{\sigma}) = \frac{1}{\Omega} \iiint_\Omega K_{ij}^{-1} \sigma_i \sigma_j d\Omega \,, \tag{2.6.17}$$

we shall now show that for any admissible $\boldsymbol{\sigma}$,

$$\boxed{\mathcal{K}_{ij}^{-1} \langle \sigma_i \rangle \langle \sigma_j \rangle = \min_{\boldsymbol{\sigma}} \mathcal{G}(\boldsymbol{\sigma})} \,. \tag{2.6.18}$$

Thus this functional gives an upper bound for $\mathcal{K}_{ij}^{-1}$, hence a lower bound of $\mathcal{K}_{ij}$; this is the dual (or second) variational principle, due to Thompson.

For the proof we need first the following inequality[1]

$$\frac{1}{2} A_{ij} \xi_i \xi_j \geq \xi_i \sigma_i - \frac{1}{2} A_{ij}^{-1} \sigma_i \sigma_j \,, \tag{2.6.19}$$

which holds for any symmetric and positive-definite matrix A_{ij} and any vectors $\boldsymbol{\xi} = (\xi_i)$ and $\boldsymbol{\sigma} = (\sigma_i)$. Moreover, the equality holds if and only if $\sigma_j = A_{ij} \xi_i$.

[1]This result is an extension of the scalar inequality: $(A\xi - \sigma)^2 = A(A\xi^2 - 2\xi\sigma + \sigma^2/A) \geq 0$ for any ξ and σ, which implies for $A \geq 0$ that $(1/2)A\xi^2 \geq \xi\sigma - (1/2A)\sigma^2$. Equality holds if and only if $\sigma = A\xi$.

To verify the above inequality, let $\xi_i = \delta_i + A_{ij}^{-1}\sigma_j$ then,

$$\begin{aligned}\frac{1}{2}A_{ij}\xi_i\xi_j &= \frac{1}{2}A_{ij}\left(\delta_i + A_{i\ell}^{-1}\sigma_\ell\right)\left(\delta_j + A_{jk}^{-1}\sigma_k\right)\\ &= \sigma_i\xi_i - \frac{1}{2}A_{ij}^{-1}\sigma_j\sigma_i + \frac{1}{2}A_{ij}\delta_j\delta_i\,.\end{aligned}$$

The last term above is positive, so the inequality (Eq. (2.6.19)) is established. Moreover, because A_{ij} is positive definite, equality holds if and only if $\delta_i = 0$ which implies $\xi_i = A_{ij}^{-1}\sigma_j$ or $\sigma_j = A_{ij}\xi_i$. Another way of stating Eq. (2.6.19) is,

$$\frac{1}{2}A_{ij}\xi_i\xi_j = \max_{\boldsymbol{\sigma}}\left(\xi_i\sigma_i - \frac{1}{2}A_{ij}^{-1}\sigma_i\sigma_j\right). \tag{2.6.20}$$

Recalling the effective conductivity formula (Eq. (2.4.9)), and applying the above inequality with $A_{ij} = K_{ij}$ and $\xi_i = \partial(W_k\xi_k)/\partial x_i$, we get:

$$\begin{aligned}\frac{1}{2}\mathcal{K}_{ij}\xi_i\xi_j &= \frac{1}{2}\left\langle K_{ij}\frac{\partial(W_k\xi_k)}{\partial x_i}\frac{\partial(W_k\xi_k)}{\partial x_j}\right\rangle\\ &\geq \left\langle\frac{\partial(W_k\xi_k)}{\partial x_i}\sigma_i\right\rangle - \frac{1}{2}\left\langle K_{ij}^{-1}\sigma_i\sigma_j\right\rangle\\ &= \left\langle\frac{\partial(w_k\xi_k)}{\partial x_i}\sigma_i\right\rangle + \langle\xi_i\sigma_i\rangle - \frac{1}{2}\left\langle K_{ij}^{-1}\sigma_i\sigma_j\right\rangle.\end{aligned} \tag{2.6.21}$$

We now choose $\boldsymbol{\sigma}$ to be an admissible vector according to Eq. (2.6.16), then the first term in the last equality is zero. Indeed by partial integration and applying Gauss' theorem the first term can be transformed to

$$\begin{aligned}&\iiint_\Omega \frac{\partial(w_k\xi_k\sigma_i)}{\partial x_i}d\Omega - \iiint_\Omega w_k\xi_k\frac{\partial\sigma_i}{\partial x_i}d\Omega\\ &= \iint_{\partial\Omega} w_k\xi_k\sigma_i n_i dS - \iiint_\Omega w_k\xi_k\nabla\cdot\boldsymbol{\sigma}d\Omega\,.\end{aligned}$$

On the right-hand side of the preceding equation, the surface integral vanishes by the periodicity of w_k and σ_k, while the second volume integral vanishes since $\boldsymbol{\sigma}$ is solenoidal (cf. Eq. (2.6.16)). Therefore

$$\frac{1}{2}\mathcal{K}_{ij}\xi_i\xi_j \geq \langle\xi_i\sigma_i\rangle - \frac{1}{2}\left\langle K_{kl}^{-1}\sigma_k\sigma_l\right\rangle, \tag{2.6.22}$$

where $\boldsymbol{\sigma}$ ranges over all Ω-periodic, divergence-free functions. Equality holds if and only if

$$\xi_i = K_{ij}^{-1}\sigma_j\,, \tag{2.6.23}$$

i.e., if

$$\sigma_i = K_{ij}\frac{\partial(W_k\xi_k)}{\partial x_j}\,. \tag{2.6.24}$$

Thus the maximum is attained for $\boldsymbol{\sigma}$ with the average

$$\langle\sigma_i\rangle = \left\langle K_{ij}\frac{\partial(W_k\xi_k)}{\partial x_j}\right\rangle = \left\langle K_{ij}\frac{\partial W_k}{\partial x_j}\right\rangle\xi_k = \mathcal{K}_{ik}\xi_k\,, \tag{2.6.25}$$

where the definition (Eq. (2.2.29)) was used.

Equation (2.6.22) can be written as

$$\begin{aligned}\frac{1}{2}\mathcal{K}_{ij}\xi_i\xi_j &= \max_{\boldsymbol{\sigma}}\left\{\langle\xi_i\sigma_i\rangle - \frac{1}{2}\left\langle K_{ij}^{-1}\sigma_i\sigma_j\right\rangle\right\}\\ &= \max_{\boldsymbol{\sigma}}\left\{\xi_i\langle\sigma_i\rangle - \frac{1}{2}\left\langle K_{ij}^{-1}\sigma_i\sigma_j\right\rangle\right\}.\end{aligned} \tag{2.6.26}$$

In view of Eq. (2.6.25), Eq. (2.6.26) becomes

$$\frac{1}{2}\mathcal{K}_{ij}\xi_i\xi_j = \mathcal{K}_{ij}\xi_i\xi_j - \min_{\boldsymbol{\sigma}}\frac{1}{2}\left\langle K_{kl}^{-1}\sigma_k\sigma_l\right\rangle. \tag{2.6.27}$$

Thus

$$\mathcal{K}_{ij}\xi_i\xi_j = \min_{\boldsymbol{\sigma}}\left\langle K_{kl}^{-1}\sigma_k\sigma_l\right\rangle = \min_{\boldsymbol{\sigma}}\mathcal{G}(\boldsymbol{\sigma})\,. \tag{2.6.28}$$

Using Eq. (2.6.25) to substitute $\xi_i = \mathcal{K}_{ij}{}^{-1}\langle\sigma_j\rangle$ we obtain the dual variational principle (Eq. (2.6.18)).

As an application, we choose a constant vector $\sigma_i = \langle\sigma_i\rangle$ in the dual variational principle and get

$$\mathcal{K}_{ij}{}^{-1}\langle\sigma_i\rangle\langle\sigma_j\rangle \le \frac{1}{\Omega}\left(\iiint_\Omega K_{ij}^{-1}d\Omega\right)\langle\sigma_i\rangle\langle\sigma_j\rangle\,. \tag{2.6.29}$$

It follows that

$$\mathcal{K}_{ij}\langle\sigma_i\rangle\langle\sigma_j\rangle \ge \left(\frac{1}{\Omega}\left(\iiint_\Omega K_{ij}^{-1}d\Omega\right)\right)^{-1}\langle\sigma_i\rangle\langle\sigma_j\rangle\,. \tag{2.6.30}$$

Thus the bilinear form of the harmonic mean $1/\langle K_{ij}^{-1}\rangle$ is a lower bound of the bilinear form of $\mathcal{K}_{ij}$.

In the special case where both component materials are isotropic, i.e., $K_{ij} = K_1\delta_{ij}$ in material 1 and $K_{ij} = K_2\delta_{ij}$ in material 2, Eq. (2.6.30) reduces to

$$\mathcal{K}_{ij}\langle\sigma_i\rangle\langle\sigma_j\rangle \geq \left(\frac{\theta_1}{K_1} + \frac{\theta_2}{K_2}\right)^{-1} \langle\sigma_i\rangle\langle\sigma_j\rangle\,, \tag{2.6.31}$$

where the volume fractions of the two materials were denoted by θ_1 and θ_2, respectively. Now by choosing $\langle\boldsymbol{\sigma}\rangle$ to be any eigenvector of $(\mathcal{K}_{ij})$, we get from Eq. (2.6.13) that

$$\lambda_{(n)} \geq \left(\frac{\theta_1}{K_1} + \frac{\theta_2}{K_2}\right)^{-1}, \quad n = 1, 2, 3\,. \tag{2.6.32}$$

Thus the harmonic average is a lower bound for the smallest of the eigenvalues $\lambda_{(n)}$.

In summary, we have from Eqs. (2.6.14) and (2.6.32),

$$\left(\frac{\theta_1}{K_1} + \frac{\theta_2}{K_2}\right)^{-1} \leq \lambda_{(n)} \leq \theta_1 K_1 + \theta_2 K_2\,, \quad n = 1, 2, 3\,. \tag{2.6.33}$$

This means that the eigenvalues of the effective conductivity tensor are bounded below by the harmonic average of the two conductivities and above by their arithmetic (or volume) average.

2.7. Hashin–Shtrikman Bounds

2.7.1. *Results and Implications*

In addition to the theory of Sec. 2.6, there are the tighter Hashin–Shtrikman bounds (Hashin and Shtrikman, 1962), which are the best that can be obtained without requiring more information about the microstructure than simply the volume fractions. We first state the bounds and explain their implications before presenting the intricate mathematical derivation.

For simplicity we limit our discussions to two homogeneous and isotropic phases, $(K_1)_{ij} = K_1\delta_{ij}, (K_2)_{ij} = K_2\delta_{ij}$, with $K_2 > K_1$. The Hashin–Shtrikman lower bound for the eigenvalues λ_i of $\mathcal{K}_{ij}$ is

$$\boxed{\frac{1}{\lambda_1 - K_1} + \frac{1}{\lambda_2 - K_1} + \frac{1}{\lambda_3 - K_1} \leq \frac{2}{m - K_1} + \frac{1}{h - K_1}}\,, \tag{2.7.1}$$

where m and h are, respectively, the arithmetic and harmonic means of the two conductivities

$$m = \theta_1 K_1 + \theta_2 K_2\,, \quad h = \left(\frac{\theta_1}{K_1} + \frac{\theta_2}{K_2}\right)^{-1}\,. \tag{2.7.2}$$

On the other hand the Hashin–Shtrikman upper bound for the eigenvalues λ_i is

$$\boxed{\frac{1}{K_2-\lambda_1}+\frac{1}{K_2-\lambda_2}+\frac{1}{K_2-\lambda_3}\le\frac{2}{K_2-m}+\frac{1}{K_2-h}}\,. \tag{2.7.3}$$

For a two-dimensional material the lower and upper bounds are, respectively,

$$\boxed{\frac{1}{\lambda_1-K_1}+\frac{1}{\lambda_2-K_1}\le\frac{1}{m-K_1}+\frac{1}{h-K_1}}\,, \tag{2.7.4}$$

$$\boxed{\frac{1}{K_2-\lambda_1}+\frac{1}{K_2-\lambda_2}\le\frac{1}{K_2-m}+\frac{1}{K_2-h}}\,. \tag{2.7.5}$$

For a two-dimensional material the elementary bounds (Eq. (2.6.33)) and the Hashin–Shtrikman bounds (Eqs. (2.7.4) and (2.7.5)) are plotted for comparison in Fig. 2.3. The elementary bounds are represented by the square since each eigenvalue λ_1, λ_2 has to be in the interval $[h, m]$. From the computations in Sec. 2.5 we know that the upper left corner and lower right corner of the square in Fig. 2.3 are attainable by laminates, since for laminates one eigenvalue equals m and the other h. It can be seen that the

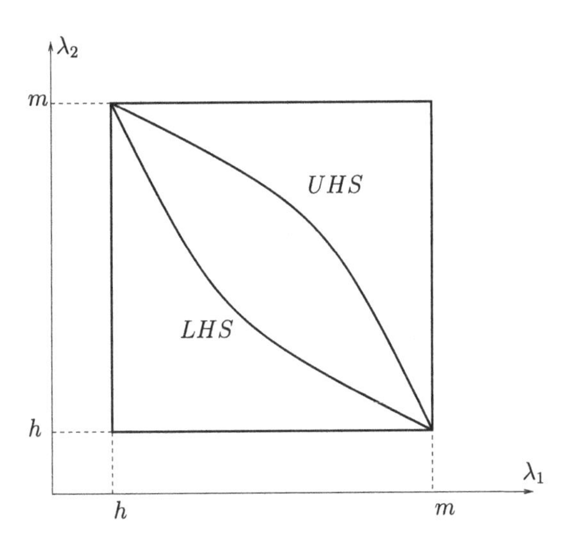

Fig. 2.3. The upper (UHS) and lower (LHS) Hashin–Shtrikman bounds for two-dimensional media, where m and h are defined in Eq. (2.7.2). Elementary bounds correspond to the sides of the square.

Hashin–Shtrikman bounds determine a *lens-like* region determined by the curves

$$\frac{1}{\lambda_1 - K_1} + \frac{1}{\lambda_2 - K_1} = \frac{1}{m - K_1} + \frac{1}{h - K_1}, \tag{2.7.6}$$

$$\frac{1}{K_2 - \lambda_1} + \frac{1}{K_2 - \lambda_2} = \frac{1}{K_2 - m} + \frac{1}{K_2 - h}. \tag{2.7.7}$$

This region is strictly included inside the square and hence the bounds are closer to each other than on the elementary bounds. Given two materials with conductivities K_1 and K_2 and volume fractions θ_1 and θ_2, we cannot obtain any composite that has the eigenvalues of the effective conductivity outside the lens-like region in Fig. 2.3, whatever the microscale geometry (particles, fibers, laminates, etc.) is.

2.7.2. Derivation of Hashin–Shtrikman Bounds

We now derive the *lower* Hashin–Shtrikman bound stated in Eq. (2.7.1).

Starting from the variational principle (Eq. (2.6.2))

$$\mathcal{K}_{ij}\xi_i\xi_j = \min_{\varphi} \mathcal{F}(\varphi) \tag{2.7.8}$$

with $\mathcal{F}$ given by Eq. (2.6.1), we add and subtract the bilinear product corresponding to a homogeneous material of conductivity K_0, with $K_0 < K_1 < K_2$,

$$\mathcal{K}_{ij}\xi_i\xi_j = \min_{\varphi} \frac{1}{\Omega}\left(\iiint_{\Omega} (K(\boldsymbol{x}) - K_0)(\nabla\varphi + \boldsymbol{\xi})^2 d\Omega \right.$$
$$\left. + \iiint_{\Omega} K_0(\nabla\varphi + \boldsymbol{\xi})^2 d\Omega \right). \tag{2.7.9}$$

In Eq. (2.6.20) let A_{ij} be replaced by the scalar $(K(\boldsymbol{x}) - K_0)$ and $\boldsymbol{\xi}$ be replaced by $\nabla\varphi + \boldsymbol{\xi}$. We then have

$$(K(\boldsymbol{x}) - K_0)(\nabla\varphi + \boldsymbol{\xi})^2 = \max_{\boldsymbol{\sigma}} \left(2\boldsymbol{\sigma} \cdot (\nabla\varphi + \boldsymbol{\xi}) - (K(\boldsymbol{x}) - K_0)^{-1}\boldsymbol{\sigma}^2\right), \tag{2.7.10}$$

which can be substituted in Eq. (2.7.9) to give

$$\mathcal{K}_{ij}\xi_i\xi_j = \min_{\varphi} \max_{\boldsymbol{\sigma}} \frac{1}{\Omega} \iiint_{\Omega} (2\boldsymbol{\sigma} \cdot (\nabla\varphi + \boldsymbol{\xi})$$
$$- (K(\boldsymbol{x}) - K_0)^{-1}\boldsymbol{\sigma}^2 + K_0(\nabla\varphi + \boldsymbol{\xi})^2) d\Omega$$

$$\geq \max_{\boldsymbol{\sigma}} \frac{1}{\Omega} \left\{ \min_{\varphi} \iiint_{\Omega} (2\boldsymbol{\sigma} \cdot \nabla\varphi + K_0(\nabla\varphi)^2) d\Omega \right.$$
$$\left. + \iiint_{\Omega} \left(2\boldsymbol{\sigma} \cdot \boldsymbol{\xi} - (K(\boldsymbol{x}) - K_0)^{-1}\boldsymbol{\sigma}^2 + K_0\boldsymbol{\xi}^2\right) d\Omega \right\}. \quad (2.7.11)$$

We note here that in general when switching min and max only an inequality is true.[2]

Let us first find the minimum of the first line above

$$\mathcal{H}(\varphi) = \iiint_{\Omega} (2\boldsymbol{\sigma} \cdot \nabla\varphi + K_0(\nabla\varphi)^2) d\Omega\,, \quad (2.7.12)$$

which, after using Gauss' theorem, can also be written as

$$\mathcal{H}(\varphi) = \iiint_{\Omega} (-2\varphi\nabla \cdot \boldsymbol{\sigma} + K_0(\nabla\varphi)^2) d\Omega\,. \quad (2.7.13)$$

The first variation of the integral in Eq. (2.7.12) is

$$\begin{aligned}
\delta\mathcal{H} &= 2 \iiint_{\Omega} (\boldsymbol{\sigma} \cdot \nabla\delta\varphi + K_0\nabla\delta\varphi \cdot \nabla\varphi) d\Omega \\
&= 2 \iiint_{\Omega} (\nabla \cdot (\boldsymbol{\sigma}\delta\varphi) - \delta\varphi\nabla \cdot \boldsymbol{\sigma} + \nabla \cdot (K_0\delta\varphi\nabla\varphi) - \delta\varphi K_0\nabla^2\varphi) d\Omega \\
&= 2 \iiint_{\Omega} (\nabla \cdot (\boldsymbol{\sigma}\delta\varphi) + \nabla \cdot (K_0\delta\varphi\nabla\varphi)) d\Omega \\
&\quad - 2 \iiint_{\Omega} \delta\varphi(\nabla \cdot \boldsymbol{\sigma} + K_0\nabla^2\varphi) d\Omega\,. \quad (2.7.14)
\end{aligned}$$

In the last equality of Eq. (2.7.14) the first integral vanishes after using Gauss' theorem and the periodicity conditions, while the second integral vanishes for any $\delta\varphi$ if and only if $\varphi = \psi$, where ψ is the solution to the Poisson problem

$$-K_0\nabla^2\psi = \nabla \cdot \boldsymbol{\sigma}\,, \quad (2.7.15)$$
$$\psi \text{ is } \Omega\text{-periodic}\,. \quad (2.7.16)$$

[2]Indeed for a function of two variables we have $f(x,y) \geq \min_x f(x,y)$ which implies $\max_y f(x,y) \geq \max_y \min_x f(x,y)$ and therefore $\min_x \max_y f(x,y) \geq \max_y \min_x f(x,y)$. In particular, in this case however one could get equality since the function in Eq. (2.7.11) is convex in φ and concave in $\boldsymbol{\sigma}$.

To get the value of the minimum in Eq. (2.7.12) we take $\varphi = \psi$, and note that from Eq. (2.7.15)

$$-\iiint_\Omega \psi \nabla \cdot \boldsymbol{\sigma} d\Omega = \iiint_\Omega K_0 \psi \nabla^2 \psi d\Omega$$
$$= \iiint_\Omega (K_0 \nabla \cdot (\psi \nabla \psi) - K_0 (\nabla \psi)^2) d\Omega$$
$$= -\iiint_\Omega K_0 (\nabla \psi)^2 d\Omega .$$

The minimum is obtained from Eq. (2.7.12)

$$\mathcal{H}(\psi) = \iiint_\Omega (2\boldsymbol{\sigma} \cdot \nabla \psi + \psi \nabla \cdot \boldsymbol{\sigma}) d\Omega = \iiint_\Omega (2\boldsymbol{\sigma} \cdot \nabla \psi - \boldsymbol{\sigma} \cdot \nabla \psi) d\Omega$$
$$= \iiint_\Omega \nabla \psi \cdot \boldsymbol{\sigma} d\Omega .$$

Since the solution ψ of problems (Eqs. (2.7.15) and (2.7.16)) depends on the vector field $\boldsymbol{\sigma}$ it will be denoted as $\psi(\boldsymbol{\sigma})$. Thus Eq. (2.7.11) becomes

$$(\mathcal{K}_{ij} - K_0 \delta_{ij}) \xi_i \xi_j$$
$$\geq \max_{\boldsymbol{\sigma}} \frac{1}{\Omega} \iiint_\Omega (-(K(\boldsymbol{x}) - K_0)^{-1} \boldsymbol{\sigma}^2 + 2\boldsymbol{\sigma} \cdot \xi + \nabla \psi(\boldsymbol{\sigma}) \cdot \boldsymbol{\sigma}) d\Omega . \tag{2.7.17}$$

We now define a piecewise constant vector $\boldsymbol{\sigma} = \chi_2 \boldsymbol{\eta}$ in terms of the characteristic function of material 2,

$$\chi_2 = \begin{cases} 0 & \text{in } \Omega_1 , \\ 1 & \text{in } \Omega_2 . \end{cases}$$

Thus

$$(\mathcal{K}_{ij} - K_0 \delta_{ij}) \xi_i \xi_j \geq \max_{\boldsymbol{\eta}} \Bigg(- \theta_2 (K_2 - K_0)^{-1} \boldsymbol{\eta} \cdot \boldsymbol{\eta} + 2\theta_2 \boldsymbol{\eta} \cdot \boldsymbol{\xi}$$
$$+ \iiint_\Omega \nabla \psi(\chi_2 \boldsymbol{\eta}) \cdot (\chi_2 \boldsymbol{\eta}) d\Omega \Bigg) . \tag{2.7.18}$$

To compute the last integral, we first remark that $\psi(\chi_2 \boldsymbol{\eta})$ defined by Eq. (2.7.15) is linear with respect to $\boldsymbol{\eta}$, indeed

$$\psi(\chi_2 \boldsymbol{\eta}) = \eta_i \Psi_i \tag{2.7.19}$$

with Ψ_i being the solutions of

$$-K_0 \nabla^2 \Psi_i = \frac{\partial \chi_2}{\partial x_i} , \tag{2.7.20}$$
$$\Psi_i \text{ is } \Omega\text{-periodic} . \tag{2.7.21}$$

Next, we define ϕ to be the solution of

$$\nabla^2\phi = \chi_2 - \theta_2\,, \tag{2.7.22}$$

$$\phi \text{ is } \Omega\text{-periodic}. \tag{2.7.23}$$

Thus the integral term in Eq. (2.7.18) becomes

$$\begin{aligned}\iiint_\Omega \nabla\psi(\chi_2\boldsymbol{\eta})\cdot(\chi_2\boldsymbol{\eta})d\Omega &= \iiint_\Omega \nabla\psi(\chi_2\boldsymbol{\eta})\cdot\boldsymbol{\eta}(\nabla^2\phi+\theta_2)d\Omega \\ &= \iiint_\Omega \eta_k\frac{\partial\Psi_k}{\partial x_i}\eta_i(\nabla^2\phi+\theta_2)d\Omega\,, \end{aligned}\tag{2.7.24}$$

where the term in θ_2 is zero after using Gauss' theorem and periodicity. Denoting

$$F_{ik} = \iiint_\Omega \frac{\partial\Psi_k}{\partial x_i}\nabla^2\phi \tag{2.7.25}$$

using Gauss' formula and canceling the boundary terms by periodicity we get

$$\begin{aligned}F_{ik} &= -\iiint_\Omega \nabla^2\Psi_k\frac{\partial\phi}{\partial x_i}d\Omega \\ &= \frac{1}{K_0}\iiint_\Omega\frac{\partial\chi_2}{\partial x_k}\frac{\partial\phi}{\partial x_i}d\Omega = -\frac{1}{K_0}\iiint_\Omega \chi_2\frac{\partial^2\phi}{\partial x_i\partial x_k}d\Omega\,.\end{aligned}\tag{2.7.26}$$

The above formula shows that (F_{ik}) is symmetric and has the trace

$$F_{ii} = -\frac{1}{K_0}\iiint_\Omega \chi_2(\chi_2-\theta_2)d\Omega = -\frac{1}{K_0}\theta_2(1-\theta_2) = -\frac{1}{K_0}\theta_1\theta_2\,. \tag{2.7.27}$$

The inequality (Eq. (2.7.18)) becomes

$$\begin{aligned}(\mathcal{K}_{ij}-K_0\delta_{ij})\xi_i\xi_j &\geq \max_{\boldsymbol{\eta}}\left(-\left(\frac{\theta_2\delta_{ij}}{K_2-K_0}-F_{ij}\right)\eta_i\eta_j + 2\theta_2\eta_i\xi_i\right) \\ &= \frac{\theta_2^2\xi_i\xi_j}{\theta_2(K_2-K_0)^{-1}\delta_{ij}-F_{ij}}\,,\end{aligned}\tag{2.7.28}$$

which is an inequality between symmetric matrices. Thus

$$(\mathcal{K}_{ij}-K_0\delta_{ij})^{-1} \leq \theta_2^{-2}(\theta_2(K_2-K_0)^{-1}\delta_{ij}-F_{ij})\,. \tag{2.7.29}$$

The trace of the left-hand side of the above equation, when $K_0 = K_1$, is in three dimensions

$$\frac{1}{\lambda_1-K_1}+\frac{1}{\lambda_2-K_1}+\frac{1}{\lambda_3-K_1}\,. \tag{2.7.30}$$

Computing the trace of the right-hand side and using the trace of F obtained above, we get

$$\frac{1}{\lambda_1 - K_1} + \frac{1}{\lambda_2 - K_1} + \frac{1}{\lambda_3 - K_1} \leq \frac{2}{m - K_1} + \frac{1}{h - K_1} \tag{2.7.31}$$

with λ_1, λ_2, and λ_3 denoting the eigenvalues of $\mathcal{K}_{ij}$. This is the lower bound of Hashin–Shtrikman.

Similarly, in two dimensions we get

$$\frac{1}{\lambda_1 - K_1} + \frac{1}{\lambda_2 - K_1} \leq \frac{1}{m - K_1} + \frac{1}{h - K_1}. \tag{2.7.32}$$

The *upper* Hashin–Shtrikman bound can be obtained by a similar procedure. Starting from Eq. (2.7.8), we add and subtract a bilinear product corresponding to a homogeneous material of conductivity K_0, with $K_1 < K_2 < K_0$. The result of the upper bound is Eq. (2.7.3) in three dimensions, and Eq. (2.7.5) in two dimensions.

2.8. Other Approximate Results for Dilute Inclusions

We mention without giving details a number of known approximate analytical theories for composites with dilute inclusions.

Maxwell (1881) was the first to consider the heat conduction through a composite made of a cubic array of spherical particles of conductivity K_2 embedded in a matrix of conductivity K_1. Approximating the array of spheres by a dipole distribution he derived an expression for the conductivity of a dilute distribution of particles, correct to order $\mathcal{O}(\theta_2)$ where θ_2 is the volume fraction of the particles. Rayleigh (1892) replaced the spherical particles in a simple cubic array by higher-order multipoles and obtained higher-order corrections

$$\frac{\mathcal{K}}{K_1} = 1 - 3\theta_2 \left(\frac{2 + K_2/K_1}{1 - K_2/K_1} + \theta_2 - \frac{1 - K_2/K_1}{4 + 3K_2/k_1} d\theta_2^{10/3} + \mathcal{O}\left(\theta_2^{14/3}\right) \right)^{-1}, \tag{2.8.1}$$

where the correct value of $d = 1.57$ was given later by Runge (1925). The method of Rayleigh was extended by Meredith and Tobias (1960), McPhedran and McKenzie (1978), and McKenzie *et al.* (1978) to higher orders of volume fractions for cubic arrays of particles.

An alternative method which uses the periodicity of the potential field for a cubic array was introduced by Zuzovski and Brenner (1977) and simplified by Sangani and Acrivos (1983). This method avoids the convergence problems of the initial method of Rayleigh and gives the same result as in McPhedran and McKenzie (1978).

Sangani and Acrivos (1983) have determined the conductivity for simple cubic, body-centered cubic, and face-centered cubic arrays of spheres. The mathematics is quite lengthy and will not be recounted here. The basic idea is to first find a family of solutions for the cell problems (Eqs. (2.2.18)–(2.2.20)), without the periodicity condition, by matching an inner to an outer expansion of spherical harmonics. To impose the periodicity, the spherical particle is replaced by multipoles and the Fourier expansion of the fundamental solution is expressed in terms of spherical harmonics using ideas from Hasimoto (1959). By matching this expansion to the outer expansion the solution to the cell problem can be found. The effective conductivity for an isotropic composite formed by a periodic array of homogeneous and isotropic spheres in a homogeneous and isotropic matrix is found to be

$$\frac{\mathcal{K}}{K_1} = 1 - 3\theta_2\bigg(- L_1^{-1} + \theta_2 + c_1 L_2 \theta_2^{10/3} \frac{1 + c_2 L_3 c^{11/3}}{1 - c_3 L_2 c^{7/3}} + c_4 L_3 \theta_2^{14/3} + c_5 L_4 \theta_2^6 + c_6 L_5 \theta_2^{22/3} + \mathcal{O}(\theta_2^{25/3})\bigg)^{-1}, \quad (2.8.2)$$

where the L_n are related to the conductivity ratio K_2/K_1 by

$$L_n = \frac{((K_2/K_1) - 1)(2n - 1)}{(K_2/K_1) + 2n}. \quad (2.8.3)$$

The coefficients c_j's depend on constants characterizing the array as given in Eq. (30) of Sangani and Acrivos (1983) and are listed in Table 2.1 for various geometries.

It is interesting to compare the variational bounds on the effective conductivity with the numerical values calculated from Eq. (2.8.2). Figure 2.4 shows the dependence of the effective conductivity on the particle volume

Table 2.1. Coefficients $c_1, c_2, \ldots, c_6$ in Eq. (2.8.2).

	Simple Cubic	Body-Centered Cubic	Face-Centered Cubic
c_1	1.3047	1.29×10^{-1}	7.529×10^{-2}
c_2	2.305×10^{-1}	-4.1286×10^{-1}	6.9657×10^{-1}
c_3	4.054×10^{-1}	7.6421×10^{-1}	-7.41×10^{-1}
c_4	7.231×10^{-2}	2.569×10^{-1}	0.4195×10^{-1}
c_5	1.526×10^{-1}	1.13×10^{-2}	2.31×10^{-2}
c_6	1.05×10^{-2}	5.62×10^{-3}	9.14×10^{-7}

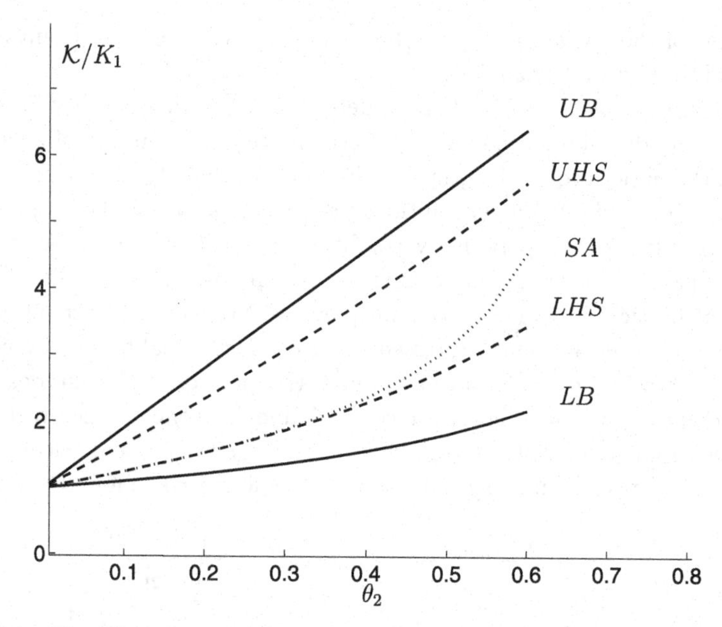

Fig. 2.4. The effective conductivity of a simple cubic array at fixed conductivity ratio ($K_2/K_1 = 10$), given by Eq. (2.8.2) by Sangani and Acrivos (SA), the classical bounds (UB, LB) and the Hashin–Shtrikman bounds (UHS, LHS). From Lipton and Vernescu (1996b), *J. Appl. Phys.*

fraction θ_2 according to Eq. (2.8.2), for a simple cubic array of spherical particles of conductivity K_2 in a matrix of conductivity K_1 with $K_2/K_1 = 10$. The conductivity increases with the volume fraction θ_2 of the better conducting material.

In Fig. 2.5 the values of the effective conductivity are shown for a fixed volume fraction, $\theta_2 = 0.3$. It can be seen that the values from Eq. (2.8.2) are very close to the Hashin–Shtrikman lower bound.

In Fig. 2.6 the lower elementary bound and the lower Hashin–Shtrikman bound are seen as an increasing function of the conductivity K_2 for a fixed volume fraction.

2.9. Thermal Resistance at the Interface

In the previous sections we have considered a model of heat conduction with perfect contact at the material interfaces. Experiments for some composites have shown that imperfect contact leads to an interfacial thermal resistance which can affect the effective conductivity of the composite significantly.

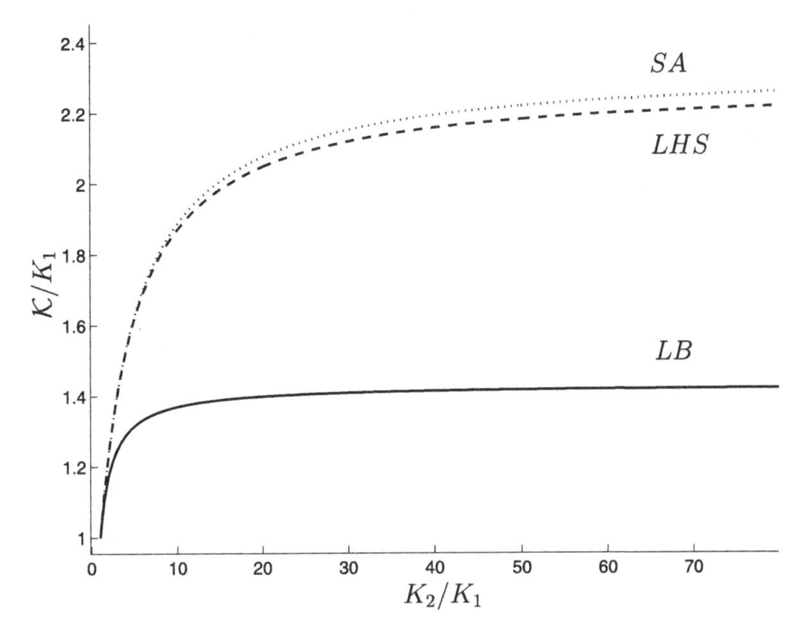

Fig. 2.5. The Sangani and Acrivos (1982) approximation (SA) for the effective conductivity of a simple cubic lattice composite at fixed volume fraction ($\theta_2 = 0.3$), compared to the lower classical bounds (LB) and the lower Hashin–Shtrikman bounds (LHS). From Lipton and Vernescu (1996b), *J. Appl. Phys.*

Thermal resistance can also be due to mismatch of thermal expansion coefficients or the presence of a thin coating. At low temperatures the interfacial thermal resistance is also due to an increased scattering mechanism of the phonons at the boundary which separates one phase from another. Kapitza (1941) was the first to observe a temperature discontinuity at metal–liquid interfaces. Hence the interfacial thermal resistance is sometimes referred to as *Kapitza resistance.* More recently, the presence of oxides as in metal powder–gas systems has been found to give rise to thermal resistance (Swift, 1966). By oxidizing uranium, zirconium, and various metal powders the effective thermal conductivity was decreased. In silicon nitride matrix–silicon carbide fibers composites, Bhatt *et al.* (1990) observed that the thermal conductivity of the composite was significantly influenced by the oxidation treatment of the carbon-rich surface of the fibers. In their experiments the oxidation decreased the transverse conductivity by a factor as high as 2. Garret and Rosenberg (1974) have studied the conductivity of epoxy–resin/powder composite materials and have observed that, at liquid helium temperatures, composites of corundum particles in

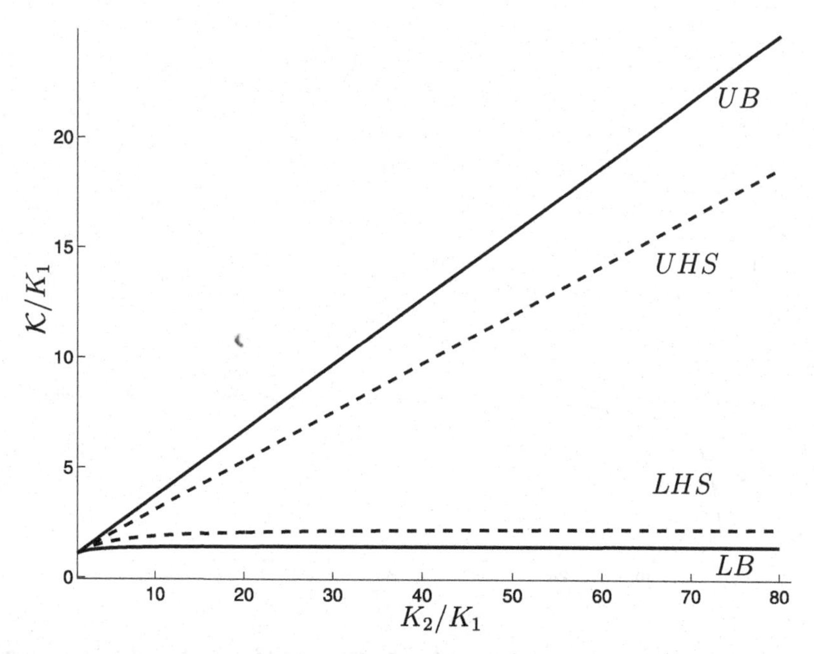

Fig. 2.6. The elementary upper and lower bounds (UB, LB) and the upper (UHS) and lower (LHS) Hashin–Shtrikman bounds for fixed volume fractions ($\theta_2 = 0.3$) as functions of the conductivity K_2/K_1. From Lipton and Vernescu (1996b), *J. Appl. Phys.*

an epoxy matrix have thermal conductivities below the conductivity of the pure epoxy (although the particles' conductivity was higher than that of the matrix). In these cases an increase in the volume fraction of corundum particles reduces the conductivity of the composite. Moreover, at constant volume fractions the conductivity decreases with particle size. A similar behavior was observed in zinc-sulfide/diamond composites. Every *et al.* (1992) observed that the thermal conductivity of zinc-sulfide is increased by adding large particles of diamond, but lowered by the addition of sub-micron size particles. For this type of problems the homogenization theory has been developed by Auriault and Ene (1984). Related references on imperfect contact include Benveniste (1986), Chiew and Glandt (1987), Hasselman and Johnson (1987), Dunn and Taya (1993), and Hasselman, Donaldson and Thomas (1993).

In Secs. 2.9–2.11 we shall extend the theory in Sec. 2.2 by introducing a model which takes into account the interfacial thermal resistance. Physical effects and comparison of theoretical and numerical results with experimental data will be discussed. On the other hand, bounds on the effective conductivity together with measurements of the effective thermal conductivity of composites can provide a simple method of estimating the interfacial thermal resistance.

In terms of the same symbols defined in previous sections, the conduction equation (2.1.2) as well as the flux continuity condition (Eq. (2.1.4)) still apply. However, temperature is no longer continuous and Eq. (2.1.3) is replaced by

$$-\beta(T_1 - T_2) = (K_1)_{ij}\frac{\partial T_1}{\partial x_i}n_j\,, \tag{2.9.1}$$

where β^{-1} is the coefficient of interfacial thermal resistance. This condition expresses the fact that there is a temperature jump across the interface which is proportional to the heat flux, and can be justified by a model with a thin insulating layer of thickness ℓ_3 and conductivity K_3, sandwiched between two thick layers of thicknesses ℓ_1, ℓ_2 and conductivities K_1, K_2. Since the diffusion times in these layers are, respectively, $T_1 = \mathcal{O}(\ell_1^2/K_1), T_3 = \mathcal{O}(\ell_3^2/K_3)$, and $T_2 = \mathcal{O}(\ell_2^2/K_2)$, the ratios

$$\frac{T_3}{T_1} = \mathcal{O}\left(\frac{\ell_3^2}{\ell_1^2}\frac{K_1}{K_3}\right) \ll 1\,, \quad \frac{T_3}{T_2} = \mathcal{O}\left(\frac{\ell_3^2}{\ell_2^2}\frac{K_2}{K_3}\right) \ll 1$$

are very small. Temperature in the insulator is practically at steady state and varies linearly with the transverse coordinate, i.e.,

$$\frac{\partial T_3}{\partial n} \cong \frac{T_2 - T_1}{\ell_3}\,.$$

Continuity of heat flux at the interface between layer 1 and layer 3 requires that

$$K_1\frac{\partial T_1}{\partial n} = \frac{K_3}{\ell_3}(T_2 - T_1)\,,$$

which is Eq. (2.9.1) with $\beta = K_3/\ell_3$.

As in Sec. 2.2 we consider a spatially periodic microstructure with the periodicity cell Ω of typical dimension ℓ, which is much smaller than the typical dimension ℓ' of the domain occupied by the composite. Introducing the same normalization as in Eqs. (2.2.2) and (2.2.6) takes the following dimensionless form

$$-\frac{\beta l}{K}(T_1^\dagger - T_2^\dagger) = (K_1^\dagger)_{ij}\frac{\partial T_1^\dagger}{\partial x_i^\dagger}n_j\,. \tag{2.9.2}$$

The dimensionless ratio

$$\beta^\dagger = \frac{\beta l}{K} = \mathcal{O}(1) \tag{2.9.3}$$

is a measure of thermal resistance; the larger the resistance, the smaller the ratio.

The multiple-scale analysis proceeds as before. The only difference is associated with the above jump condition (Eq. (2.9.1)) which gives at different orders of ϵ

$$-\beta(T_1^{(0)} - T_2^{(0)}) = (K_1)_{ij}\frac{\partial T_1^{(0)}}{\partial x_j}n_i\,, \quad \boldsymbol{x} \in \Gamma\,, \tag{2.9.4}$$

$$-\beta(T_1^{(1)} - T_2^{(1)}) = (K_1)_{ij}\left(\frac{\partial T_1^{(0)}}{\partial x'_j} + \frac{\partial T_1^{(1)}}{\partial x_j}\right)n_i\,, \quad \boldsymbol{x} \in \Gamma\,, \tag{2.9.5}$$

$$-\beta(T_1^{(2)} - T_2^{(2)}) = (K_1)_{ij}\left(\frac{\partial T_1^{(1)}}{\partial x'_j} + \frac{\partial T_1^{(2)}}{\partial x_j}\right)n_i\,, \quad \boldsymbol{x} \in \Gamma\,. \tag{2.9.6}$$

The leading-order solution is still

$$T_1^{(0)} = T_2^{(0)} = T^{(0)}(\boldsymbol{x}', t)\,. \tag{2.9.7}$$

The $\mathcal{O}(\epsilon)$ problem is linear and the unknown $T^{(1)}$ can be expressed as

$$T_\alpha^{(1)} = w_j\frac{\partial T^{(0)}}{\partial x'_j}\,, \tag{2.9.8}$$

with the vector $\boldsymbol{w} = (w_l(x_j))$ satisfying the following cell problem

$$\frac{\partial}{\partial x_i}\left((K_\alpha)_{ij}\left(\delta_{jl} + \frac{\partial w_{\alpha_l}}{\partial x_j}\right)\right) = 0\,, \quad x_i \in \Omega_\alpha\,, \tag{2.9.9}$$

$$-\beta(w_{1_l} - w_{2_l}) = (K_1)_{ij}\left(\delta_{jl} + \frac{\partial w_{1_l}}{\partial x_j}\right)n_i\,, \quad x_i \in \Gamma\,, \tag{2.9.10}$$

$$(K_1)_{ij}\left(\delta_{jl} + \frac{\partial w_{1_l}}{\partial x_j}\right)n_i = (K_2)_{ij}\left(\delta_{jl} + \frac{\partial w_{2_l}}{\partial x_j}\right)n_i\,, \quad x_i \in \Gamma\,, \tag{2.9.11}$$

$$w_\alpha \text{ is } \Omega\text{-periodic}\,. \tag{2.9.12}$$

From these conditions the vector $\boldsymbol{w}$ is determined uniquely up to an additive constant.

At the next order $\mathcal{O}(\epsilon^2)$ the problem for $T_\alpha^{(2)}$ is governed by Eqs. (2.2.24), (2.2.26), (2.2.27), and (2.9.6). Upon integrating Eq. (2.2.24) over the volumes Ω_α, $\alpha = 1, 2$ and using the divergence theorem and the boundary conditions, we obtain the effective equation governing the macroscale behavior. The results are of the same form as Eqs. (2.2.31) and (2.2.29). However, the interfacial thermal resistance is present in the effective conductivity via the solution for $\boldsymbol{w}$ of the cell problems (Eqs. (2.9.9)–(2.9.12)).

By following the procedure in Sec. 2.3, the symmetry and positive-definiteness of the effective conductivity tensor $(\mathcal{K}_{ij})$ can be proven. In particular the reader can verify readily that Eq. (2.4.5) is changed to

$$\iiint_{\Omega_1} (K_1)_{ij} \frac{\partial W_l}{\partial x_j} \frac{\partial \Phi}{\partial x_i} d\Omega + \iiint_{\Omega_2} (K_2)_{ij} \frac{\partial W_l}{\partial x_j} \frac{\partial \Phi}{\partial x_i} d\Omega = -\beta \iint_\Gamma [w_l][\Phi] dS\,, \tag{2.9.13}$$

where the square brackets denote the jump (discontinuity) across the interface Γ, i.e.,

$$[w] \equiv (w_1)_\Gamma - (w_2)_\Gamma\,. \tag{2.9.14}$$

The effective conductivity defined in Eq. (2.2.29) can be written as

$$\mathcal{K}_{ij} = \left\langle K_{il} \frac{\partial W_j}{\partial x_l} \right\rangle = \left\langle K_{kl} \frac{\partial W_j}{\partial x_l} \frac{\partial x_i}{\partial x_k} \right\rangle = \left\langle K_{kl} \frac{\partial W_j}{\partial x_l} \frac{\partial (W_i - w_i)}{\partial x_k} \right\rangle. \tag{2.9.15}$$

After expanding the above average and using Eq. (2.9.13) for $\Phi = w_i$ we get

$$\mathcal{K}_{ij} = \left\langle K_{kl} \frac{\partial W_i}{\partial x_k} \frac{\partial W_j}{\partial x_l} \right\rangle + \frac{\beta}{\Omega} \iint_\Gamma [w_i][w_j] dS\,. \tag{2.9.16}$$

Since $W_i = w_i + x_i$, another equivalent form can be written by noting that $[w_j] = [W_j]$,

$$\mathcal{K}_{ij} = \left\langle K_{kl} \frac{\partial W_j}{\partial x_l} \frac{\partial W_i}{\partial x_k} \right\rangle + \frac{\beta}{\Omega} \iint_\Gamma [W_j][W_i] dS\,. \tag{2.9.17}$$

From this form of $(\mathcal{K}_{ij})$, the symmetry of the effective conductivity tensor is evident.

To prove the positiveness of $(\mathcal{K}_{ij})$ we take a constant vector $\xi = (\xi_i)$ and form a bilinear product with Eq. (2.9.17):

$$\mathcal{K}_{ij}\xi_i\xi_j = \left\langle K_{kl} \frac{\partial (\boldsymbol{W} \cdot \boldsymbol{\xi})}{\partial x_l} \frac{\partial (\boldsymbol{W} \cdot \boldsymbol{\xi})}{\partial x_k} \right\rangle + \frac{\beta}{\Omega} \iint_\Gamma [\boldsymbol{W} \cdot \boldsymbol{\xi}]^2 dS\,. \tag{2.9.18}$$

The positiveness of (K_{ij}) guarantees the same for $(\mathcal{K}_{ij})$. To see if $(\mathcal{K}_{ij})$ is positive definite we first assume that $\mathcal{K}_{ij}\xi_i\xi_j = 0$, then

$$K_{kl} \frac{\partial (\boldsymbol{W} \cdot \boldsymbol{\xi})}{\partial x_l} \frac{\partial (\boldsymbol{W} \cdot \boldsymbol{\xi})}{\partial x_k} = 0 \quad \text{and} \quad [\boldsymbol{W} \cdot \boldsymbol{\xi}] = 0 \quad \text{on } \Gamma\,. \tag{2.9.19}$$

Since (K_{kl}) is positive definite it is necessary that

$$\frac{\partial(\boldsymbol{W}\cdot\boldsymbol{\xi})}{\partial x_l}=0$$

in each phase, which implies, since $W_k = w_k + x_k$, that

$$\frac{\partial(\boldsymbol{w}\cdot\boldsymbol{\xi})}{\partial x_l}+\xi_l=0\,.$$

Integrating the above equation over Ω and using $[\boldsymbol{w}\cdot\boldsymbol{\xi}]=0$ from Eq. (2.9.19) and the Ω-periodicity, the integral of the first term vanishes, hence $\xi_l=0$. It follows that $(\mathcal{K}_{ij})\xi_i\xi_j$ vanishes only when $\xi_i=0$, guaranteeing that $(\mathcal{K}_{ij})$ is positive definite.

2.10. Laminated Composites with Thermal Resistance

As a special example we consider a laminated composite by bonding alternate sheets of two materials periodically. Heat conduction in each component layer of a typical period is described by the one-dimensional equations

$$\rho_\alpha c_\alpha\frac{\partial T_\alpha}{\partial t}=K_\alpha\frac{\partial^2 T_\alpha}{\partial x^2}\,,\qquad \alpha=1,2 \tag{2.10.1}$$

in each component layer α, subject to the interface conditions

$$-\beta(T_1-T_2)=K_1\frac{\partial T_1}{\partial x}\,, \tag{2.10.2}$$

$$K_1\frac{\partial T_1}{\partial x}=K_2\frac{\partial T_2}{\partial x}\,, \tag{2.10.3}$$

where x is the coordinate transverse to the layers. T_α, ρ_α, c_α, and K_α denote, respectively, the temperature, density, specific heat, and heat conductivity in the component layer α.

2.10.1. *Effective Coefficients*

Since the problem is one-dimensional, the key task is to solve the following cell problem for the function $w_\alpha = w_\alpha(x)$:

$$\frac{\partial}{\partial x}\left(K_\alpha\left(1+\frac{\partial w_\alpha}{\partial x}\right)\right)=0\,,\quad x\in\Omega_\alpha\,,\ \alpha=1,2 \tag{2.10.4}$$

$$-\beta(w_1-w_2)=K_2\left(1+\frac{\partial w_2}{\partial x}\right),\qquad x\in\Gamma \tag{2.10.5}$$

$$K_1\left(1+\frac{\partial w_1}{\partial x}\right)=K_2\left(1+\frac{\partial w_2}{\partial x}\right), \quad x\in\Gamma. \tag{2.10.6}$$

$$w_\alpha \text{ is } x\text{-periodic.} \tag{2.10.7}$$

Afterward the macroscale diffusion equation for the composite is

$$\langle\rho c\rangle\frac{\partial T^{(0)}}{\partial t}=\mathcal{K}\frac{\partial^2 T^{(0)}}{\partial x'^2}, \tag{2.10.8}$$

where $x'=\epsilon x$ with $\epsilon=\ell/\ell'\ll 1$, and

$$\langle f\rangle=\frac{1}{\ell}\left(\iiint_{\Omega_1} f_1 dx+\iiint_{\Omega_2} f_2 dx\right). \tag{2.10.9}$$

The effective conductivity $\mathcal{K}$ is:

$$\mathcal{K}=\left\langle K\left(1+\frac{\partial w}{\partial x}\right)\right\rangle. \tag{2.10.10}$$

The solution w to the ordinary differential equation (2.2.18) can again be explicitly computed. Referring to Fig. 2.7, the boundaries of the periodic cell must lie in the same material. The interface between the two constituents within the cell has two components Γ_1 and Γ_2 and material 1 has two components Ω_{11} and Ω_{12}. Then w is a linear function in x in each domain:

$$w_1=Ax+a \quad \text{in } \Omega_{11}, \quad w_2=Bx \quad \text{in } \Omega_2, \quad w_1=Cx+c \quad \text{in } \Omega_{12}. \tag{2.10.11}$$

Imposing the boundary conditions (Eqs. (2.10.5)–(2.10.7)) we get five equations for A, B, C, a, and c,

$$K_1(A+1)=-\beta(A-B)\lambda\theta_1\ell-\beta a, \tag{2.10.12}$$

$$K_2(B+1)=\beta(C-B)(\lambda\theta_1\ell+\theta_2\ell)+\beta c, \tag{2.10.13}$$

$$K_1(A+1)=K_2(B+1)=K_1(C+1), \tag{2.10.14}$$

$$a=C\ell+c. \tag{2.10.15}$$

The system (Eqs. (2.10.12)–(2.10.15)) yields

$$\begin{aligned} A=C&=\frac{(K_2-K_1)\,\theta_2-2(K_1K_2/\beta\ell)}{K_1\theta_2+K_2\theta_1+2(K_1K_2/\beta\ell)},\\ B&=-\frac{(K_2-K_1)\,\theta_1+2(K_1K_2/\beta\ell)}{K_1\theta_2+K_2\theta_1+2(K_1K_2/\beta\ell)}. \end{aligned} \tag{2.10.16}$$

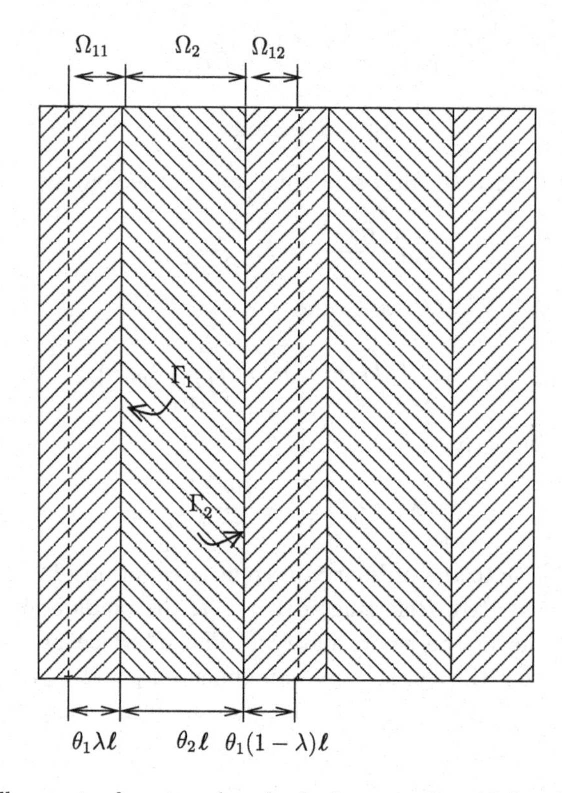

Fig. 2.7. Cell geometry for a two-phase laminate composite with interface resistance.

Thus the effective conductivity (Eq. (2.10.10)) is

$$\mathcal{K} = \theta_1 K_1(A+1) + \theta_2 K_2(B+1) = \left(\frac{\theta_1}{K_1} + \frac{\theta_2}{K_2} + \frac{2}{\beta\ell}\right)^{-1}. \qquad (2.10.17)$$

Due to the presence of the factor $\beta\ell$ the effect of interface resistance is greater for thinner layers (smaller ℓ). In particular for sufficiently small ℓ, $\mathcal{K}$ can be smaller than both K_1 and K_2. Thus a composite made of very thin layers can be a better insulator than a layer of pure material. This interesting property called the *size effect* is at the base of modern nanotechnology.

The theoretical result above can also be derived from a model where the thermal resistance is the consequence of another thin layer with low conductivity. Let the nominal thicknesses of the two coatings in a bilayer be ℓ_1 and ℓ_2 with $\ell_1 + \ell_2 = \ell$. Let there be a very thin layer with very low conductivity (i.e., $\ell_3 \ll \ell_1, \ell_2$, $K_3 \ll K_1, K_2$). From the well-known result for a periodic laminate composed of three homogeneous layers in perfect thermal contact, the effective conductivity $\mathcal{K}$ is the *harmonic* average of the component conductivities,

$$\begin{aligned}\frac{1}{\mathcal{K}} &= \frac{(\ell_1 - (1/2)\ell_3)/\ell}{K_1} + \frac{(\ell_2 - (1/2)\ell_3)/\ell}{K_2} + \frac{2\ell_3/\ell}{K_3} \\ &= \frac{\ell_1/\ell}{K_1} + \frac{\ell_2/\ell}{K_2} + \frac{\ell_3}{\ell}\left(\frac{2}{K_3} - \frac{1}{2K_1} - \frac{1}{2K_2}\right). \end{aligned} \tag{2.10.18}$$

Assume $\ell_3, K_3 \to 0$, but

$$\frac{\ell_3}{K_3} \to \frac{1}{\beta}. \tag{2.10.19}$$

We define β to be the thermal resistance, thus

$$\frac{1}{\mathcal{K}} = \frac{\ell_1/\ell}{K_1} + \frac{\ell_2/\ell}{K_2} + \frac{2}{\beta\ell}, \tag{2.10.20}$$

which confirms Eq. (2.10.17).

We now describe an application.

2.10.2. *Application to Thermal Barrier Coatings*

A laminate made of multilayer coatings can be used as thermal barriers, since interface contact is often imperfect, due either to manufacturing difficulties or oxidation of materials, resulting in thermal resistance, which can lower the thermal conductivity of the laminate below the thermal conductivities of the constituent materials. After the laminate is formed, it is necessary to perform experiments to help determine the effective conductivity. Originally designed for a layer of pure material (Parker *et al.*, 1961), one method requires that one face (say $x = 0$) of the material be heated up quickly initially. Then the transient temperature $T(a, t)$ on the opposite face at $x = a$ is measured, while both surfaces are well insulated for all $t > 0$. Let us define $t_{1/2}$ to be the time required for temperature to reach half of its maximum, $T(a, t_{1/2}) = T_{\max}/2$. By equating the predicted and the measured $t_{1/2}$, one can determine the laminate conductivity.

For thermal barriers made of many thin layers of metal coatings Josell *et al.* (1997a, b, 1998) performed heat-pulse experiments and developed a theory to determine the effective thermal properties. In one of these experiments the composite was made of ten periodic bilayers of iron/copper coatings of known thicknesses and thermal properties. In order to avoid vibrations induced by the initial heat pulse, the composite was deposited on a thick molybdenum substrate. Initially the outer face of the substrate was quickly heated. Temperature transient on the face of the laminate was recorded. From the recorded temperature at $t = t_{1/2}$, the effective heat conductivity was obtained. Since the conductivities and thicknesses of iron and copper coating can be measured, finding the effective conductivity $\mathcal{K}$ is equivalent to finding the resistance coefficient β.

The theory by Josell *et al.* (1998) was based on the solution of coupled heat conduction problem for 2N layers. The computational task is relatively laborious. On the other hand, the effective conductivity of the multilayer system can be quickly obtained by the theory in the previous section. The combination with the substrate is then reduced to an initial-boundary-value problem for just two layers, which is easily solved, as given in Appendix 2A. We summarize the physical results here.

The properties for the materials used the cited experiment are as follows. From the recorded time history of $T(a,t)/T_{\max}$ plotted in Fig. 2.8, we get $t_{1/2} = 53.4 \pm 0.8\ \mu\text{s}$.

With the known conductivities of copper and iron listed in Table 2.2, we first use different trial values of β to get the effective conductivity of the laminate. By solving the two-layer problem for the laminate/substrate composite as described in Appendix 2A, we calculate $T(a,t)$. It is found that the best value is $\beta = 5\times 10^4\,\text{K}^{-1}\cdot\text{cm}^{-2}\cdot\text{W}$ which gives $t_{1/2} = 52.5\ \mu\text{s}$ so that $T(a, t_{1/2}) \approx T_{\max}/2$. The corresponding effective conductivity is $\mathcal{K} = 0.3757\,\text{W}\cdot\text{cm}^{-1}\cdot\text{s}$. Based on these best values, the predicted $T(a,t)/T_{\max}$ for all t is shown in Fig. 2.8 and agrees with the data for all $t > 0$.

Table 2.2. Material properties and thicknesses in experiment by Josell *et al.* (1997a, b).

Material	Thermal Diffusivity ($\text{cm}^2\cdot\text{s}^{-1}$)	Thermal Conductivity ($\text{W}\cdot\text{cm}^{-1}\cdot\text{K}^{-1}$)	Layer Thickness (μm)	Layer Number
Copper	0.838	3.42	2/3	10
Iron	0.0605	0.282	4/3	10
Molybdenum	0.339	1.05	75	1

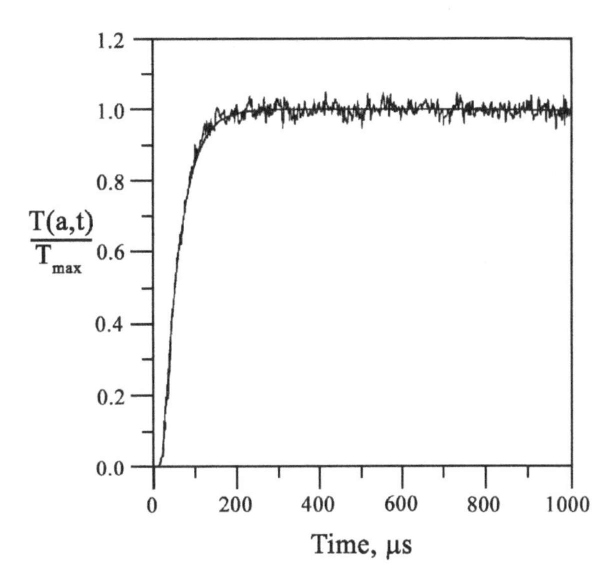

Fig. 2.8. Comparison of transient temperature $T(a,t)/T_{\max}$ at the rear surface of a laminate composed of ten iron/copper bilayers. Jagged curve: data by Josell *et al.* (1997a); smooth curve: prediction.

2.11. Bounds for the Effective Conductivity

2.11.1. *Variational Principles and Bounds*

As an extension of the earlier results, we shall show first that the variational principle for an upper bound is

$$\boxed{\mathcal{K}_{ij}\xi_i\xi_j = \min_{\varphi}\left\{\frac{1}{\Omega}\iiint_\Omega K_{ij}\left(\frac{\partial\varphi}{\partial x_i}+\xi_i\right)\left(\frac{\partial\varphi}{\partial x_j}+\xi_j\right)d\Omega + \frac{\beta}{\Omega}\iint_\Gamma [\varphi]^2 dS\right\}}, \tag{2.11.1}$$

for any constant vectors $\boldsymbol{\xi} = (\xi_i)$, where the minimum is taken over all Ω-periodic functions φ, possibly discontinuous on Γ.

We shall also show that the dual variational principle for the lower bound is

$$\boxed{{\mathcal{K}_{ij}}^{-1}\langle\sigma_i\rangle\langle\sigma_j\rangle = \min_{\boldsymbol{\sigma}}\left\{\frac{1}{\Omega}\iiint_\Omega K_{ij}^{-1}\sigma_i\sigma_j d\Omega + \frac{1}{\beta\Omega}\iint_\Gamma (\boldsymbol{\sigma}\cdot\boldsymbol{n})^2 dS\right\}}, \tag{2.11.2}$$

where the minimum is taken over all Ω-periodic vector functions $\boldsymbol{\sigma}$ that have given average $\langle\boldsymbol{\sigma}\rangle$ and satisfy:

$$\nabla\cdot\boldsymbol{\sigma}=0\,,\quad \boldsymbol{\sigma}\cdot\boldsymbol{n}\text{ continuous on }\Gamma\,. \tag{2.11.3}$$

First, from Eq. (2.9.17)

$$\begin{aligned}\mathcal{K}_{ij}\xi_i\xi_j=\frac{1}{\Omega}\iiint_\Omega K_{ij}\left(\frac{\partial(\boldsymbol{w}\cdot\boldsymbol{\xi})}{\partial x_i}+\xi_i\right)\left(\frac{\partial(\boldsymbol{w}\cdot\boldsymbol{\xi})}{\partial x_j}+\xi_j\right)d\Omega\\ +\frac{\beta}{\Omega}\iint_\Gamma[\boldsymbol{w}\cdot\boldsymbol{\xi}]^2dS\,,\end{aligned} \tag{2.11.4}$$

and since $\boldsymbol{w}\cdot\boldsymbol{\xi}$ is a Ω-periodic function we get

$$\begin{aligned}\mathcal{K}_{ij}\xi_i\xi_j\geq\min_\varphi\left\{\frac{1}{\Omega}\iiint_\Omega K_{ij}\left(\frac{\partial\varphi}{\partial x_i}+\xi_i\right)\left(\frac{\partial\varphi}{\partial x_j}+\xi_j\right)d\Omega\right.\\ \left.+\frac{\beta}{\Omega}\iiint_\Gamma[\varphi]^2dS\right\}.\end{aligned} \tag{2.11.5}$$

We shall now show the converse inequality by considering a Ω-periodic φ, possibly discontinuous on Γ and denote $\delta=\boldsymbol{w}\cdot\boldsymbol{\xi}-\varphi$,

$$\begin{aligned}&\left\langle K_{ij}\left(\frac{\partial\varphi}{\partial x_i}+\xi_i\right)\left(\frac{\partial\varphi}{\partial x_j}+\xi_j\right)\right\rangle+\frac{\beta}{\Omega}\iint_\Gamma[\varphi]^2dS\\ &\quad=\left\langle K_{ij}\left(\frac{\partial(\boldsymbol{w}\cdot\boldsymbol{\xi})}{\partial x_i}+\xi_i\right)\left(\frac{\partial(\boldsymbol{w}\cdot\boldsymbol{\xi})}{\partial x_j}+\xi_j\right)\right\rangle+\frac{\beta}{\Omega}\iint_\Gamma[\boldsymbol{w}\cdot\boldsymbol{\xi}]^2dS\\ &\quad+\left\langle K_{ij}\frac{\partial\delta}{\partial x_i}\frac{\partial\delta}{\partial x_j}\right\rangle+\frac{\beta}{\Omega}\iint_\Gamma[\delta]^2dS-2\left\langle K_{ij}\frac{\partial\delta}{\partial x_i}\left(\frac{\partial(\boldsymbol{w}\cdot\boldsymbol{\xi})}{\partial x_j}+\xi_j\right)\right\rangle\\ &\quad-2\frac{\beta}{\Omega}\iint_\Gamma[\delta][\boldsymbol{w}\cdot\boldsymbol{\xi}]dS\,.\end{aligned} \tag{2.11.6}$$

The sum of the two last terms is zero. Indeed by using Gauss' formula and the cell problems (Eqs. (2.9.9)–(2.9.12)), the term before the last becomes

$$\frac{2}{\Omega}\iint_\Gamma K_{ij}[\delta]\left(\frac{\partial(\boldsymbol{w}\cdot\boldsymbol{\xi})}{\partial x_j}+\xi_j\right)n_idS=-\frac{2}{\Omega}\beta\iint_\Gamma[\delta][\boldsymbol{w}\cdot\boldsymbol{\xi}]dS\,. \tag{2.11.7}$$

Since the second, third and fourth terms in Eq. (2.11.6) are positive we obtain

$$\mathcal{K}_{ij}\xi_i\xi_j\leq\left\{\frac{1}{\Omega}\iiint_\Omega K_{ij}\left(\frac{\partial\varphi}{\partial x_i}+\xi_i\right)\left(\frac{\partial\varphi}{\partial x_j}+\xi_j\right)d\Omega+\frac{\beta}{\Omega}\iint_\Gamma[\varphi]^2dS\right\} \tag{2.11.8}$$

for any φ a Ω-periodic function and therefore

$$\mathcal{K}_{ij}\xi_i\xi_j \leq \min_{\varphi}\left\{\frac{1}{\Omega}\iiint_{\Omega} K_{ij}\left(\frac{\partial\varphi}{\partial x_i}+\xi_i\right)\left(\frac{\partial\varphi}{\partial x_j}+\xi_j\right)d\Omega+\frac{\beta}{\Omega}\iint_{\Gamma}[\varphi]^2 dS\right\}. \tag{2.11.9}$$

The inequalities (Eqs. (2.11.5) and (2.11.9)) can both be true only if equality holds. This implies the variational principle (Eq. (2.11.1)).

The dual variational principle will be deduced by applying the inequality (Eq. (2.6.19)) to the effective conductivity formula (Eq. (2.9.18)) for an admissible vector $\boldsymbol{\sigma}$ with $\boldsymbol{\sigma}\cdot\boldsymbol{n}$ continuous on Γ

$$\begin{aligned}\frac{1}{2}\mathcal{K}_{ij}\xi_i\xi_j &= \frac{1}{2}\left\langle K_{kl}\frac{\partial(\boldsymbol{W}\cdot\boldsymbol{\xi})}{\partial x_l}\frac{\partial(\boldsymbol{W}\cdot\boldsymbol{\xi})}{\partial x_k}\right\rangle+\frac{\beta}{2\Omega}\iint_{\Gamma}[\boldsymbol{W}\cdot\boldsymbol{\xi}]^2 dS\\ &\geq \left\langle\frac{\partial(\boldsymbol{w}\cdot\boldsymbol{\xi})}{\partial x_k}\sigma_k\right\rangle+\langle\boldsymbol{\xi}\cdot\boldsymbol{\sigma}\rangle-\frac{1}{2}\left\langle K^{-1}kl\sigma_k\sigma_l\right\rangle\\ &\quad+\frac{1}{\Omega}\iint_{\Gamma}[\boldsymbol{w}\cdot\boldsymbol{\xi}]\boldsymbol{\sigma}\cdot\boldsymbol{n}\,dS-\frac{1}{2\beta\Omega}\iint_{\Gamma}(\boldsymbol{\sigma}\cdot\boldsymbol{n})^2 dS.\end{aligned} \tag{2.11.10}$$

After partial integration and use of Gauss' formula, the first term in the last equality above cancels with the fourth term since $\boldsymbol{\sigma}$ satisfies Eq. (2.11.3). Thus

$$\frac{1}{2}\mathcal{K}_{ij}\xi_i\xi_j \geq \langle\boldsymbol{\xi}\cdot\boldsymbol{\sigma}\rangle-\frac{1}{2}\langle K_{kl}^{-1}\sigma_k\sigma_l\rangle-\frac{1}{2\beta\Omega}\iint_{\Gamma}(\boldsymbol{\sigma}\cdot\boldsymbol{n})^2 dS. \tag{2.11.11}$$

Equality holds for

$$\sigma_i = K_{ij}\frac{\partial(\boldsymbol{W}\cdot\boldsymbol{\xi})}{\partial x_j} \quad \text{and} \quad \boldsymbol{\sigma}\cdot\boldsymbol{n}=\beta[\boldsymbol{W}\cdot\boldsymbol{\xi}], \tag{2.11.12}$$

as can be seen in the first line of Eq. (2.11.10). Note that the first condition in Eq. (2.11.12) and the boundary condition (Eq. (2.9.10)) of the cell problem imply the second. Now we can write

$$\frac{1}{2}\mathcal{K}_{ij}\xi_i\xi_j = \max_{\boldsymbol{\sigma}}\left\{\langle\boldsymbol{\xi}\cdot\boldsymbol{\sigma}\rangle-\frac{1}{2}\langle K_{kl}^{-1}\sigma_k\sigma_l\rangle-\frac{1}{2\beta\Omega}\iint_{\Gamma}(\boldsymbol{\sigma}\cdot\boldsymbol{n})^2 dS\right\}, \tag{2.11.13}$$

where $\boldsymbol{\sigma}$ ranges over all Ω-periodic, divergence-free functions with $\boldsymbol{\sigma}\cdot\boldsymbol{n}$ continuous on Γ.

The maximum is attained for $\boldsymbol{\sigma}$ given by Eq. (2.11.12) which has the average

$$\langle\sigma_i\rangle = \left\langle K_{ij}\frac{\partial(\boldsymbol{W}\cdot\boldsymbol{\xi})}{\partial x_j}\right\rangle = \mathcal{K}_{ij}\xi_j. \tag{2.11.14}$$

With Eq. (2.11.13), it becomes

$$\frac{1}{2}\mathcal{K}_{ij}\xi_i\xi_j = \mathcal{K}_{ij}\xi_i\xi_j - \min_{\boldsymbol{\sigma}}\left\{\frac{1}{2}\langle K_{kl}^{-1}\sigma_k\sigma_l\rangle + \frac{1}{2\beta\Omega}\iint_\Gamma (\boldsymbol{\sigma}\cdot\boldsymbol{n})^2 dS\right\}. \tag{2.11.15}$$

Let us denote $\langle\sigma_i\rangle = \mathcal{K}_{ij}\xi_j$, or equivalently $\mathcal{K}_{ij}{}^{-1}\langle\sigma_j\rangle = \xi_i$. Then Eq. (2.11.15) gives the variational principle (Eq. (2.11.2)).

Having established the variational principles (Eqs. (2.11.1) and (2.11.2)), we can deduce, as before, in Sec. 2.6, simple bounds by choosing suitable test functions φ and $\boldsymbol{\sigma}$ in Eqs. (2.11.1) and (2.11.2). For example, an upper bound can be obtained by making the simplest choice for a Ω-periodic function in Eq. (2.11.1): $\varphi = 0$. This yields

$$\mathcal{K}_{ij}\xi_i\xi_j \leq \langle K_{ij}\rangle\xi_i\xi_j\,, \tag{2.11.16}$$

i.e., the arithmetic mean is an upper bound. This result is unfortunately too crude since it does not include the influence of the interfacial thermal resistance. In the next section for particulate composites, we shall illustrate that by having more information about the geometry of the interface a more precise upper bound can be found.

For the lower bound we take $\boldsymbol{\sigma}$ to be a constant vector and get

$$\mathcal{K}_{ij}{}^{-1}\langle\sigma_i\rangle\langle\sigma_j\rangle \leq \left(\langle K_{ij}^{-1}\rangle + \frac{1}{\beta\Omega}\iint_\Gamma n_i n_j dS\right)\langle\sigma_i\rangle\langle\sigma_j\rangle\,. \tag{2.11.17}$$

As before, information on the geometry of the interface Γ will enable us compute the surface integral above.

2.11.2. *Application to a Particulate Composite*

Let us consider the example of a composite with spherical particles embedded in a matrix (Lipton and Vernescu, 1996a). Specifically in a cell Ω with the matrix phase Ω_1, there are n spherical particles $\Omega_{2_1}, \Omega_{2_2}, \ldots \Omega_{2_n}$. The mth sphere has the radius R_m, $m = 1, 2, \ldots n$, the specific volume θ_{2_m}, and is centered at $\boldsymbol{r}^{(m)}$. The total volume fraction is $\theta_2 = \theta_{2_1} + \theta_{2_2} + \cdots + \theta_{2_n}$. If we denote the position of a point in the mth sphere by $\boldsymbol{x}$, its position with respect to the center of the mth sphere by $\boldsymbol{x}^{(m)}$ and the unit normal to the surface of the sphere by $\boldsymbol{n}$ then

$$\boldsymbol{x}^{(m)} = \boldsymbol{x} - \boldsymbol{r}^{(m)}, \quad n_i = \left(\frac{x_i^{(m)}}{R_m}\right)_{\Gamma_m}. \tag{2.11.18}$$

The surface integral in Eq. (2.11.17) can be explicitly computed; indeed by using Eq. (2.11.18)

$$\frac{1}{\Omega}\sum_{m=1}^{n}\iint_{\Gamma_m} n_i n_j dS = \frac{1}{\Omega}\sum_{m=1}^{n}\iint_{\Gamma_m}\frac{x_i^{(m)}}{R_m} n_j dS$$
$$= \frac{1}{\Omega}\sum_{m=1}^{n}\iint_{\Gamma_m}\frac{x_i - r_i^{(m)}}{R_m} n_j dS\,.$$

Next, using Gauss's theorem:

$$\frac{1}{\Omega}\sum_{m=1}^{n}\iint_{\Gamma_m} n_i n_j dS = \frac{1}{\Omega}\sum_{m=1}^{n}\iiint_{\Omega_{2m}}\frac{1}{R_m}\frac{\partial(x_i - r_i^{(m)})}{\partial x_j} d\Omega$$
$$= \left(\sum_{m=1}^{n}\frac{\theta_{2_m}}{R_m}\right)\delta_{ij}\,. \tag{2.11.19}$$

The sum above represents the total volume of particles times the inverse of the harmonic mean of the particle radii.

In the case of monodisperse particles of same radius R the above sum becomes $(\theta_2/R)\delta_{ij}$. The lower bound can be readily obtained from Eq. (2.11.17)

$$\mathcal{K}_{ij}{}^{-1}\langle\sigma_i\rangle\langle\sigma_j\rangle \le \left(\frac{\theta_1}{K_1} + \frac{\theta_2}{K_2} + \frac{\theta_2}{\beta R}\right)\langle\boldsymbol{\sigma}\rangle\cdot\langle\boldsymbol{\sigma}\rangle\,. \tag{2.11.20}$$

To get an upper bound we choose in Eq. (2.11.1) the following admissible functions

$$\varphi_{\boldsymbol{\eta}}(x) = \begin{cases} 0 & \text{in the matrix}\,,\\ \boldsymbol{\eta}\cdot\boldsymbol{x}^{(m)} & \text{in the } m\text{th sphere}\,,\end{cases} \tag{2.11.21}$$

where $\boldsymbol{\eta}$ is any constant vector. This corresponds to discontinuous, piecewise constant, temperature gradients. Then the upper bound becomes

$$\min_{\boldsymbol{\eta}}\left\{\frac{1}{\Omega}\left(\theta_1(K_1)_{ij}\xi_i\xi_j + \theta_2(K_2)_{ij}(\eta_i+\xi_i)(\eta_j+\xi_j)\right)\right.$$
$$\left. + \frac{\beta}{\Omega}\left(\iint_{\Gamma} x_i^{(m)}x_j^{(m)} dS\right)\eta_i\eta_j\right\}. \tag{2.11.22}$$

Since the particles are spherical the surface integral above is the same as Eq. (2.11.19). For particles of equal radius, the bound becomes

$$\min_{\boldsymbol{\eta}}\{\theta_2\left((K_2)_{ij} + \beta R\delta_{ij}\right)\eta_i\eta_j + 2\theta_2(K_2)_{ij}\eta_i\xi_j$$
$$+ (\theta_1(K_1)_{ij} + \theta_2(K_2)_{ij})\xi_i\xi_j\}\,. \tag{2.11.23}$$

To fix ideas we restrict our attention to isotropic materials, then

$$\mathcal{K}_{ij}\xi_i\xi_j \leq \min_{\boldsymbol{\eta}}\{\theta_2(K_2+\beta R)\boldsymbol{\eta}^2 + 2\theta_2 K_2 \boldsymbol{\eta}\cdot\boldsymbol{\xi} + (\theta_1 K_1 + \theta_2 K_2)\boldsymbol{\xi}\cdot\boldsymbol{\xi}\}\,. \tag{2.11.24}$$

After computing the minimum of the right-hand side, we get

$$\mathcal{K}_{ij}\xi_i\xi_j \leq \left(\theta_1 K_1 + \theta_2 K_2 - \frac{\theta_2 K_2^2}{K_2+\beta R}\right)\boldsymbol{\xi}\cdot\boldsymbol{\xi}\,. \tag{2.11.25}$$

Let $\lambda_{(n)}$ be the eigenvalues of $(\mathcal{K}_{ij})$ we get by combining Eqs. (2.11.20) and (2.11.25),

$$\left(\frac{\theta_1}{K_1}+\frac{\theta_2}{K_2}+\frac{\theta_2}{\beta R}\right)^{-1} \leq \lambda_{(n)} \leq \left(\theta_1 K_1 + \theta_2 K_2 - \frac{\theta_2 K_2^2}{K_2+\beta R}\right). \tag{2.11.26}$$

The above bounds depend not only on the phase conductivities and volume fractions but also on the specific surface area Γ/θ_2 of the inclusions. Note also that the upper bound is less than the arithmetic mean of the two conductivities and thus improves Eq. (2.11.16).

An interesting consequence of the interface resistance is the existence of a special particle size for which better conducting particles have no effect on the thermal conductivity of the composite. No matter how large the volume fraction of particles is, the composite has the same conductivity as the matrix. Indeed it can be seen from Eq. (2.11.26) that for $K_1 < K_2$ the upper and lower bounds coincide

$$\left(\frac{\theta_1}{K_1}+\frac{\theta_2}{K_2}+\frac{\theta_2}{\beta R_{cr}}\right)^{-1} = \left(\theta_1 K_1 + \theta_2 K_2 - \frac{\theta_2 K_2^2}{K_2+\beta R_{cr}}\right), \tag{2.11.27}$$

for a critical particle radius R_{cr}

$$R_{cr} = \frac{1}{\beta}\frac{K_1 K_2}{K_2 - K_1}\,. \tag{2.11.28}$$

The larger the interface resistance β^{-1}, the greater the critical radius. For commonly used composites this critical particle size is of the order of microns.

The interface resistance model can also explain why in some experiments composites with a higher volume fraction of the better conducting material can have a lower effective thermal conductivity. In the experiments of Every *et al.* (1992) for zinc-sulfide/diamond composites the effect of the particle size on the thermal conductivity was studied. They observed that the conductivity of the composite can be increased by adding large diamond particles to a ZnS matrix, but decreased by adding sub-micron size particles. Thus, although diamond has the highest known thermal conductivity of all materials, small diamond particles can lower the conductivity of the composite below the conductivity of the matrix. This result is in sharp contrast to the case of perfect interface, since the harmonic mean lower bound (Eq. (2.6.30)) is always greater than the lowest of the component conductivities.

In Fig. 2.9 we plot the normalized effective conductivity $\mathcal{K}/K_1$ bounds (Eq. (2.11.26)) for ZnS/diamond composites, for different particle sizes R, as functions the particle volume fraction. As in Every *et al.* (1992) we take the particle conductivity $K_2 = 1000\,\mathrm{W/(mK)}$, the matrix conductivity $K_1 = 17.4\,\mathrm{W/(mK)}$ and the interface conductivity to be $\beta = 0.167\times 10^8\,\mathrm{W/(m^2K)}$ (conductivity is measured in Watts/meter/Kelvin and the interface conductivity in Watts/meter square/Kelvin). For the critical size $R_{cr} = 1.06\,\mu\mathrm{m}$ the upper and lower bounds coincide with the matrix conductivity. Below this critical size the bounds decrease by increasing the particle volume fraction, and above the critical size the trend is reversed.

In many cases the interfacial thermal resistance is difficult to measure experimentally. Thus, it is of practical interest to estimate the resistance

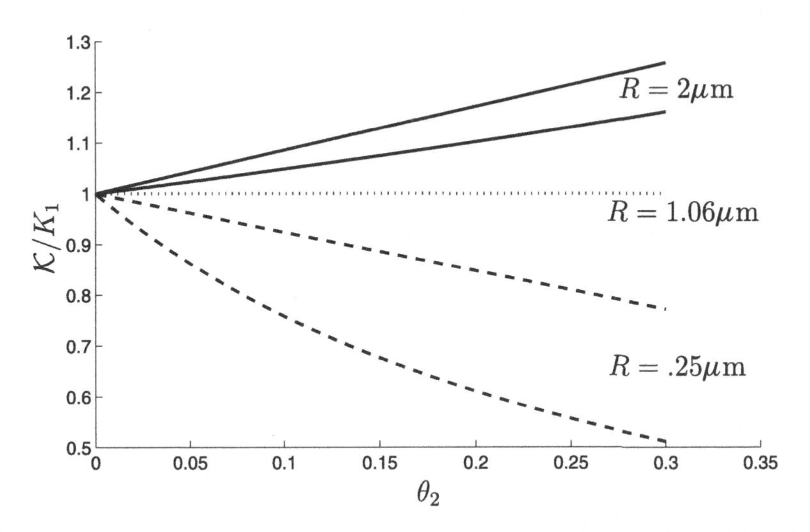

Fig. 2.9. Upper and lower bounds for monodisperse suspensions of diamond in a ZnS matrix as functions of volume fraction θ_2, for different values of particle radius R; at $R = R_{cr}$ the upper and lower bounds are equal (dotted line). From Lipton and Vernescu (1996a), *Proc. R. Soc. Lond. A.*

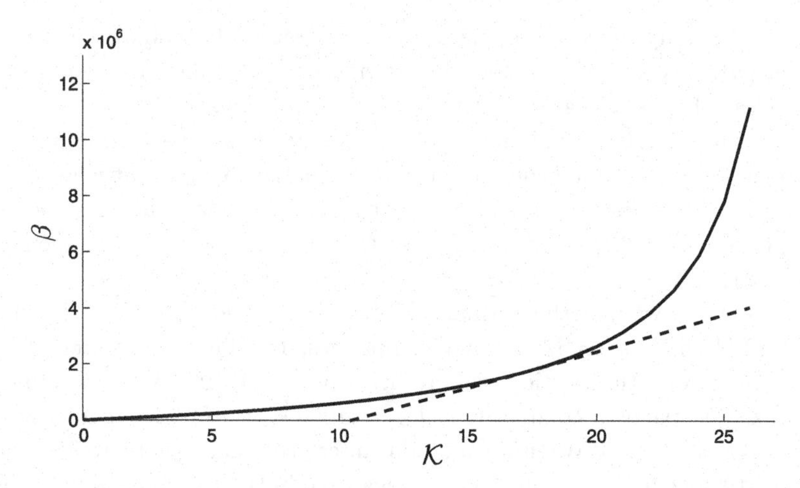

Fig. 2.10. Bounds on the interfacial resistance coefficient β for periodic three-dimensional monodisperse suspensions of diamonds in a ZnS matrix with $K_1 = 17.4\,\mathrm{W/(mK)}$, $K_2 = 1000\,\mathrm{W/(mK)}$, $\theta_2 = 0.4$, and radii $R = 10\,\mu\mathrm{m}$. From Lipton and Vernescu (1996a), *Proc. R. Soc. Lond. A.*

coefficient β in terms of the measured effective conductivity of the composite. Indeed the bounds (Eq. (2.11.26)) are monotonically increasing function of β. By inverting the relationships one can get bounds on the interfacial resistance coefficient β, in terms of the measured effective property, the volume fractions, and particle size

$$\frac{1}{R}\frac{(\theta_1 K_1 - \mathcal{K})K_2}{\mathcal{K} - (\theta_1 K_1 + \theta_2 K_2)} \leq \beta \leq \frac{\theta_2}{R}\frac{\mathcal{K}K_1K_2}{K_1K_2 - \mathcal{K}(\theta_1 K_2 + \theta_2 K_1)}. \tag{2.11.29}$$

The bounds are plotted in Fig. 2.10.

In Lipton and Vernescu (1996a) tighter bounds than those in Eq. (2.11.26) are obtained by relating the effective conductivity for the imperfect interface case with the one for a material with the same geometry but perfect contact between phases and by using the results developed in Bergman (1978) and Bruno (1991).

2.12. Chemical Transport in Aggregated Soil

Due to spills from filling stations, airports, petroleum refineries, and oil fields, volatile organic compounds (VOC) stay in wet soils both as a solute in the water phase and as sorbate on the surface of soil grains. One popular method of removal, called *Soil–Vapor Extraction* or *Air Venting*, is to inject compressed air through a well into the groundwater (aquifer). A bubble plume is created which rises above the water table by buoyancy. In this process VOC can be diffused from the water phase and the adsorbed phase into the moving air and carried into the unsaturated zone. It is then easier to pump and transfer the contaminated air to a treatment plant above the ground. Prediction of the transport process is therefore an important task in both design and operation of this technology.

Soils are often composed of aggregates with micropores which are generally much finer than the macropores between the aggregates. Water captured inside the aggregates is quite immobile. Air flow is significant only through the macropores. The transport process is a combination of convective diffusion in the moving air and static diffusion in and from the aggregates. Usually the diffusivity D_g in gas-filled macropores is much greater than D_w in the liquid-filled aggregates. Despite this sharp contrast, the diffusion time scale through air over a macroscale length of the soil ℓ'^2/D_g is comparable to the diffusion time scale ℓ^2/D_w through water in the small aggregates. It is this time ratio that results in some distinguishing features in the overall diffusion process. The macroscale equation of diffusion through soils has been derived by heuristic arguments in the chemical engineering literature of packed beds. For periodic aggregates Auriault (1983) and Hornung (1991) have used the theory of homogenization to get similar governing equations. Here we describe a related work of Ng and Mei (1996b)[3] for modeling air-venting of volatile organic compounds (VOC).

We model the soil as a periodic array of packed spherical aggregates of uniform size. Water with dissolved and sorbed chemicals is held immobile in the aggregates, while air with chemical vapors can flow in the pore space between aggregates. Thus a unit microcell Ω consists of a spherical aggregate Ω_w of radius a, surrounded by air-filled void Ω_g, and the surface of the aggregate is denoted by Γ_{gw}, as sketched in Fig. 2.11. For simplicity, the aggregate geometry is assumed to be uniform on the large scale. The volume ratios are denoted by $\theta_g = \Omega_g/\Omega$ and $\theta_w = \Omega_w/\Omega$, while the volume-averages of a quantity f over an Ω-cell are denoted by angle brackets,

$$\langle f \rangle_g \equiv \frac{1}{\Omega} \iiint_{\Omega_g} f d\Omega \,, \quad \langle f \rangle_w \equiv \frac{1}{\Omega} \iiint_{\Omega_w} f d\Omega \,.$$

In air-venting operations, the pressure variation can be large enough so that air compressibility is important. The theory of seepage flow in Sec. 3.5

[3]Material of the present section is essentially taken from this article.

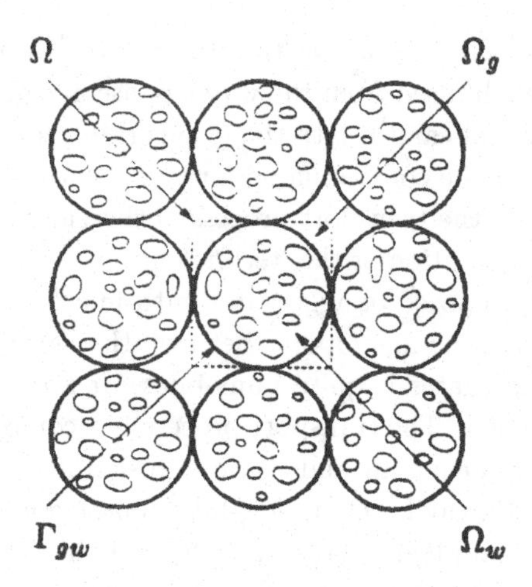

Fig. 2.11. A cubic array of spherical aggregates. Ω: the unit cell; Ω_g: gas-filled macropores in a unit cell; Ω_w: water-saturated aggregate in a unit cell; Γ_{ga}: air/aggregate interface.

of the next chapter applies. We shall suppose that the interstitial gas flow is known.

Let C_g and C_w be the concentrations of the chemical compound in the pore air and in the immobile water, respectively. The variation in the gaseous pores is governed by the law of convective diffusion

$$\frac{\partial C_g}{\partial t} + \frac{\partial}{\partial x_i}(u_i C_g) = D_g \frac{\partial^2 C_g}{\partial x_i \partial x_i}, \tag{2.12.1}$$

and by the simple diffusion law in the immobile water-saturated aggregates,

$$\frac{\partial C_w}{\partial t} = D_e \frac{\partial^2 C_w}{\partial x_i \partial x_i}. \tag{2.12.2}$$

D_e represents the effective diffusivity accounting for adsorption of the chemical on the surface of the solid. Because the aggregate is a mixture of solid and water the effective diffusivity is different from that of pure water. Let ϕ be the porosity of the aggregate, and assume linear isotherm, then

$$D_e = \frac{\phi D_w}{K_d(1-\phi)\rho_s + \phi}, \tag{2.12.3}$$

where K_d is the empirical sorption coefficient.

In addition, we require two matching conditions on the air–water interface Γ_{gw} which is a part of the surface of the aggregate: (i) continuity of flux:

$$D_g \frac{\partial C_g}{\partial x_i} n_i = D_e \frac{\partial C_w}{\partial x_i} n_i \,, \tag{2.12.4}$$

where n_i is the unit normal vector directing into the aggregate and (ii) local chemical equilibrium so that Henry's law of concentration partition applies

$$H = \frac{C_g}{C_w} \,, \tag{2.12.5}$$

where H denotes Henry's law constant. Along the cell boundaries, C_g and C_w are Ω-periodic.

If dimensionless variables are introduced as before, a number of dimensionless ratios appear. For air-venting operations, the following numerical values are typical: $U_g = \mathcal{O}(0.1)\,\text{mm/s}$, $\ell = \mathcal{O}(1-10)\,\text{mm}$, $\nu_g = 0.15\,\text{cm}^2/\text{s}$, $D_g = \mathcal{O}(0.1)\,\text{cm}^2/\text{s}$, $D_w = \mathcal{O}(10^{-7})\,\text{cm}^2/\text{s}$. Based on these, we make the following order estimates of the dimensionless parameters,

(i) Péclét number $Pe = U_g\ell/D_g = Re(\nu_g/D_g) = \mathcal{O}(Re) = \mathcal{O}(\epsilon)$.
(ii) Ratio of diffusivities $D_e/D_g = \mathcal{O}(\epsilon^2)$.
(iii) Time scale T is controlled by the diffusion time in the immobile aggregate, i.e., $\ell^2/TD_e = \mathcal{O}(1)$. It follows that $\ell^2/TD_g = \mathcal{O}(\epsilon^2)$.

Returning to physical variables but retaining the order symbols according to the estimates, the vapor transport equation is

$$\epsilon^2 \frac{\partial C_g}{\partial t} + \epsilon \frac{\partial}{\partial x_i}(u_i C_g) = D_g \frac{\partial^2 C_g}{\partial x_i \partial x_i} \,. \tag{2.12.6}$$

Unlike the dispersion example in Chapter 1 (cf. Eq. (1.3.7)) where convective transport dominates, here molecular diffusion prevails. Introducing two scale expansions $C_g = C_g^{(0)} + \epsilon C_g^{(1)} + \cdots$, we obtain

$$\mathcal{O}(1)\colon 0 = D_g \frac{\partial^2 C_g^{(0)}}{\partial x_i \partial x_i} \,, \tag{2.12.7}$$

$$\mathcal{O}(\epsilon)\colon \frac{\partial}{\partial x_i}(u_i^{(0)} C_g^{(0)}) = D_g \left(\frac{\partial^2 C_g^{(1)}}{\partial x_i^2} + 2\frac{\partial^2 C_g^{(0)}}{\partial x_i' \partial x_i} \right), \tag{2.12.8}$$

$$\mathcal{O}(\epsilon^2)\colon \frac{\partial C_g^{(0)}}{\partial t} + \frac{\partial}{\partial x_i}(u_i^{(0)} C_g^{(1)} + u_i^{(1)} C_g^{(0)}) + \frac{\partial}{\partial x_i'}(u_i^{(0)} C_g^{(0)})$$

$$= D_g \frac{\partial}{\partial x_i}\left(\frac{\partial C_g^{(2)}}{\partial x_i} + \frac{\partial C_g^{(1)}}{\partial x_i'} \right) + D_g \frac{\partial}{\partial x_i'}\left(\frac{\partial C_g^{(1)}}{\partial x_i} + \frac{\partial C_g^{(0)}}{\partial x_i'} \right). \tag{2.12.9}$$

Similarly, diffusion in the immobile water Ω_w in the aggregates obeys

$$\epsilon^2 \frac{\partial C_w}{\partial t} = \epsilon^2 D_e \frac{\partial^2 C_w}{\partial x_i \partial x_i}, \tag{2.12.10}$$

where C_w is the chemical concentration in the immobile water inside an aggregate. Expanding $C_w = C_w^{(0)} + \cdots$, the leading-order equation is

$$\frac{\partial C_w^{(0)}}{\partial t} = D_e \frac{\partial^2 C_w^{(0)}}{\partial x_i \partial x_i}. \tag{2.12.11}$$

By expanding the boundary conditions similarly, we obtain, for the continuity of flux along Γ_{gw},

$$\mathcal{O}(1)\colon D_g \frac{\partial C_g^{(0)}}{\partial x_i} n_i = 0, \tag{2.12.12}$$

$$\mathcal{O}(\epsilon)\colon D_g \left(\frac{\partial C_g^{(1)}}{\partial x_i} + \frac{\partial C_g^{(0)}}{\partial x_i'} \right) n_i = 0, \tag{2.12.13}$$

$$\mathcal{O}(\epsilon^2)\colon D_g \left(\frac{\partial C_g^{(2)}}{\partial x_i} + \frac{\partial C_g^{(1)}}{\partial x_i'} \right) n_i = D_e \frac{\partial C_w^{(0)}}{\partial x_i} n_i, \tag{2.12.14}$$

where n_i is the unit normal vector directing into the aggregate. From Henry's law of chemical partition,

$$H = \frac{C_g^{(m)}}{C_w^{(m)}} \quad m = 0, 1, 2, \ldots. \tag{2.12.15}$$

Along the cell boundaries, $C_g^{(m)}$ and $C_w^{(m)}$, $m = 0, 1, \ldots$ are Ω-periodic.

From Eqs. (2.12.7) and (2.12.12), it is obvious that

$$C_g^{(0)} = C_g^{(0)}(x_i', t). \tag{2.12.16}$$

At the order $\mathcal{O}(\epsilon)$, we make use of Eq. (3.5.21) to be derived in Chapter 3 to reduce Eq. (2.12.8) to

$$0 = D_g \frac{\partial^2 C_g^{(1)}}{\partial x_i \partial x_i}.$$

In view of Eq. (2.12.13) we let

$$C_g^{(1)} = -N_l \frac{\partial C_g^{(0)}}{\partial x_\ell'}. \tag{2.12.17}$$

N_ℓ must then be the solution of a canonical boundary-value cell problem defined by Laplace's equation

$$\frac{\partial^2 N_\ell}{\partial x_i \partial x_i} = 0, \quad x_i \in \Omega_g, \tag{2.12.18}$$

and the interface condition

$$\frac{\partial N_\ell}{\partial x_i} n_i = n_\ell \quad \text{along } \Gamma_{gw} . \tag{2.12.19}$$

Together with the requirement that N_ℓ must be Ω-periodic, Eqs. (2.12.18) and (2.12.19) define the cell problem for N_ℓ, which is the same as that for heat conduction in a composite with nonconducting inclusions. Indeed the cell problem for N_ℓ is a limiting case of Eqs. (2.2.19)–(2.2.22).

At order $\mathcal{O}(\epsilon^2)$: we average Eqs. (2.12.9) and (2.12.14) over an Ω-cell, and add the results to get, for the chemical concentration in the gaseous phase,

$$\theta_g \frac{\partial C_g^{(0)}}{\partial t} + \frac{\partial \langle C_w^{(0)} \rangle}{\partial t} + \frac{\partial}{\partial x_i'} \left[\langle u_i^{(0)} \rangle C_g^{(0)} \right] = \frac{\partial}{\partial x_i'} \left[\{ \mathcal{D}_{il} + n_g D_g \delta_{il} \} \frac{\partial C_g^{(0)}}{\partial x_\ell'} \right] , \tag{2.12.20}$$

where $\theta_g = \Omega_g / \Omega$ is the volume fraction of the gaseous phase. The tensor

$$\mathcal{D}_{il} = -D_g \left\langle \frac{\partial N_\ell}{\partial x_i} \right\rangle_g \tag{2.12.21}$$

represents the contribution of microstructure to the effective diffusivity, and is symmetric and positive as shown in Sec. 2.4. The source-like term $-\partial \langle C_w^{(0)} \rangle / \partial t$ represents the rate of chemical transfer from immobile water to the gaseous phase, and can be evaluated more explicitly as follows. From Eq. (2.12.16) $C_g^{(0)}$ is uniform in a local cell. In view of the boundary condition on the spherical surface of an aggregate,

$$C_w^{(0)}(r = a, t) = \frac{1}{H} C_g^{(0)}(x_i', t) , \tag{2.12.22}$$

$C_w^{(0)}$ must be radially symmetric and a function of the local radial position r and time only. Hence the governing equation is simply

$$\frac{\partial C_w^{(0)}}{\partial t} = \frac{D_e}{r^2} \frac{\partial}{\partial r} \left(r^2 \frac{\partial C_w^{(0)}}{\partial r} \right) \tag{2.12.23}$$

subject to the initial condition

$$C_w^{(0)}(r, t = 0) \, (\equiv C_{w0}) = \frac{C_{g0}}{H} . \tag{2.12.24}$$

Thus we have a classical diffusion problem in a sphere whose solution can be found by Laplace transform (Carlslaw and Jaeger, 1959):

$$C_w^{(0)} = -\frac{2}{ar} \sum_{n=1}^{\infty} (-1)^n e^{-\lambda_n t} \sin \left(\frac{n \pi r}{a} \right) \left\{ \frac{a^2}{n\pi} C_{w0} + n \pi D_e \int_0^t e^{\lambda_n \tau} \frac{C_g^{(0)}}{H} d\tau \right\} , \tag{2.12.25}$$

where

$$\lambda_n = \frac{D_e n^2 \pi^2}{a^2}. \tag{2.12.26}$$

Upon averaging over Ω_w and differentiating with respect to time, the source term is calculated as

$$\frac{\partial \langle C_w^{(0)} \rangle}{\partial t} = \frac{6\theta_w D_e}{a^2} \sum_{n=1}^{\infty} \left[e^{-\lambda_n t} \left(\frac{C_{g0}}{H} - C_{w0} \right) + \frac{e^{-\lambda_n t}}{H} \int_0^t e^{\lambda_n \tau} \frac{\partial C_g^{(0)}}{\partial \tau} d\tau \right], \tag{2.12.27}$$

where $C_{g0} \equiv C_g^{(0)}(t = 0)$. The volume fraction of the water phase θ_w is related to the aggregate volume fraction by $\theta_w = \phi\theta_a$, where $\theta_a = 1 - \theta_g$. If initially the partition between air and immobile water is in equilibrium, the two terms inside the parentheses cancel each other. It follows by substitution of Eq. (2.12.27) into Eq. (2.12.20) that

$$\theta_g \frac{\partial C_g^{(0)}}{\partial t} + \frac{6\theta_w D_e}{a^2} \sum_{n=1}^{\infty} \left[\frac{e^{-\lambda_n t}}{H} \int_0^t e^{\lambda_n \tau} \frac{\partial C_g^{(0)}}{\partial \tau} d\tau \right] + \frac{\partial}{\partial x_i'} [\langle u_i^{(0)} \rangle_g C_g^{(0)}]$$
$$= \frac{\partial}{\partial x_i'} \left[\{\mathcal{D}_{il} + n_g D_g \delta_{il}\} \frac{\partial C_g^{(0)}}{\partial x_l'} \right]. \tag{2.12.28}$$

Equation (2.12.28) is an integral–differential equation governing the macroscale transport of the vapor phase solute $C_g^{(0)}$. With given flow velocity distribution, it can be solved numerically under appropriate boundary conditions on the ground surface, at the well casing, at the water table below, if any, and at infinity. Afterward $C_w^{(0)}$ can be computed from Eq. (2.12.25).

Based on Eq. (2.12.28) Ng and Mei (1996a) have described the simulation of a one-dimensional laboratory experiment by Gierke *et al.* (1990) and Gierke *et al.* (1992) for the feeding and cleansing of methane in a soil column. Results compare well with the measurements by a chromatograph. Ng and Mei (1996a) have also studied the effect of air venting from a vertical well through a contaminated soil layer. In air-venting operations it is useful to know how fast the pumping rate should be to achieve a desired rate of chemical removal. They idealize the problem by considering the radial flow into an infinitely long vertical well or radius r_w, at the center of a circular zone of contamination with radius r_c. If the pressure P_w of the pore air in the pumping well ($r = r_w$) and the pressure P_m in a monitoring well ($r = r_m$) are known, the seepage velocity in the radial direction is easily found by using the theory for compressible air to be derived in Sec. 3.5,

$$\langle u_r^{(0)}(r)\rangle = -\frac{KP_w}{2\mu\theta_g r}\frac{[(P_m/P_w)^2-1]}{[\ln(r_m/r_w)+((P_m/P_w)^2-1)\ln(r/r_w)]^{1/2}}\,. \quad (2.12.29)$$

An initial-boundary-value problem can be formulated to simulate the extraction from a cylindrical domain of contamination $r_w < r < r_c$ where r_w is the radius of the well. Initially the gaseous concentration is constant (C_{g0}) in the cylindrical domain outside the well. Using r_c as the length scale, the magnitude of air seepage velocity $|u_r(r_c)| = U_g$ as the velocity scale, and r_c/U_g as the time scale, we can

$$C_g^{(0)} = C_{g0}C_g^\dagger\,,\quad \langle C_w^{(0)}\rangle = \left(\frac{C_{g0}}{H}\right)\langle C_w^\dagger\rangle,\quad r = r_c r^\dagger\,,$$
$$t = \left(\frac{r_c}{U_g}\right)t^\dagger\,,\quad u = U_g u^\dagger, \quad (2.12.30)$$

normalize the governing equation as follows:

$$\frac{\partial C_g^\dagger}{\partial t^\dagger} + \frac{1}{r^\dagger}\frac{\partial(r^\dagger u_r^\dagger C_g^\dagger)}{\partial r^\dagger} - \frac{1}{Pe}\frac{1}{r^\dagger}\frac{\partial}{\partial r^\dagger}\left(r^\dagger\frac{\partial C_g^\dagger}{\partial r^\dagger}\right)$$
$$= -6\xi\sigma\sum_{n=1}^{\infty}\int_0^{t^\dagger} e^{-\lambda_n^\dagger(t^\dagger-\tau^\dagger)}\frac{\partial C_g^\dagger}{\partial \tau^\dagger}d\tau^\dagger\,, \quad (2.12.31)$$

where

$$Pe = \frac{U_g r_c}{D_g}\,,\quad \lambda_n^\dagger = n^2\pi^2\sigma, \quad (2.12.32)$$

and

$$\sigma = \frac{D_e r_c}{a^2 U_g} \quad (2.12.33)$$

is the ratio of the convection time r_c/U_g to the aggregate diffusion time a^2/D_e. Finally,

$$\xi = \frac{D_w\theta_w}{HD_e\theta_g} = \frac{[K_d(1-\phi)\rho_s + \phi]\theta_a}{H\theta_g} \quad (2.12.34)$$

is the ratio of aggregate to vapor properties.

Since $C_g^\dagger = 0$ for all $r^\dagger > 1$, continuity of flux along $r^\dagger = 1$ requires that the total flux is zero

$$u_r^\dagger C_g^\dagger - \frac{1}{Pe}\frac{\partial C_g^\dagger}{\partial r^\dagger} = 0 \quad \text{at } r^\dagger = 1\,,\quad t^\dagger > 0\,. \quad (2.12.35)$$

Inside the well free of grains, diffusion is rapid. A good approximation is that $C_g^\dagger$ is independent of $r^\dagger$. At the well $C_g^\dagger$ must be continuous. Since the radial velocities are the same: $(u_g^\dagger)_- = \theta_g(u_g^\dagger)_+$, continuity of fluxes requires that

$$\frac{\partial C_g^\dagger}{\partial r^\dagger} = 0 \quad \text{at } r^\dagger = r_w^\dagger = \frac{r_w}{r_c}, \quad t^\dagger > 0 . \tag{2.12.36}$$

Equations (2.12.35) and (2.12.36) correspond to Danckwerts conditions given first for one-dimensional problems. At the initial instant we have

$$C_g^\dagger = \langle C_w^\dagger \rangle = 1 \quad \text{at } t^\dagger = 0, \; r_w^\dagger \le r^\dagger \le 1 . \tag{2.12.37}$$

Let P_m and P_w be the pore pressures at the monitoring station ($r = r_m$) and at the well ($r = r_w$), respectively. The normalized radial seepage velocity can be easily found by the one-dimensional theory involving Darcy's law,

$$u_r^\dagger = -\frac{1}{r^\dagger}\left[\frac{\ln(r_m/r_w) + \left((P_m/P_w)^2 - 1\right)\ln(r_c/r_w)}{\ln(r_m/r_w) + \left((P_m/P_w)^2 - 1\right)\ln(r/r_w)}\right]^{1/2} . \tag{2.12.38}$$

In Fig. 2.12 we show the time evolution of both $C_g^\dagger$ and $C_w^\dagger$ at the well. Pumping is first maintained for a finite duration, then halted for certain period, and is resumed indefinitely afterward. During the shut-down, $C_g^\dagger$ rebounds due to the slow release of chemical from the aggregates. For

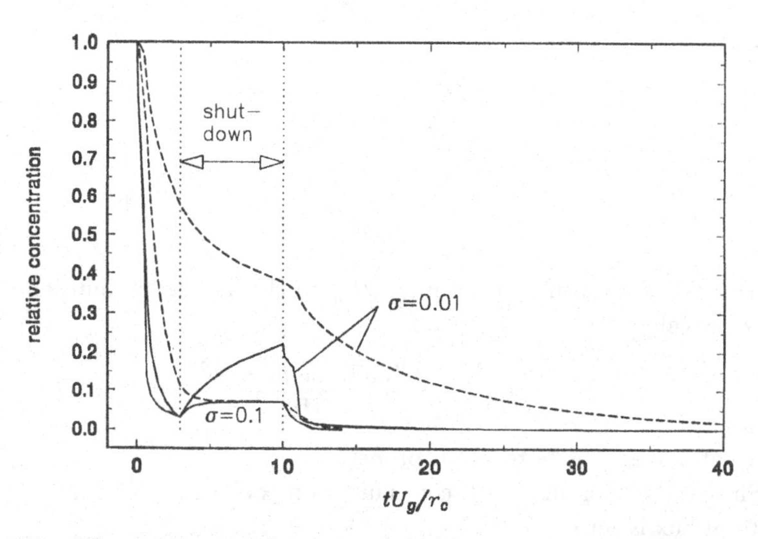

Fig. 2.12. Effects of interrupted pumping and of aggregate properties for fixed $Pe = U_gL/D_g = 500$. Solid curve: $C_g^\dagger(r_w^\dagger, t^\dagger)$. Dashed curves: $\langle C_w^\dagger(r_w^\dagger, t^\dagger)\rangle$. From Ng and Mei (1996b), *Water Resour. Res.*

smaller σ (larger aggregates or smaller D_e), the effluent vapor concentration is depleted more slowly during the first stage of pumping and rebounds more. The aqueous concentration stays higher throughout the operation.

Appendix 2A. Heat Transfer in a Two-Slab System

We will now solve the macroscale problem of heat conduction in a two-slab system, with one slab ($-b < x < 0$) to be a single-phase substrate and the other ($0 < x < a$) to be a multilayer laminate.

The heat transfer problem in two slabs is described by

$$\frac{\partial T_b}{\partial t} = k_b \frac{\partial^2 T_b}{\partial x^2}, \quad -b < x < 0, \tag{A.1}$$

$$\frac{\partial T_a}{\partial t} = k_a \frac{\partial^2 T_a}{\partial x^2}, \quad 0 < x < a. \tag{A.2}$$

Assuming perfect contact between the two slabs, the boundary conditions are:

$$\frac{\partial T_b}{\partial x} = 0, \quad \text{at } x = -b, \quad \frac{\partial T_a}{\partial x} = 0, \quad \text{at } x = a, \tag{A.3}$$

$$T_b = T_a, \quad K_b \frac{\partial T_b}{\partial x} = K_a \frac{\partial T_a}{\partial x}, \quad \text{at } x = 0. \tag{A.4}$$

The conductivities and diffusivities are related by

$$k_a = \frac{K_a}{\rho_a c_a}, \quad k_b = \frac{K_b}{\rho_b c_b}, \tag{A.5}$$

where c_a and c_b are the specific heats. Corresponding to an impulsive heat flux on the side of $x = -b$, we prescribed the initial condition (Parker *et al.*, 1961):

$$T_b(x,0) = \begin{cases} \dfrac{Q}{\rho_b c_b \epsilon}, & -b < x < -b + \epsilon \\ 0, & -b + \epsilon < x < 0; \end{cases} \tag{A.6}$$

$$T_a(x,0) = 0, \quad 0 < x < a. \tag{A.7}$$

The depth ϵ is very small compared to a or b and Q is the total heat energy in the initial pulse.

A similar two-slab problem has been solved by Laplace transform in (Carlslaw and Jaeger, 1959). An equivalent but simpler alternative is to use the method of separation of variables

$$T(x,t) = \chi(x)\theta(t)\,, \tag{A.8}$$

where

$$\chi(x) = \begin{cases} \chi_b(x)\,, & -b < x < 0\,; \\ \chi_a(x)\,, & 0 < x < a\,. \end{cases} \tag{A.9}$$

From Eqs. (A.1) and (A.2) and the separation of variables (Eq. (A.8)) we get the differential equations,

$$\chi_b'' + \alpha_b^2 \chi_b = 0\,, \quad \theta' = k_b \alpha_b^2 \theta\,, \tag{A.10}$$

$$\chi_a'' + \alpha_a^2 \chi_a = 0\,, \quad \theta' = k_a \alpha_a^2 \theta\,. \tag{A.11}$$

We require that

$$k_b \alpha_b^2 = k_a \alpha_a^2\,, \tag{A.12}$$

so that the time dependence is the same in both slabs. By integrating Eqs. $(A.10)_1$ and $(A.11)_1$ and from the boundary conditions (Eq. (A.3)) we get

$$\chi_b = A_b \cos \alpha_b (x + b)\,, \quad \chi_a = A_a \cos \alpha_a (x - a)\,. \tag{A.13}$$

The continuity condition (Eq. (A.4)) on the interface $x = 0$ gives

$$A_b \cos \alpha_b b - A_a \cos \alpha_a a = 0\,, \tag{A.14}$$

$$A_b K_b \alpha_b \sin(\alpha_b b) + A_a K_a \alpha_a \sin(\alpha_a a) = 0\,. \tag{A.15}$$

The two equations above give us the eigenvalue condition

$$K_b \alpha_b \sin(\alpha_b b) \cos(\alpha_a a) + K_a \alpha_a \sin(\alpha_a a) \cos(\alpha_b b) = 0\,, \tag{A.16}$$

or with Eq. (A.12)

$$\frac{K_b}{\sqrt{k_b}} \sin(\alpha_b b) \cos(\alpha_a a) + \frac{K_a}{\sqrt{k_a}} \sin(\alpha_a a) \cos(\alpha_b b) = 0\,. \tag{A.17}$$

Equations (A.17) and (A.12) determine the eigenvalues α_{bn}, with $n = 0, 1, 2, 3 \ldots$. Note that $\alpha_{b0} = 0$.

Using the first of the continuity conditions (Eq. (A.4)), we define the eigenfunctions

$$\chi_n = \begin{cases} \chi_{bn} = \dfrac{\cos(\alpha_{bn}(x + b))}{\cos(\alpha_{bn} b)}\,, & -b < x < 0\,, \\[2ex] \chi_{an} = \dfrac{\cos(\alpha_{an}(x - a))}{\cos(\alpha_{an} a)}\,, & 0 < x < a\,. \end{cases} \tag{A.18}$$

Note that the first eigenfunction for $n = 0$ is $\chi_0 = \chi_{b0} = \chi_{a0} = 1$.

For each $n \neq 0$, the corresponding time factor is

$$\theta_n = e^{-\lambda_n t}, \quad \text{where } \lambda_n = k_b \alpha_{bn}^2 = k_a \alpha_{an}^2, \tag{A.19}$$

for $n \neq 0$ and $\theta_0(t) = 1$, so that $n = 0$ corresponds to the final temperature.

For different eigenvalues, the eigenfunctions χ_n and χ_m are orthogonal with weighting function K/k which is K_b/k_b in $(-b, 0)$ and K_a/k_a in $(0, a)$. Indeed by integrating Eqs. (A.10) and (A.11)

$$[\chi_{bm}\chi_{bn}' - \chi_{bn}\chi_{bm}']_{-b}^{0} + (\alpha_{bn}^2 - \alpha_{bm}^2) \int_{-b}^{0} \chi_{bn}\chi_{bm} dx = 0, \tag{A.20}$$

$$[\chi_{am}\chi_{an}' - \chi_{an}\chi_{am}']_{0}^{a} + (\alpha_{an}^2 - \alpha_{am}^2) \int_{0}^{a} \chi_{an}\chi_{am} dx = 0. \tag{A.21}$$

Adding Eq. (A.20) multiplied by K_b with Eq. (A.21) multiplied by K_a, we get after using Eq. (A.4),

$$(\alpha_{bn}^2 - \alpha_{bm}^2) \left\{ \frac{K_b}{k_b} \int_{-b}^{0} \chi_{bn}\chi_{bm} dx + \frac{K_a}{k_a} \int_{0}^{a} \chi_{an}\chi_{am} dx \right\} = 0, \tag{A.22}$$

which implies the desired orthogonality.

For later use let us compute the norm of the eigenvectors χ_n, for $n \neq 0$:

$$\int_{-b}^{a} \frac{K}{k} \chi_n^2 dx = \frac{bK_b}{2k_b} \frac{1}{\cos^2 \alpha_b b} + \frac{aK_a}{2k_a} \frac{1}{\cos^2 \alpha_a a}. \tag{A.23}$$

For $n = 0$ we get

$$\int_{-b}^{a} \frac{K}{k} \chi_0^2 dx = \frac{K_b b}{k_b} + \frac{K_a a}{k_a}. \tag{A.24}$$

Let

$$T(x,t) = \sum_{n=0}^{\infty} C_n e^{-\lambda_n t} \chi_n(x), \quad -b < x < a. \tag{A.25}$$

Apply the initial condition:

$$T(x,0) = T_0(x) = \sum_{n=0}^{\infty} C_n \chi_n(x). \tag{A.26}$$

Using the orthogonality the Fourier coefficients can be determined:

$$C_n = \frac{\int_{-b}^{a} (K/k) T_0(x) \chi_n(x) dx}{\int_{-b}^{a} (K/k) \chi_n^2(x) dx}, \tag{A.27}$$

where the denominator is given by Eq. (A.23). In particular, for $n = 0$ the first term in the series (Eq. (A.25)) is

$$C_0 = T_{max} = \frac{\int_{-b}^{a} T_0(x)(K/k)\chi_0 dx}{\int_{-b}^{a}(K/k)\chi_0^2 dx} = \frac{(Q/\rho_b c_b)(K_b/k_b)}{(K_b b/k_b) + (K_a a/k_a)}. \tag{A.28}$$

Thus the solution of the heat conduction problem in a two-slab composite is given by Eq. (A.25) with the coefficients given by Eqs. (A.27) and (A.28).

References

Auriault, J. L. (1983). Effective macroscopic description for heat conduction in periodic composites. *Int. J. Heat Mass Transfer* **26**: 861–869.

Auriault, J.-L. and H. I. Ene (1994). Macroscopic modelling of heat transfer in composites with interfacial barrier. *Int. J. Heat Mass Transfer* **37**(18): 2885–2892.

Benveniste, Y. (1986). Effective thermal conductivity of composites with a thermal contact resistance between the constituents: Nondilute case. *J. Appl. Phys.* **61**: 2840–2844.

Bergman, D. (1978). The dielectric constant of a composite material — A problem in classical physics. *Phys. Rep.* **43**: 377–407.

Bhatt, H., K. Y. Donaldson and D. P. H. Hasselman (1990). Role of the interfacial thermal barrier in the effective thermal diffusivity/conductivity of the SiC-fiber-reinforced reaction-bonded silicon nitride. *J. Am. Ceram. Soc.* **73**(2): 312–316.

Bruno, O. P. (1991). The effective conductivity of strongly heterogeneous composites. *Proc. R. Soc. Lond. A* **433**: 353–381.

Carlslaw, H. S. and J. C. Jaeger (1959). *Conduction of Heat in Solids*, Clarendon Press, Oxford, pp. 109–112.

Chiew, Y. C. and E. D. Glandt (1987). Effective conductivity of dispersions: The effect of resistance at the particle surfaces. *Chem. Eng. Sci.* **42**: 2677–2685.

Dunn, M. L. and M. Taya (1993). The effective thermal conductivity of composites with coated reinforcement and the application to imperfect interfaces. *J. Appl. Phys.* **73**: 1711.

Every, A. G., Y. Tzou, D. P. H. Hasselman and R. Raj (1992). The effect of particle size on the thermal conductivity of ZnS/diamond composites. *Acta Metall. Matter* **40**: 123–129.

Garret, K. W. and H. M. Rosenberg (1974). The thermal conductivity of epoxy-resin powder composite materials. *J. Phys. D Appl. Phys.* **7**: 1247–1258.

Gierke, J. S., N. J. Hutzler and J. C. Crittenden (1990). Modeling the movement of volatile organic chemicals in columns of unsaturated soil. *Water Resour. Res.* **26**(7): 1529–1547.

Gierke, J. S., N. J. Hutzler and D. B. McKenzie (1992). Vapor transport in unsaturated soil columns. Implications for vapor extraction. *Water Resour. Res.* **28**(2): 323–335.

Hashin, Z. and S. Shtrikman (1962). A variational approach to the theory of the effective magnetic permeability of multiphase materials. *J. Appl. Phys.* **33**: 3125–3131.

Hasimoto, H. (1959). On the periodic fundamental solutions of the Stokes equation and their application to viscous flow past a cubic array of spheres. *J. Fluid Mech.* **5**: 317–328.

Hasselman, D. P. H. and L. F. Johnson (1987). Effective thermal conductivity of composites with interfacial thermal barrier resistance. *J. Composite Materials* **21**: 508–515.

Hasselman, D. P. H., K. Y. Donaldson and J. R. Thomas, Jr. (1993). Effective thermal conductivity of uniaxial composite with cylindrically orthotropic carbon fibers and interfacial thermal barrier. *J. Composite Materials* **27**: 637–644.

Hornung, U. (ed.) (1997). *Homogenization and Porous Media*, Springer-Verlag.

Josell, D., A. Cezairliyan, D. van Heerden and B. T. Murray (1997a). An integral solution for thermal diffusion in periodic multilayer materials: Application to iron/copper multilayers. *Int. J. Thermophysics* **18**(3): 865–885.

Josell, D., A. Cezairliyan, D. van Heerden and B. T. Murray (1997b). Thermal diffusion through multilayer coatings: Theory and experiment. *Nanostructured Materials* **9**: 727–736.

Josell, D., A. Cezairliyan and J. E. Bonevich (1998). Thermal diffusion through multilayer coatings: Theory and experiment. *Int. J. Thermophysics* **19**(2): 525–535.

Kapitza, P. L. (1941). The study of heat transfer in helium II. *J. Phys. (USSR)* **60**: 354.

Lipton, R. and B. Vernescu (1996a). Composites with imperfect interface. *Proc. R. Soc. Lond.* **452**(329): 329–358.

Lipton, R. and B. Vernescu (1996b). Critical radius, size effects and inverse problems for composites with imperfect interface. *J. Appl. Phys.* **79**(12): 8964–8966.

Maxwell, J. C. (1881). *A Treatise on Electricity and Magnetism*, Clarendon Press, Oxford, **1**: 314.

McKenzie, D. R., R. C. McPhedran and G. H. Derrick (1978). The conductivity of lattices of spheres. II. The body-centered and face-centered cubic lattices. *Proc. R. Soc. Lond. A* **362**: 211.

McPhedran, R. C. and D. R. McKenzie (1978). The conductivity of lattices of spheres I. The simple cubic lattice. *Proc. R. Soc. Lond. A* **359**: 45.

Meredith, R. E. and C. W. Tobias (1960). Resistance to potential flow through a cubical array of spheres. *J. Appl. Phys.* **31**: 1270-3.

Ng, C. O. and C. C. Mei (1996a). Homogenization theory applied to soil vapor extraction in aggregated soils. *Phys. Fluids* **8**(9): 2298–2306.

Ng, C. O. and C. C. Mei (1996b). Aggregate diffusion model applied to soil vapor extraction in unidirectional and radial flows. *Water Resour. Res.* **32**: 1289–1297.

Parker, W. J., R. J. Jenkins, C. P. Butler and G. L. Abbot (1961). Flash method of determining thermal diffusivity, heat capacity and thermal conductivity. *J. Appl. Phys.* **32**(9): 1679–1684.

Prat, M. (1989). On the boundary conditions at the macroscopic level. *Transport in Porous Media* **4**: 259–280.

Prat, M. (1992). Some refinements concerning the boundary conditions at the macroscopic level. *Transport in Porous Media* **7**: 147–161.

Rayleigh, Lord (1892). On the influence of obstacles arranged in rectangular order upon the properties of a medium. *Philos. Mag.* **34**: 481–507.

Runge, I. (1925). On the electrical conductivity of metallic aggregates. *Z. Tech. Physik.* **6**: 61–68.

Sangani, A. S. and A. Acrivos (1983). The effective conductivity of a periodic array of spheres. *Proc. R. Soc. Lond. A* **386**: 263–275.

Swift, D. L. (1966). The thermal conductivity of spherical metal powders including the effect of an oxide coating. *Int. J. Heat Mass Transfer* **9**: 1061–1074.

Zuzovski, M. and H. Brenner (1977). Effective conductivities of composite materials composed of cubic arrangements of spherical particles embedded in an isotropic matrix. *J. Appl. Math. Phys.* (ZAMP) **28**: 979.

Seepage in Rigid Porous Media

3

Seepage flow through porous media is one of the earliest topics in multi-phase mechanics to which the method of homogenization was applied. In this chapter, we shall first show how the empirical law of Darcy can be theoretically derived, and how the hydraulic conductivity can be computed from a cell problem. In addition, certain important properties of the hydraulic conductivity can be proven from the conditions governing the cell problem. In most of this chapter the solid matrix is supposed to be non-deformable. A minor exception is an application to biomechanics where the solid deformation is simply related to fluid flow by an algebraic relation. Here we illustrate how a two-dimensional flow on the macroscale depends on the three-dimensional microstructure. Extensions to media with three scales is also described. A theoretical justification of Brinkman's law for high porosity is also given. Finally weak nonlinear effects of inertia are discussed by a higher-order analysis.

A more mathematical treatment of homogenization analysis of flow in porous media can be found in Hornung (1997).

3.1. Equations for Seepage Flow and Darcy's Law

Let a rigid porous medium be saturated by an incompressible Newtonian fluid of constant density. Driven by ambient pressure gradient, the steady flow velocity u_i and pressure p in the pores are governed by Navier–Stokes equations

$$\frac{\partial u_i}{\partial x_i} = 0, \tag{3.1.1}$$

$$\rho u_j \frac{\partial u_i}{\partial x_j} = -\frac{\partial p}{\partial x_i} + \mu \nabla^2 u_i . \tag{3.1.2}$$

On the wetted surface of the solid matrix Γ, there is no slip,

$$u_i = 0, \quad \boldsymbol{x} \in \Gamma. \tag{3.1.3}$$

In most porous media problems the pores are small and the flow is slow. It is well known in the fluid dynamics of slow viscous flows that the two terms on the right-hand side of Eq. (3.1.2), representing respectively the pressure gradient and the viscous force, are dominant over the convective inertia on the left-hand side. In principle both pressure and velocity vary according to two length scales: the microscale ℓ characteristic of the size of pores and grains (millimeters to centimeters for natural soils) and the macroscale ℓ' associated with the global pressure gradient (meters or greater for soils) which drives the seepage flow. Anticipating that the driving global pressure gradient $\Delta P/\ell'$ is balanced by viscous resistance we have

$$\mathcal{O}\left(\frac{\Delta P}{\ell'}\right) = \mathcal{O}\left(\frac{\mu U}{\ell^2}\right), \tag{3.1.4}$$

which defines the velocity scale U. Let us normalize the space coordinates by the microscale length, and the unknowns according to the estimates just found,

$$x_i = \ell x_i^\dagger, \quad p = \Delta P p^\dagger, \quad u_i = \frac{\ell^2 \Delta P}{\mu \ell'} u_i^\dagger, \tag{3.1.5}$$

where the primed quantities are dimensionless and ΔP here stands for the scale of pressure variation. Equation (3.1.2) becomes formally,

$$Re\, u_j^\dagger \frac{\partial u_i^\dagger}{\partial x_j^\dagger} = -\frac{1}{\epsilon}\frac{\partial p^\dagger}{\partial x_i^\dagger} + \nabla^{\dagger 2} u_i^\dagger, \tag{3.1.6}$$

where

$$Re \equiv \epsilon \frac{\rho \ell^2 \Delta P}{\mu^2} \tag{3.1.7}$$

is just the Reynolds number. The dimensionless continuity equation remains in the form of Eq. (3.1.1).

Let us assume the Reynolds number to be small, say of order $\mathcal{O}(\epsilon)$, i.e.,

$$Re = \mathcal{O}(\epsilon), \quad \text{or} \quad \frac{\rho \ell^2 \Delta P}{\mu^2} = \mathcal{O}(1). \tag{3.1.8}$$

Note that the pressure gradient term appears formally dominant over all other terms in Eq. (3.1.6). This is because $\boldsymbol{x}$ is normalized by the micro-length scale ℓ in every term.

Again we return to dimensional variables but retain the ordering symbol ϵ to help indicate the relative magnitude of each term

$$\epsilon^2 \rho u_j \frac{\partial u_i}{\partial x_j} = -\frac{\partial p}{\partial x_i} + \epsilon \mu \nabla^2 u_i \,. \tag{3.1.9}$$

The boundary condition on the wetted surface Γ of the pores is already given by Eq. (3.1.3).

Consider a periodic geometry on the microscale, as depicted in Fig. 3.1. Each periodic cell Ω is a box of dimension $\mathcal{O}(\ell)$. We then expect u_i and p to be spatially periodic from cell to cell, while changing slowly over the macroscale.

Let us now introduce the multiple-scale coordinates $\boldsymbol{x}$ and $\boldsymbol{x}' = \epsilon \boldsymbol{x}$ and the perturbation expansions

$$u_i = u_i^{(0)} + \epsilon u_i^{(1)} + \epsilon^2 u_i^{(2)} + \cdots \,, \tag{3.1.10}$$

$$p = p^{(0)} + \epsilon p^{(1)} + \epsilon^2 p^{(2)} + \cdots \,, \tag{3.1.11}$$

where $u_i^{(n)}, p^{(n)}$ are functions of $\boldsymbol{x}$ and $\boldsymbol{x}'$.

From Eq. (3.1.1) perturbation equations at the first three orders $\mathcal{O}(\epsilon^0)$, $\mathcal{O}(\epsilon)$ and $\mathcal{O}(\epsilon^2)$ are obtained:

$$\frac{\partial u_i^{(0)}}{\partial x_i} = 0 \,, \tag{3.1.12}$$

$$\frac{\partial u_i^{(1)}}{\partial x_i} + \frac{\partial u_i^{(0)}}{\partial x_i'} = 0 \,, \tag{3.1.13}$$

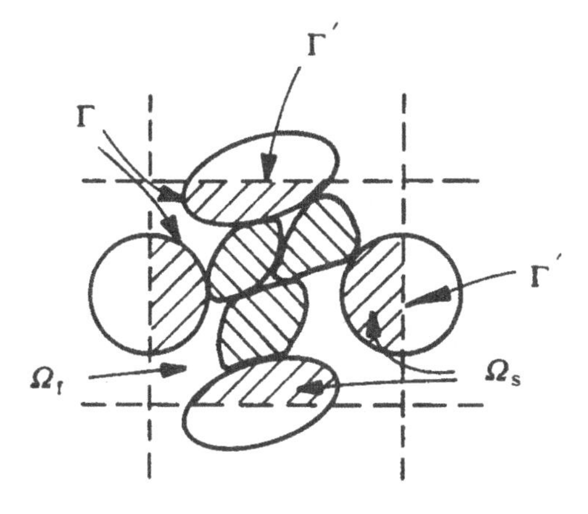

Fig. 3.1. Typical geometry in a periodic cell.

$$\frac{\partial u_i^{(2)}}{\partial x_i} + \frac{\partial u_i^{(1)}}{\partial x_i'} = 0, \tag{3.1.14}$$

$$\vdots$$

Similarly we get from Eq. (3.1.9)

$$0 = -\frac{\partial p^{(0)}}{\partial x_i}, \tag{3.1.15}$$

$$0 = -\frac{\partial p^{(0)}}{\partial x_i'} - \frac{\partial p^{(1)}}{\partial x_i} + \mu \nabla^2 u_i^{(0)}, \tag{3.1.16}$$

$$\rho u_j^{(0)} \frac{\partial u_i^{(0)}}{\partial x_j} = -\frac{\partial p^{(1)}}{\partial x_i'} - \frac{\partial p^{(2)}}{\partial x_i} + \mu \nabla^2 u_i^{(1)} + 2\mu \nabla \cdot \nabla' u_i^{(0)}, \tag{3.1.17}$$

$$\vdots$$

where $\nabla' = \{\partial/\partial x_j'\}$. On the fluid/grain interface Γ the velocity vanishes, hence

$$u_i^{(0)} = u_i^{(1)} = u_i^{(2)} = \cdots = 0 \quad x_i \in \Gamma. \tag{3.1.18}$$

In a typical Ω cell the flow must be periodic,

$$u_i^{(0)}, u_i^{(1)}, u_i^{(2)}, \cdots \quad p^{(0)}, p^{(1)}, p^{(2)}, \cdots \quad \text{are } \Omega\text{-periodic}. \tag{3.1.19}$$

From Eq. (3.1.15) it is clear that

$$p^{(0)} = p^{(0)}(\boldsymbol{x}'), \tag{3.1.20}$$

i.e., the leading-order pressure does not fluctuate on the pore scale. Now Eqs. (3.1.12) and (3.1.16) constitute a Stokes problem for $u_i^{(0)}$ and $p^{(1)}$ forced by the macroscale pressure gradient, yet unknown. Thanks to linearity the solution can be formally represented by

$$u_i^{(0)} = -K_{ij} \frac{\partial p^{(0)}}{\partial x_j'}, \quad p^{(1)} = -A_j \frac{\partial p^{(0)}}{\partial x_j'} + p_o^{(1)}(\boldsymbol{x}'), \tag{3.1.21}$$

where $p_o^{(1)}$ is an integration "constant". It then follows that $K_{ij}(\boldsymbol{x}, \boldsymbol{x}')$ and $A_j(\boldsymbol{x}, \boldsymbol{x}')$ must satisfy

$$\frac{\partial K_{ij}}{\partial x_i} = 0, \tag{3.1.22}$$

$$-\frac{\partial A_j}{\partial x_i} + \mu \nabla^2 K_{ij} = -\delta_{ij}, \tag{3.1.23}$$

where

$$K_{ij} = 0 \quad \text{on } \Gamma, \tag{3.1.24}$$

$$K_{ij} \text{ and } A_j \text{ are } \Omega\text{-periodic}. \tag{3.1.25}$$

These conditions define a Stokes problem of low Reynolds number flow in a unit cell.

Let us define the volume (or geometrical) average over a Ω-cell by

$$\langle f \rangle = \frac{1}{\Omega} \iiint_{\Omega_f} f \, d\Omega \,, \tag{3.1.26}$$

where Ω is also used to denote the gross volume of the Ω-cell and Ω_f is the fluid volume inside. Because only the gradient of A_j appears in the governing equations we impose a constraint to ensure its uniqueness,

$$\langle A_j \rangle = 0 \,. \tag{3.1.27}$$

In short for each $j = 1, 2$, or 3, we must solve a boundary value problem in a unit cell for the Stokes flow forced by a unit pressure gradient; there being four scalar unknowns K_{ij} and A_j. This cell problem can in principle be solved numerically for any prescribed microstructure.

By taking the cell average of Eq. (3.1.21) we get

$$\left\langle u_i^{(0)} \right\rangle = -\mathcal{K}_{ij} \frac{\partial p^{(0)}}{\partial x_j'} \,, \quad \text{where } \mathcal{K}_{ij} = \langle K_{ij} \rangle \,, \tag{3.1.28}$$

and

$$\langle p^{(1)} \rangle = \theta p_o^{(1)} \,, \tag{3.1.29}$$

where $\left\langle u_i^{(0)} \right\rangle$ is called the seepage velocity and $\mathcal{K}_{ij} = \langle K_{ij} \rangle$ the hydraulic conductivity, and

$$\theta = \frac{\Omega_f}{\Omega} \tag{3.1.30}$$

is the porosity. Equation (3.1.28) is the celebrated Darcy's law which was first established empirically. Theoretical derivation in the present manner was first given by Ene and Sanchez-Palencia (1975), Bensoussan *et al.* (1978), and Keller (1980).

The Ω-average of Eq. (3.1.13) gives

$$\left\langle \frac{\partial u_i^{(0)}}{\partial x_i'} \right\rangle + \frac{1}{\Omega} \iiint_{\Omega_f} \frac{\partial u_i^{(1)}}{\partial x_i} d\Omega = 0 \,. \tag{3.1.31}$$

By using the Gauss theorem and the boundary conditions, the second term above vanishes. In Appendix 3A, we shall show the *spatial averaging theorem* which permits the interchange of integration with respect to $\boldsymbol{x}$ and differentiation with respect to $\boldsymbol{x}'$. It follows that

$$\frac{\partial \langle u_i^{(0)} \rangle}{\partial x_i'} = 0 \,, \tag{3.1.32}$$

which implies, in turn,

$$\frac{\partial}{\partial x_i'}\left(\mathcal{K}_{ij}\frac{\partial p^{(0)}}{\partial x_j'}\right) = 0\,. \tag{3.1.33}$$

Equations (3.1.28) and (3.1.32), or (3.1.33) alone govern the seepage flow in a rigid porous medium on the macroscale.

If the medium is isotropic and homogeneous on the macroscale, we have

$$\mathcal{K}_{ij} = \mathcal{K}\delta_{ij}\,, \tag{3.1.34}$$

where $\mathcal{K}$ is a constant. Equation (3.1.33) reduces further to Laplace's equation

$$\frac{\partial^2 p^{(0)}}{\partial x_k'\partial x_k'} = 0\,. \tag{3.1.35}$$

With proper boundary conditions on the macroscale, $p^{(0)}$ can then be found.

It may be pointed out that it is the assumption (Eq. (3.1.8)) that renders the perturbation equations linear at each order. If the Reynolds number is finite, i.e., $Re = \mathcal{O}(1)$, then the convective inertia would be of the order $\mathcal{O}(\epsilon)$ and Eq. (3.1.16) must be replaced by

$$\rho u_j \frac{\partial u_i^{(0)}}{\partial x_j} = -\frac{\partial p^{(0)}}{\partial x_j'} - \frac{\partial p^{(1)}}{\partial x_i} + \mu\nabla^2 u_i^{(0)}\,. \tag{3.1.36}$$

Together with Eq. (3.1.12) the cell problem is fully nonlinear (Ene and Sanchez-Palencia, 1975). Equation (3.1.33) and Darcy's law no longer hold and must be replaced by a nonlinear relation between $\langle \boldsymbol{u}^{(0)}\rangle$ and $\nabla' p^{(0)}$.

3.2. Uniqueness of the Cell Boundary-Value Problem

We shall now show that if $\left(K_{ij}^{(1)}, A_j^{(1)}\right)$ and $\left(K_{ij}^{(2)}, A_j^{(2)}\right)$ are two solutions to the same cell boundary problem, the two K_{ij}s and the two A_js must be the same.

Due to linearity, the two differences $\widehat{K}_{ij} = K_{ij}^{(1)} - K_{ij}^{(2)}$ and $\widehat{A}_j = A_j^{(1)} - A_j^{(2)}$ must be the solution to the homogeneous cell problem, defined by Eqs. (3.1.22)–(3.1.25) with zero on the right-hand side of Eq. (3.1.23), i.e.,

$$-\frac{\partial \widehat{A}_j}{\partial x_i} + \mu\nabla^2 \widehat{K}_{ij} = 0\,. \tag{3.2.1}$$

Taking the scalar product of $\widehat{K}_{ij}$ and Eq. (3.2.1), we get, after partial integration

$$-\frac{\partial}{\partial x_i}\left(\widehat{K}_{ij}\widehat{A}_j\right)+\mu\frac{\partial}{\partial x_k}\left(\widehat{K}_{ij}\frac{\partial \widehat{K}_{ij}}{\partial x_k}\right)-\mu\left(\frac{\partial \widehat{K}_{ij}}{\partial x_k}\frac{\partial \widehat{K}_{ij}}{\partial x_k}\right)=0\,.$$

If the entire equation is integrated over Ω_f, the first two volume integrals can be turned to surface integrals by Gauss' theorem, and be shown to vanish because of the boundary conditions, therefore,

$$-\mu\iiint_{\Omega}\frac{\partial \widehat{K}_{ij}}{\partial x_k}\frac{\partial \widehat{K}_{ij}}{\partial x_k}d\Omega=0\,.$$

This equality is possible if and only if

$$\widehat{K}_{ij}\equiv 0\,,\quad \forall\,\boldsymbol{x}\in\Omega_f\,, \tag{3.2.2}$$

which ensures uniqueness. From Eqs. (3.2.1) and (3.1.27), it is obvious that $\widehat{A}_j\equiv 0$.

3.3. Symmetry and Positiveness of Hydraulic Conductivity

As in Chapter 2 on the diffusion of heat, the hydraulic conductivity tensor can be shown to be symmetric and positive. Since there is only one phase, the proof is simpler.

Let us form the scalar product of K_{ip} and Eq. (3.1.23)

$$-K_{ip}\frac{\partial A_j}{\partial x_i}+\mu K_{ip}\frac{\partial^2 K_{ij}}{\partial x_k\partial x_k}=-\delta_{ij}K_{ip}=-K_{jp}\,.$$

Taking the volume average over Ω and integrating by parts, we get

$$-\iiint_{\Omega_f}\frac{\partial}{\partial x_i}(K_{ip}A_j)d\Omega+\mu\iiint_{\Omega_f}\frac{\partial}{\partial x_k}\left(K_{ip}\frac{\partial K_{ij}}{\partial x_k}\right)d\Omega$$
$$-\mu\iiint_{\Omega_f}\frac{\partial K_{ip}}{\partial x_k}\frac{\partial K_{ij}}{\partial x_k}d\Omega=-\Omega\langle K_{jp}\rangle\,.$$

By Gauss' theorem and the boundary conditions, the first two integrals vanish, hence,

$$\mu\left\langle\frac{\partial K_{ip}}{\partial x_k}\frac{\partial K_{ij}}{\partial x_k}\right\rangle=\langle K_{jp}\rangle\,.$$

Since the left-hand side is symmetric with respect to the interchange of subscripts j and p, the symmetry of hydraulic conductivity follows,

$$\langle K_{jp}\rangle=\langle K_{pj}\rangle\,,\quad\text{or}\quad \mathcal{K}_{jp}=\mathcal{K}_{pj}\,. \tag{3.3.1}$$

To prove the positive-definiteness of the effective permeability let us consider the quadratic scalar product

$$\mu \left\langle \frac{\partial K_{ip}}{\partial x_k} \frac{\partial K_{ij}}{\partial x_k} \right\rangle \xi_p \xi_j = \langle K_{jp} \rangle \xi_j \xi_p \,, \tag{3.3.2}$$

where $\boldsymbol{\xi} = (\xi_p)$ is an arbitrary vector independent of the short scale coordinates. Since the left-hand side can be written as

$$\mu \left\langle \frac{\partial K_{ip}\xi_p}{\partial x_k} \frac{\partial K_{ij}\xi_j}{\partial x_k} \right\rangle,$$

which is clearly positive, the right-hand side of Eq. (3.3.2) must also be positive for any ξ_i.

Next, if the right-hand side in Eq. (3.3.2) is zero, then $K_{ip}\xi_p = 0$. Let us multiply Eq. (3.1.23) by a vector function $\boldsymbol{w}$ such that

$$\frac{\partial w_i}{\partial x_i} = 0, \text{in } \Omega_f, w_i = 0 \text{ on } \Gamma, \langle w_i \rangle = \xi_i$$

and integrate on Ω_f. The cell problem (Eqs. (3.1.22)–(3.1.25)) implies

$$-\mu \iiint_{\Omega_f} \frac{\partial (K_{ij}\xi_j)}{\partial x_k} \frac{\partial w_i}{\partial x_k} d\Omega = - \iiint_{\Omega_f} w_i \xi_j \delta_{ij} d\Omega = -\Omega \xi_i \xi_i$$

and thus if the left-hand side is zero, then $\xi_i = 0$. Therefore the effective permeability tensor $\mathcal{K}$ is positive definite.

3.4. Numerical Computation of the Permeability Tensor

Numerical solution of the Stokes flow through a bed of uniform spheres in cubic packing has been described by Snyder and Stewart (1966) by a Galerkin method with trigonometric series as trial functions. Later, Zick and Homsy (1980) applied the technique of integral equations, and obtained more accurate solutions for six different arrays of uniform spheres. The cases of contacting spheres are taken as special limits, corresponding to just six different values of porosity. Lee *et al.* (1996) have chosen the Wigner–Seitz grain[1] which is a polyhedron with 14 sides as shown in Fig. 3.2. For a

[1]This is a well-known model in solid state physics (Burns, 1985).

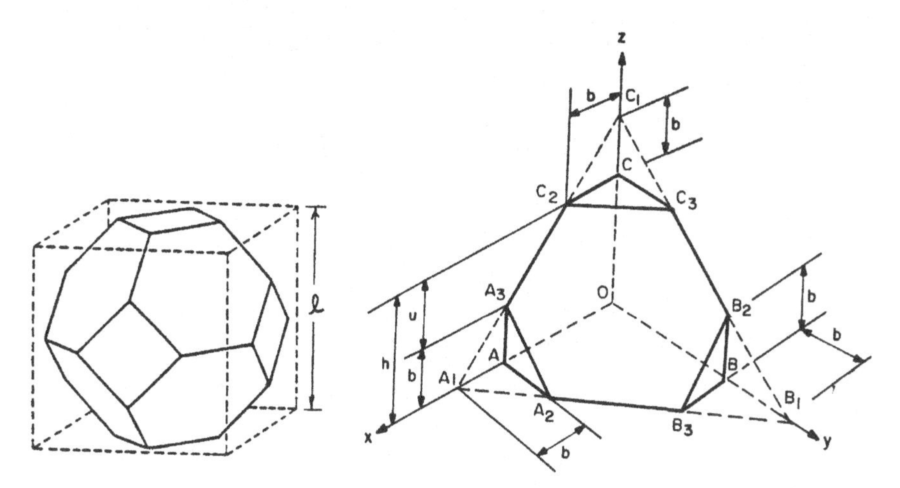

Fig. 3.2. Left: A unit cell containing a Wigner–Seitz grain. Right: Geometrical parameters of a Wigner–Seitz grain in one-eighth of a cell.

cubically packed array, each grain is in contact with six neighbors. It can be shown that the porosity is given by

$$\theta = 1 - \frac{1}{6}\left[\left(1 + \frac{b}{h}\right)^3 - 3\left(\frac{b}{h}\right)^3\right]. \tag{3.4.1}$$

At the limit of $b/h = 0$, the porosity is the greatest $\theta = 5/6$ while neighboring grains are still in point contact. The minimum porosity is $\theta = 1/6$ corresponding to $b/h = 1$; fluid cannot flow from pore to pore. Beyond the higher limit ($\theta = 5/6$), the grains lose contact and are suspended in fluid.

The Ω-cell is a unit cube containing just one grain, and the cell problem can be solved by the finite element method. First let us recast the Stokes problem in the cell as a variational principle, i.e., its solution is equivalent to the minimization of the following functional over all symmetric (K_{ij}) with square integrable gradients, and square-integrable (A_j):

$$\mathcal{F} = \frac{\mu}{2}\iiint_{\Omega_f} \frac{\partial K_{ij}}{\partial x_m}\frac{\partial K_{ij}}{\partial x_m} d\Omega$$

$$+ \iiint_{\Omega_f} K_{ij}\left(\frac{\partial A_j}{\partial x_i} - \delta_{ij}\right) d\Omega + \lambda_j \iiint_{\Omega_f} A_j d\Omega\,, \tag{3.4.2}$$

where $\lambda_j, j = 1, 2, 3$ are Lagrange's multipliers. By taking the first variation and performing partial integration, we get

$$\delta\mathcal{F} = \mu \iiint_{\Omega_f} \frac{\partial}{\partial x_m}\left(\delta K_{ij}\frac{\partial K_{ij}}{\partial x_m}\right) d\Omega + \iiint_{\Omega_f} \frac{\partial}{\partial x_i}(K_{ij}\delta A_j) d\Omega$$
$$- \iiint_{\Omega_f} \delta K_{ij}\left(\mu\frac{\partial^2 K_{ij}}{\partial x_m \partial x_m} - \frac{\partial A_j}{\partial x_i} + \delta_{ij}\right) d\Omega$$
$$- \iiint_{\Omega_f} \delta A_j \frac{\partial K_{ij}}{\partial x_i} d\Omega + \lambda_j \iiint_{\Omega_f} \delta A_j d\Omega\,.$$

By Gauss' theorem, the first two volume integrals can be transformed to surface integrals

$$\mu \iint_{\partial\Omega_f} \delta K_{ij}\frac{\partial K_{ij}}{\partial x_m} dS + \iint_{\partial\Omega_f} K_{ij}\delta A_j dS\,,$$

where $\partial\Omega_f = \partial\Omega \cup \Gamma$; both of these vanish by virtue of the boundary conditions. If K_{ij} and A_j are the solutions of the cell boundary value problem, subjected to the constraint (3.1.27), the remaining volume integrals also vanish. Hence the first variation of $\mathcal{F}$ is zero,

$$\delta\mathcal{F} = 0\,, \tag{3.4.3}$$

and $\mathcal{F}$ is stationary. Conversely, since δK_{ij} and δA_j are arbitrary, vanishing of $\delta\mathcal{F}$ also implies all the conditions defining boundary-value problem for K_{ij} and A_j. Thus the Stokes problem in the Ω-cell is equivalent to the stationarity of the functional $\mathcal{F}$.

By using finite elements to approximate K_{ij} and A_j in Ω_f, a linear system of algebraic equations for the nodal coefficients can be obtained by extremizing $\mathcal{F}$, subject to the constraint of Eq. (3.1.27). Because of the geometrical symmetry, it is only necessary to consider $\langle K_{ij}\rangle$ induced by unit pressure gradient in the direction of x_1, and limit the discrete computation to one-eighth of the cell.

Many unidirectional-flow experiments have been performed in the past for nearly uniform spheres, powders, and natural sand with a considerable size variation and for a wide range of Reynolds numbers ($0 < Re < \mathcal{O}(1000)$). Based on averages of measured data, empirical formulas have been proposed. Among them that of Kozeny and Carman[2] (Carman, 1937)

[2]Within the measured range the data may deviate from the empirical formula by 10–20%, due to the irregularity in shape and size variations of the specimen tested (Carman, 1937).

is the best known,

$$\mathcal{K} = \frac{\mathcal{L}^2}{5\mu} \frac{\theta^3}{(1-\theta)^2}, \tag{3.4.4}$$

where $\mathcal{L}$ is the ratio of volume to surface area of the grain, a length characterizing both the size and the shape of the grain. For the Wigner–Seitz particle the value of $\mathcal{L}$ can be calculated and put in the Kozeny–Carman formula; the empirical result is compared with the finite element calculations in Fig. 3.3(a) for the normalized conductivity $\langle K_{ii}\rangle\mu/\ell^2$. Despite

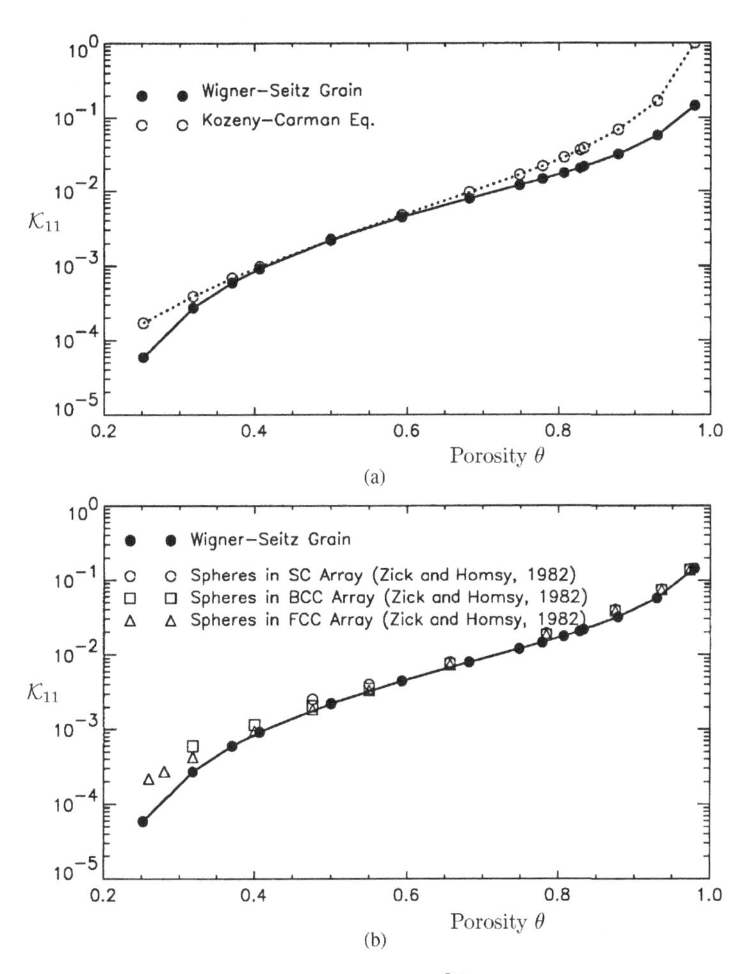

Fig. 3.3. Variation of permeability normalized by ℓ^2/μ versus porosity θ. Computed permeability is compared (a) with empirical formula by Kozeny and Carman, and (b) with computations by Zick and Homsy for uniform spheres. From Lee *et al.* (1996), *Int. J. Heat Mass Transf.*

the idealized model of periodic and uniform grains, the agreement is quite good over the range $0.3 < \theta < 0.8$, where most of the measurements were recorded. Outside this range the empirical formula is only an extrapolation of measured data. Nevertheless, the agreement indicated here suggests that the hydraulic conductivity may be rather insensitive to the irregularities in grain size or shape. Note that the normalized permeability varies over a wide range depending on the porosity, from $\mathcal{O}(1)$ at high porosity to $\mathcal{O}(10^{-4})$ at low porosity. Therefore the scale ℓ^2/μ gives only the maximum of $\mathcal{K}$.

The computed permeability is also compared with the numerical results of Zick and Homsy (1980) for uniform spheres in various arrangements in Fig. 3.3(b). The agreement is also good for a wide range of porosities.

In the following two sections, we discuss two variations of the same theme.

3.5. Seepage of a Compressible Fluid

Sometimes air or gas is forced to flow through a porous medium under very high pressure. A method for cleansing wet soils contaminated by volatile organic compounds is to inject compressed air through a nozzle into the groundwater beneath the water table, so as to create a buoyant plume of bubbles or air channels. Chemicals dissolved in groundwater can be passed to the gaseous phase and rise with the plume above the water table. Then the contaminated gas can be vacuumed to another site for treatment. Since the typical pressure at the point of injection can be a sizable fraction of an atmosphere, compressibility of the gas must be accounted for. The transport of chemicals by diffusion and convection, which was treated in Sec. 2.12, depends on the knowledge of gas flow through the pores.

For the steady flow of a compressible gas, the continuity equation is

$$\frac{\partial(\rho u_i)}{\partial x_i} = 0, \tag{3.5.1}$$

and the momentum equation is

$$\rho u_j \frac{\partial u_i}{\partial x_j} = -\frac{\partial p}{\partial x_i} + \mu \frac{\partial^2 u_i}{\partial x_j \partial x_j} + \frac{\mu}{3} \frac{\partial^2 u_j}{\partial x_i \partial x_j}, \tag{3.5.2}$$

where gravity has been ignored and the bulk viscosity is assumed to be zero. Assuming an ideal gas, the equation of state is

$$\frac{pM}{R_g T} = \rho. \tag{3.5.3}$$

In the preceding equations, u_i is the velocity component, p is the pressure, ρ is the air density, μ is the dynamic viscosity, M is the molecular weight of air mixture, R_g is the universal gas constant, and T is the absolute temperature. Under isothermal conditions the dynamic viscosity is essentially constant owing to its insensitivity to pressure variations.

As discussed before, the scale of pressure variation P is characterized by $\mu\ell' U_g/\ell^2$ over the global length scale ℓ'. In terms of the normalized variables (distinguished by daggers):

$$x_i = l x_i^\dagger\,, \quad p = P p^\dagger\,, \quad u_i = U_g u_i^\dagger\,, \quad \rho = \frac{PM}{R_g T}\rho^\dagger\,, \tag{3.5.4}$$

with $U_g = \ell^2 \Delta P/\mu\ell'$, the dimensionless Navier–Stokes equations are

$$\frac{\partial (p^\dagger u_i^\dagger)}{\partial x_i^\dagger} = 0\,, \tag{3.5.5}$$

$$\frac{\partial p^\dagger}{\partial x_i^\dagger} = \epsilon \left(\frac{\partial^2 u_i^\dagger}{\partial x_j^\dagger \partial x_j^\dagger} + \frac{1}{3}\frac{\partial^2 u_j^\dagger}{\partial x_i^\dagger \partial x_j^\dagger} \right) - \epsilon Re\, \rho^\dagger u_j^\dagger \frac{\partial u_i^\dagger}{\partial x_j^\dagger}\,. \tag{3.5.6}$$

We shall assume

$$Re = \frac{PM}{R_g T}\frac{U_g \ell}{\mu} = \mathcal{O}(\epsilon) \ll 1\,, \tag{3.5.7}$$

hence the last term in Eq. (3.5.6) is of order $\mathcal{O}(\epsilon^3)$.

From here on let us return to physical variables but keep the ordering parameter for identification,

$$\frac{\partial (p u_i)}{\partial x_i} = 0\,, \tag{3.5.8}$$

$$\frac{\partial p}{\partial x_i} = \epsilon\mu \left(\frac{\partial^2 u_i}{\partial x_j^2} + \frac{1}{3}\frac{\partial^2 u_j}{\partial x_i \partial x_j} \right) - \epsilon^2 \rho u_j \frac{\partial u_i}{\partial x_j}\,. \tag{3.5.9}$$

Let us introduce the local and global coordinates x_i and $x_i' = \epsilon x_i$, and multiple-scale expansions as in Eqs. (3.1.10) and (3.1.11), we obtain from Eqs. (3.5.8) and (3.5.9) the following perturbation equations:

$\underline{\mathcal{O}(\epsilon^0)}$:

$$\frac{\partial}{\partial x_i}\left(p^{(0)} u_i^{(0)} \right) = 0\,, \tag{3.5.10}$$

$$\frac{\partial p^{(0)}}{\partial x_i} = 0\,. \tag{3.5.11}$$

$\underline{\mathcal{O}(\epsilon)}$:

$$\frac{\partial}{\partial x_i}\left(p^{(0)}u_i^{(1)} + p^{(1)}u_i^{(0)}\right) + \frac{\partial}{\partial x_i'}\left(p^{(0)}u_i^{(0)}\right) = 0\,, \tag{3.5.12}$$

$$\frac{\partial p^{(1)}}{\partial x_i} + \frac{\partial p^{(0)}}{\partial x_i'} = \mu\frac{\partial^2 u_i^{(0)}}{\partial x_j \partial x_j} + \frac{1}{3}\mu\frac{\partial^2 u_j^{(0)}}{\partial x_i \partial x_j}\,. \tag{3.5.13}$$

The boundary conditions are

$$u_i^{(m)} = 0 \quad \text{on } \Gamma\,, \tag{3.5.14}$$

$$u_i^{(m)} \quad \text{and} \quad p^{(m)} \quad \text{are } \Omega\text{-periodic}\,, \tag{3.5.15}$$

where $m = 0, 1, \cdots$.

Clearly from Eq. (3.5.11)

$$p^{(0)} = p^{(0)}(x_i')\,, \tag{3.5.16}$$

and hence from Eq. (3.5.10)

$$\frac{\partial u_i^{(0)}}{\partial x_i} = 0\,. \tag{3.5.17}$$

That is, to the leading order the pressure depends only on the global coordinates and the air flow is incompressible on the microscale. Consequently the last term on the right-hand side of Eq. (3.5.13) vanishes. Finally the microscale cell problem for $u_i^{(0)}$ and $p^{(1)}$ is the same as that defined by Eqs. (3.1.12)–(3.1.16). Hence we can adopt Eq. (3.1.20) and solve Eqs. (3.1.22)–(3.1.26) to get Darcy's law

$$\langle u_i^{(0)}\rangle = -\langle K_{ij}\rangle\frac{\partial p^{(0)}}{\partial x_j'} = -\mathcal{K}_{ij}\frac{\partial p^{(0)}}{\partial x_j'}\,. \tag{3.5.18}$$

From the law of mass conservation (Eq. (3.5.12)) we obtain by taking the cell average and using Gauss' theorem and Eq. (3.5.14),

$$\frac{\partial}{\partial x_i'}\left(p^{(0)}\langle u_i^{(0)}\rangle\right) = 0\,, \tag{3.5.19}$$

which displays the effects of compressibility. Together Eqs. (3.5.18) and (3.5.19) are the effective flow equations for air on the macroscale.

If the porous matrix is isotropic, $\mathcal{K}_{ij} = \mathcal{K}\delta_{ij}$ and we may combine Eqs. (3.5.18) and (3.5.19) to yield the nonlinear equation for $p^{(0)}$,

$$\frac{\partial}{\partial x'_i}\left(\mathcal{K}p^{(0)}\frac{\partial p^{(0)}}{\partial x'_i}\right) = 0 . \tag{3.5.20}$$

If we further assume a homogeneous soil, $\mathcal{K}$ reduces to a constant and Eq. (3.5.20) can be simplified as

$$\frac{\partial^2 \left(p^{(0)}\right)^2}{\partial x'_i \partial x'_i} = 0 . \tag{3.5.21}$$

Subject to appropriate global boundary conditions, the nonlinear problem must be solved first; specific discharge $\langle u_i^{(0)} \rangle$ is then found from Eq. (3.5.18).

3.6. Two-Dimensional Flow Through a Three-Dimensional Matrix

In animal lungs there are many alveolar sheets surrounded by air. Each sheet is formed by two membranes held together by an array of transverse tissue posts. Driven by the pressure difference between the arteriole and the venule, blood plasma flows through the gap between the membranes and around the posts. The sheets are semipermeable so that oxygen can enter the blood from, and carbon dioxide can be expelled to, the air outside. For cats the typical post diameter is 4–4.5 μm, the spacing between posts is 12–13 μm, and the membrane thickness is 1–5 μm.

The exchange process between blood and gas depends not only on the permeability of the membranes and is enhanced on the convection and diffusion in the flowing blood which spreads the dissolved chemicals over a large area. The dynamical problem is similar to seepage flow and contaminant transport in saturated soils. An important difference is that the alveolar membranes are elastic. When blood plasma and red blood cells pass through the thin and tortuous space between the membranes and the posts, the gap between the two membranes can expand with increasing blood pressure. Based on measured evidence (Fung, 1977), the following empirical relation has been established between the mean thickness h of the sheet and the pressure difference across the membrane,

$$h = \begin{cases} h_0 + \alpha(p - p_A) , & p > p_A , \\ 0 , & p < p_A , \end{cases} \tag{3.6.1}$$

where p_A is the threshold pressure.

Fung and Sobin (1969) took the first theoretical step by modeling the flow in the alveolar sheet. Their model consists of two elastic and closely spaced membranes connected by a periodic array of transverse posts. By considering only the average quantities such as pressure and the averaged seepage velocities (U, V) parallel to the plane of the walls. Darcy's law between the local pressure gradient and the seepage velocity is first assumed as

$$U_j = -\frac{h^2}{\mu k f}\frac{\partial p}{\partial x_j}, \tag{3.6.2}$$

where k is presumed to be an empirical constant and f is the geometrical friction factor depending on the postal geometry and is a constant in the limiting case without posts (Couette flow). With Eq. (3.6.1), the law of mass conservation leads to the following differential equation of the blood pressure p,

$$\frac{1}{\mu}\frac{\partial}{\partial x_1}\left(\frac{h^3}{kf}\frac{\partial p}{\partial x_1}\right) + \frac{1}{\mu}\frac{\partial}{\partial x_2}\left(\frac{h^3}{kf}\frac{\partial p}{\partial x_2}\right) = \frac{\partial h}{\partial t} + \frac{2K_p}{\rho}(p - p_A), \tag{3.6.3}$$

where p_A is the ambient pressure outside the sheet and K_p the seepage coefficient of the membrane. If k and f are known, Eq. (3.6.1) can be used to give a nonlinear partial differential equation for p or h alone.

As a complement to the macroscale model above, Lee (1969) derived the permeability by solving a microscale problem for the Stokes flow though a periodic cell (septum) bounded by two parallel walls and a periodic array of circular posts. An approximate analytical solution is obtained for the geometrical friction factor f when the solid fraction and h/a are both small, where a is the radius of the posts. In comparison with model experiments, the approximate result is found to be limited to the range of $h/a \leq \mathcal{O}(1)$.

This problem is concerned with two-dimensional seepage through a three-dimensional porous matrix, and has been re-examined by Zhong *et al.* (2002) with a view to putting the heuristic reasoning of Fung and Sobin on a microscale foundation, and to giving accurate numerical results for a wide range of geometrical parameters. Despite the elasticity of the membranes, the passage from microscale to macroscale is very much similar to the rigid matrix cases studied in this chapter.

3.6.1. *Governing Equations*

We consider the flow of pure plasma so that Navier–Stokes equations governing Newtonian fluids are appropriate

$$\frac{\partial u_i}{\partial x_i} = 0, \tag{3.6.4}$$

$$\rho \frac{\partial u_i}{\partial t} + \rho\, u_j \frac{\partial u_i}{\partial x_j} = -\frac{\partial p}{\partial x_i} + \mu \frac{\partial^2 u_i}{\partial x_j \partial x_j}. \tag{3.6.5}$$

Let ω^{-1} be the time scale, U the velocity scale, ΔP the pressure variation scale, ℓ the typical interportal distance, and ℓ' the size of the alveolar sheet. Then for low speed flows we expect the pressure gradient to be comparable to the viscous stress,

$$\frac{\Delta P}{\ell'} \sim \frac{\mu U}{\ell^2}.$$

Compared to the viscous stress, the local acceleration is of the magnitude

$$\frac{\rho(\partial u_i/\partial t)}{\mu(\partial^2 u_i/\partial x_j \partial x_j)} \sim \frac{\rho \omega \ell^2}{\mu} = \frac{\omega \ell^2}{\nu} = W_0^2,$$

where

$$W_0 = \ell \sqrt{\frac{\omega}{\nu}} \tag{3.6.6}$$

is called the Womersley number in biomechanics which is the ratio of microscale length to the thickness of Stokes boundary layer. The local convective inertia is

$$\frac{\rho u_j (\partial u_i/\partial x_j)}{\mu(\partial^2 u_i/\partial x_i \partial x_j)} \sim \frac{\rho U \ell}{\mu} = Re,$$

where Re is the Reynolds number. Taking for estimation, $\ell' \sim 1\,\text{cm} = 10^4\,\mu\text{m}$, $\ell \sim 10^{-3}\,\text{cm} = 10\,\mu\text{m}$, $\nu \sim 10^{-2}\,\text{cm}^2/\text{s}$ $\omega \sim 2\pi\,\text{s}^{-1}$ and $U \sim 0.1\,\text{cm/s}$, we get $\ell/\ell' = 0.006 = \mathcal{O}(10^{-2}) \ll 1$, $Re \sim 10^{-2} \ll 1$ and $W_0^2 \sim 10^{-3}$ which are all small.

Normalizing all coordinates by ℓ, we then have

$$\frac{(\partial p/\partial x_i)}{\mu(\partial^2 u_i/\partial x_j \partial x_j)} \sim \frac{\Delta P l}{\mu U} \sim \frac{\ell'}{\ell} \gg 1.$$

Let us define a small parameter

$$\epsilon = \frac{\ell}{\ell'} \ll 1, \tag{3.6.7}$$

and assume

$$Re \sim W_0^2 = \mathcal{O}(\epsilon). \tag{3.6.8}$$

Using ϵ to express the relative magnitude we may rewrite Eq. (3.6.5) with dimensions as

$$\epsilon^3 \rho \frac{\partial u_i}{\partial t} + \epsilon^2 \rho\, u_j \frac{\partial u_i}{\partial x_j} = -\frac{\partial p}{\partial x_i} + \epsilon\, \mu \frac{\partial^2 u_i}{\partial x_i \partial x_j}. \tag{3.6.9}$$

We shall model the posts by circular cylinders and ignore the transverse deformation. Condition of no slippage on the walls (B) requires that

$$u_i = 0 \quad \text{on } B\,. \tag{3.6.10}$$

Let $\boldsymbol{q}$ denote the transverse membrane velocity. To account for seepage across the membranes we assume that the normal flux $(\boldsymbol{u} - \boldsymbol{q}) \cdot \boldsymbol{n}$ be proportional to the pressure difference on two sides of the alveolar membrane,

$$\rho(\boldsymbol{u} - \boldsymbol{q}) \cdot \boldsymbol{n} = \pm K_p (p - p_A)\,, \tag{3.6.11}$$

where K_p denotes the membrane seepage coefficient.

On the membrane surfaces described by

$$F(x, y, z, t) = z \mp \frac{h(x, y, t)}{2} = 0\,, \tag{3.6.12}$$

the motion of a point on the membrane must obey the kinematic condition

$$\frac{\partial F}{\partial t} + \boldsymbol{q} \cdot \nabla F = 0\,, \tag{3.6.13}$$

hence the normal velocity of the membrane is

$$\boldsymbol{q} \cdot \boldsymbol{n} = \frac{\boldsymbol{q} \cdot \nabla F}{|\nabla F|} = \frac{-(\partial F/\partial t)}{|\nabla F|}\,. \tag{3.6.14}$$

It follows from Eq. (3.6.11) that

$$\frac{(\partial F/\partial t)}{|\nabla F|} + \frac{\boldsymbol{u} \cdot \nabla F}{|\nabla F|} = \pm \frac{K_p}{\rho}(p - p_A)\,, \tag{3.6.15}$$

or

$$\mp \frac{1}{2}\left(\frac{\partial h}{\partial t} + u\frac{\partial h}{\partial x} + v\frac{\partial h}{\partial y}\right) + w = \pm \frac{K_p}{\rho}(p - p_A)\sqrt{1 + h_x^2 + h_y^2} \tag{3.6.16}$$

on $z = \pm h/2$.

In principle there can be some local indentations where the posts and the membranes are joined. To be confirmed later we assume the indentations to be of higher order so that

$$\frac{\partial h}{\partial x}, \frac{\partial h}{\partial y} = \mathcal{O}\left(\frac{\ell}{\ell'}\right) = \mathcal{O}(\epsilon)\,,$$

$$\frac{\partial h/\partial t}{w} \sim \frac{w\ell}{U} \sim \frac{w}{\nu}\frac{\ell^2 \nu}{U\ell} \sim \frac{\ell^2}{\delta^2}\frac{1}{Re} = \mathcal{O}(\epsilon),$$

and

$$\frac{u(\partial h/\partial x), v(\partial h/\partial y)}{w} = \mathcal{O}(\epsilon).$$

For generality we assume

$$K_p(p - p_A) = \mathcal{O}\left(\frac{\partial h}{\partial t}\right),$$

which determines the maximum magnitude of K_p. Finally we display the order of magnitude of each term in Eq. (3.6.16) by rewriting it as

$$w = \pm\frac{\epsilon}{2}\left(\frac{\partial h}{\partial t} + u\frac{\partial h}{\partial x} + v\frac{\partial h}{\partial y}\right) \pm \epsilon\frac{K_p}{\rho}(p - p_A)\sqrt{1 + \epsilon^2 h_x^2 + \epsilon^2 h_y^2} \tag{3.6.17}$$

on $z = \pm h/2$.

3.6.2. *Homogenization*

We now introduce fast and slow variables $x_i = (x, y, z)$ and $x_i' = (\epsilon x, \epsilon y)$ and the usual expansions

$$u_i = u_i^{(0)} + \epsilon u_i^{(1)} + \cdots, \quad p = p^{(0)} + \epsilon p^{(1)} + \cdots, \\ h = h^{(0)} + \epsilon h^{(1)} + \cdots, \tag{3.6.18}$$

where $u_i^{(n)}$, $p^{(n)}$, and $h^{(n)}$ are functions of (x_i, x_i', t). Note that the slow variables are only in two dimensions. It follows from the continuity equation that

$$\sum_{i=1}^{3} \frac{\partial u_i^{(0)}}{\partial x_i} = 0, \\ \sum_{i=1}^{3} \frac{\partial u_i^{(1)}}{\partial x_i} + \sum_{i=1}^{2} \frac{\partial u_i^{(0)}}{\partial x_i'} = 0, \tag{3.6.19}$$

and from the momentum equations that

$$0 = -\frac{\partial p^{(0)}}{\partial x_i}, \\ 0 = -\frac{\partial p^{(1)}}{\partial x_i} - \frac{\partial p^{(0)}}{\partial x_i'} + \mu \sum_{k=1}^{3} \frac{\partial^2 u_i^{(0)}}{\partial x_j \partial x_j}. \tag{3.6.20}$$

The membrane elasticity condition now implies

$$\begin{aligned} h^{(0)} - h_0 &= \alpha(p^{(0)} - p^*)\,, \\ h^{(1)} &= \alpha\, p^{(1)}\,. \end{aligned} \tag{3.6.21}$$

All three velocity components vanish on the walls of the posts B

$$u_i^{(0)} = u_i^{(1)} = \cdots 0 \quad i = 1, 2, 3 \quad \text{on } B\,. \tag{3.6.22}$$

On the membranes only the horizontal components vanish

$$u_i^{(0)} = u_i^{(1)} = \cdots 0 \quad i = 1, 2 \quad z = \pm\frac{h}{2}\,. \tag{3.6.23}$$

As for the vertical component, we expand Eq. (3.6.17) and get $z = \pm h/2$,

$$\begin{aligned} w^{(0)} &= 0\,, \\ w^{(1)} &= \mp\left[\frac{\partial h}{\partial t} + \frac{K_p}{\rho}(p^{(0)} - p_A)\right]. \end{aligned} \tag{3.6.24}$$

All unknowns are periodic from cell to cell.

We now solve the perturbation problems at the first two orders.

From the first term of Eq. (3.6.20) we get at $\mathcal{O}(\epsilon^0)$ the familiar result

$$p^{(0)} = p^{(0)}(x_i', t)\,, \tag{3.6.25}$$

so that at the leading-order pressure is just the averaged pressure independent of the microscale coordinates. From the first membrane elasticity condition (Eq. (3.6.17)), it follows that $h^{(0)} = h^{(0)}(x_i', t)$, confirming a scale estimate made earlier. At the order $\mathcal{O}(\epsilon)$, $u_i^{(0)} (i = 1, 2, 3)$ and $p^{(1)}$ are governed by the first of Eq. (3.6.19) and second of Eq. (3.6.20), subject to the boundary condition on the walls and periodicity. We now divide the sheet into periodic cells centered around each post. Within the typical cell volume Ω we let

$$u_i^{(0)} = -\sum_{j=1}^{2} K_{ij}\frac{\partial p^{(0)}}{\partial x_j'}\,, \quad p^{(1)} = -\sum_{j=1}^{2} A_j\frac{\partial p^{(0)}}{\partial x_j'} \tag{3.6.26}$$

for $i = 1, 2, 3$. The new unknowns K_{ij} and A_j are functions of $x_i, i = 1, 2, 3$. We get the familiar three-dimensional problem governing $K_{ij}(x_k)$

and $A_j(x_k)$ in a cylindrical cell,

$$\sum_{i=1}^{3} \frac{\partial K_{ij}}{\partial x_i} = 0\,, \tag{3.6.27}$$

$$\mu \sum_{k=1}^{3} \frac{\partial^2 K_{ij}}{\partial x_k \partial x_k} - \frac{\partial A_j}{\partial x_i} = -\delta_{ij} \tag{3.6.28}$$

for $i = 1, 2, 3$, but $j = 1, 2$. In addition,

$$K_{ij} = 0 \tag{3.6.29}$$

on the material boundaries, and

$$K_{ij} \text{ and } A_i \text{ are } \Omega\text{-periodic}\,. \tag{3.6.30}$$

To ensure that $p^{(0)}$ is the average pressure with at most $\mathcal{O}(\epsilon^2)$ error, we impose further

$$\langle A_j \rangle = 0\,, \tag{3.6.31}$$

where $\langle \cdot \rangle$ represents the cell average

$$\langle f \rangle = \frac{1}{S_b h} \int_{-h/2}^{h/2} dz \iint_{S_b} f\, dxdy\,. \tag{3.6.32}$$

S_b denotes the base area of the cylindrical cell space occupied by the fluid so that $S_b h$ is the fluid volume in the cell.

Equations (3.6.27) and (3.6.28), and the boundary conditions define the three-dimensional Stokes problem in a unit cell bounded by the membranes, the post, and the cell boundaries. Once solved for all K_{ij}, Darcy's law follows for the two seepage velocity components parallel to the sheet,

$$\langle u_i^{(0)} \rangle = -\sum_{j=1}^{2} \mathcal{K}_{ij} \frac{\partial p^{(0)}}{\partial x'_j}\,, \quad i = 1, 2\,, \tag{3.6.33}$$

where

$$\mathcal{K}_{ij} \equiv \langle K_{ij} \rangle, \quad i, j = 1, 2 \tag{3.6.34}$$

is the 2×2 conductivity tensor.

We next integrate the second term of Eq. (3.6.19) over the unit cell Ω,

$$\iiint_\Omega \sum_{i=1}^{3} \frac{\partial u_i^{(1)}}{\partial x_i} d\Omega + \iiint_\Omega \sum_{i=1}^{2} \frac{\partial u_i^{(0)}}{\partial x_i'} d\Omega = 0 . \tag{3.6.35}$$

By Gauss theorem and the boundary conditions including periodicity, the first term above becomes

$$S_b[w^{(1)}]_{-h/2}^{h/2} = S_b \left[\frac{\partial h^{(0)}}{\partial t} + \frac{2K_p}{\rho}(p^{(0)} - p_A) \right] . \tag{3.6.36}$$

For the second term in Eq. (3.6.35) we note first that $w^{(0)}$ does not enter. Recalling the averaging theorem, volume integration and differentiation can be interchanged so that

$$\iiint_\Omega \sum_{i=1}^{2} \frac{\partial u_i^{(0)}}{\partial x_i'} d\Omega = \sum_{i=1}^{2} \frac{\partial}{\partial x_i'} \iiint_\Omega u_i^{(0)} \, d\Omega = S_b \sum_{i=1}^{2} \frac{\partial}{\partial x_i'} (\langle u_i^{(0)} \rangle h) . \tag{3.6.37}$$

It follows that

$$\sum_{i=1}^{2} \frac{\partial}{\partial x_i'} (\langle u_i^{(0)} \rangle h^{(0)}) = \frac{\partial h^{(0)}}{\partial t} + \frac{2K_p}{\rho}(p^{(0)} - p_A) . \tag{3.6.38}$$

Since

$$\langle u_i^{(0)} \rangle = - \sum_{j=1}^{2} \mathcal{K}_{ij} \frac{\partial p^{(0)}}{\partial x_j'} , \quad i = 1, 2 , \tag{3.6.39}$$

we get

$$\sum_{i=1}^{2} \sum_{j=1}^{2} \frac{\partial}{\partial x_i'} \left(\mathcal{K}_{ij} h^{(0)} \frac{\partial p^{(0)}}{\partial x_j'} \right) = \frac{\partial h^{(0)}}{\partial t} + \frac{2K_p}{\rho}(p^{(0)} - p_A) , \tag{3.6.40}$$

which agrees formally with Eq. (3.6.3) if we let

$$f_{ij} = \frac{h^2}{\mu k} \mathcal{K}_{ij}^{-1} , \tag{3.6.41}$$

and assume $\mathcal{K}_{ij} = \mathcal{K} \delta_{ij}$.

We leave it to the reader to prove the symmetry and positive definiteness of $\mathcal{K}_{ij}$, and to derive the variational principle that is equivalent to the Stokes boundary value problem for the unit cell.

3.6.3. Numerical Results

Zhong *et al.* (2002) employed a known method of three-dimensional finite elements in elastostatics to solve the three-dimensional Stokes problem here for K_{ij} and A_j. The unit cell is hexagonal of height of h with a circular cylinder of diameter a at its center, as shown in Fig. 3.4. A local coordinate system is used where the posts are in the z-direction. The dimensions of the hexagon can be expressed in terms of the post-spacing (x_0, y_0):

$$l_1 = \frac{1}{2}\left(y_0 + \frac{x_0^2}{y_0}\right), \quad l_2 = \frac{1}{2}\left(y_0 - \frac{x_0^2}{y_0}\right). \tag{3.6.42}$$

The porosity, known also as the vascular–space tissue ratio (VSTR), is defined as the ratio of the vascular lumen volume to the circumscribing volume of the cell, and can be found to be

$$\mathrm{VSTR} = 1 - \frac{\pi a^2}{8\,(l_1 + l_2)\,x_0} = 1 - \frac{\pi a^2}{8 x_0 y_0}. \tag{3.6.43}$$

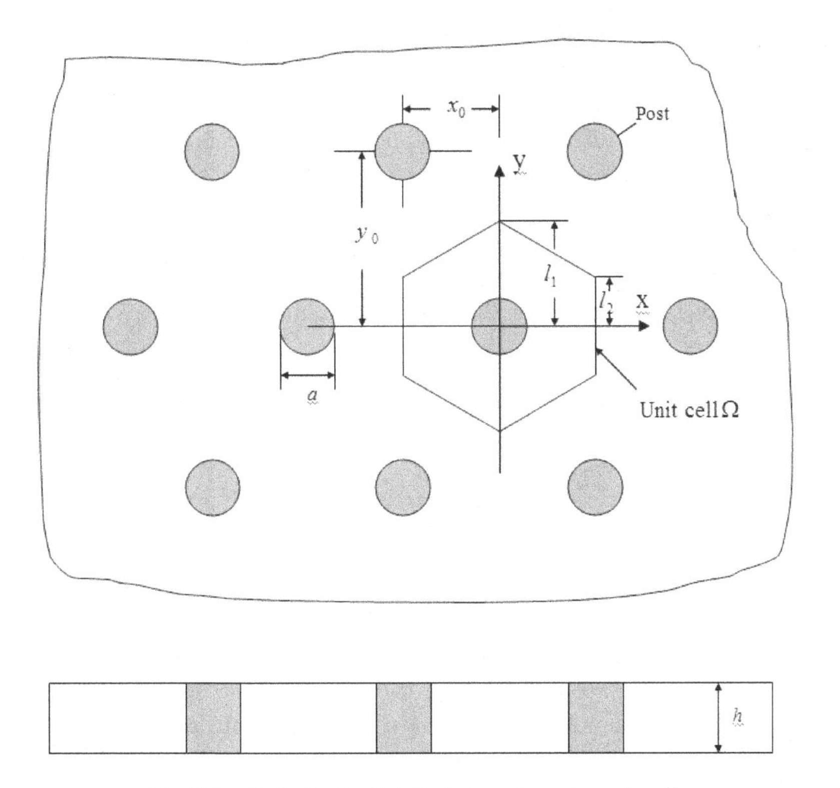

Fig. 3.4. Periodically distributed posts and a unit cell.

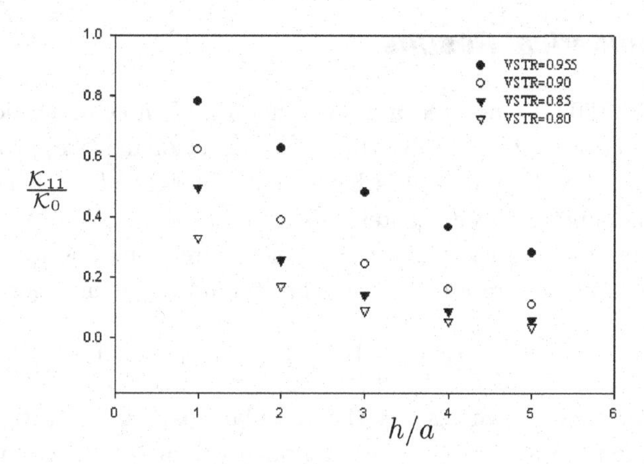

Fig. 3.5. Variation of $\mathcal{K}_{11}/\mathcal{K}_0$ for a regular hexagonal cell with $y_0/x_0 = 1.732$. From Zhong *et al.* (2002), *Comp. Model. Eng. Sci.*

For this symmetric geometry isotropy is expected, i.e., $\mathcal{K}_{12} = \mathcal{K}_{21} = 0$ and $\mathcal{K}_{11} = \mathcal{K}_{22}$. Let $\mathcal{K}_0 \left(= h^2/12\right)$ be the permeability of the classical Couette flow (no posts, i.e., VSTR = 1). The variation of the ratio $\mathcal{K}_{11}/\mathcal{K}_0$ is shown in Fig. 3.5 as a function of the ratio of sheet thickness to post-diameter h/a, for different values of porosity. The relation is nonlinear, implying that f in Eq. (3.6.2) is not a constant. For the same porosity, $\mathcal{K}_{11}/\mathcal{K}_0$ decreases with increasing h/a; flow resistance is reduced for smaller posts or wider gap. The ratio $\mathcal{K}_{11}/\mathcal{K}_0$ appears to approach an asymptotic limit for very large h/a. For a fixed h/a, a smaller porosity leads to a smaller permeability ratio, naturally. The asymptotic limit at large h/a is higher for a larger porosity.

A sample of computed results for elongated cells is shown in Fig. 3.5. From Eq. (3.6.41)

$$\begin{aligned} f_x \equiv f_{11} &= \frac{h^2}{\mu k \mathcal{K}_{11}} = \frac{12}{\mu k}\frac{\mathcal{K}_0}{\mathcal{K}_{11}}, \\ f_y \equiv f_{22} &= \frac{h^2}{\mu k \mathcal{K}_{22}} = \frac{12}{\mu k}\frac{\mathcal{K}_0}{\mathcal{K}_{22}}, \quad f_{xy} = f_{yx} = 0. \end{aligned} \tag{3.6.44}$$

Figure 3.6 shows the calculated f_x versus the ratio of sheet thickness to post-diameter h/a. For comparison, the experimental results of Yen and Fung (1973) as well as the approximate predictions of Lee (1969) are also included. Agreement with the measured data is good for $h/a < 4$. For larger values of h/a, Lee's approximate theory for small h/a loses accuracy.

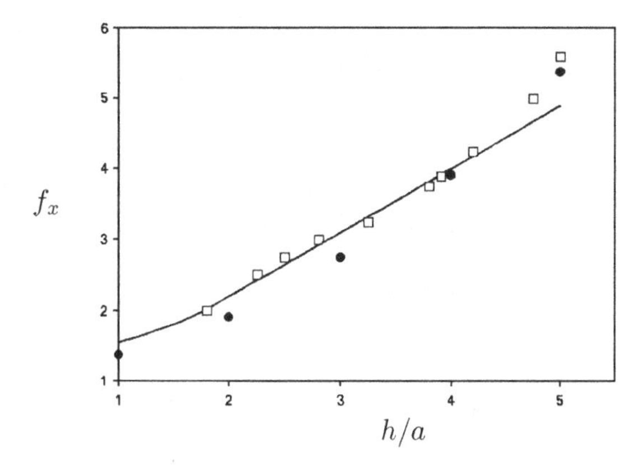

Fig. 3.6. Comparison between the present theory (solid circles) with experiments of Yen and Fung (1973) (hollow squares) and the approximate theory of Lee (1969) (line). The hexagonal cell is elongated with $y_0/x_0 = 3.99, VSTR = 0.955$. From Zhong *et al.* (2002), *Comp. Math. Eng. Sci.*

Since the oxygenation of blood and the removal of carbon dioxide may depend on the details of flow in the septa, an accurate theory of convection and diffusion based on micromechanical theory is needed. This task is mathematically similar to the transport of contaminants in soils. Related theories will be described in Chapter 4.

3.7. Porous Media with Three Scales

Stimulated by environmental concerns of contaminant transport in groundwater, a great deal of recent attention has been focused on the effects of soil heterogeneity over spatial scales much greater than the dimensions in the laboratory. In nature, geological variations cover a wide and continuous range of scales: the aquifer depth ($\mathcal{O}(10\,\text{m})$), the geological formation depth $\mathcal{O}(100\,\text{m})$, and the regional size $\mathcal{O}(1\text{–}10\,\text{km})$ (Dagan, 1989). Physics at one scale may affect the physics at another. In hydrology, statistical models have been constructed to schematize the problems, so that with a finite number of parameters to be supplied by field measurements, certain statistical information on the largest (regional) scale can be predicted (Dagan, 1989; Gelhar, 1993). In these stochastic theories, Darcy's law is taken as the starting basis and the permeability is regarded as a given random quantity. Theories of stochastic differential equations are applied. Assumptions such as stationarity and small departure from the statistical mean are often added to enable analytical progress.

While it is questionable to model earth strata in nature as periodic media, our simple model will give some qualitative guidance to the effects of heterogeneity. Many man-made materials, such as solids composed of laminates of different porous materials, can be approximately treated as periodic media with three or more sharply contrasting scales. We therefore illustrate our homogenization analysis for such composites.

3.7.1. *Effective Equations*

Figure 3.7 shows an idealized model of three-scale porous medium which is a composite consisting of a periodic array of lenses of a different permeability. Then ℓ corresponds to the granular size, ℓ' to the thickness of a period, and ℓ'' the global scale of the entire composite, with $\ell \ll \ell' \ll \ell''$.

The homogenization theory can be achieved in two equivalent ways. One is to break the homogenization process in two steps: first, obtain the mesoscale equations from the microscale, and then the macroscale equations from the mesoscale results. Another is to apply the method of multiple scales with three levels of fast and slow variables $(\boldsymbol{x}, \boldsymbol{x}' = \epsilon\boldsymbol{x}, \boldsymbol{x}'' = \epsilon^2\boldsymbol{x})$ to the basic equations. We shall illustrate the first approach and refer the readers to Mei and Auriault (1989) for the second.

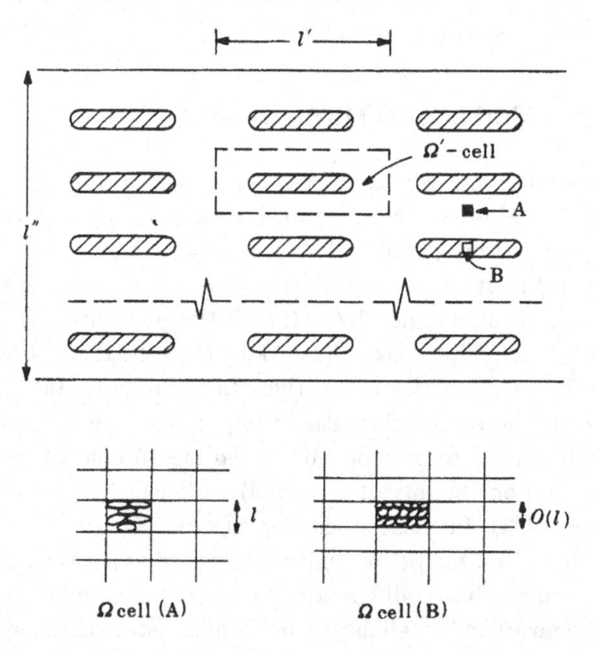

Fig. 3.7. A schematic three-scale porous medium.

We begin with the effective equations already established for the mesoscale, i.e.,

$$\frac{\partial \langle u_i \rangle}{\partial x_i'} = 0 \tag{3.7.1}$$

with Darcy's law

$$\langle u_i \rangle = -\mathcal{K}_{ij} \frac{\partial p}{\partial x_j'}, \tag{3.7.2}$$

where $\langle f \rangle$ denotes the Ω-average of f and

$$\mathcal{K}_{ij} \equiv \langle K_{ij} \rangle. \tag{3.7.3}$$

To respond to the global pressure gradient of the order P/ℓ'', the velocity scale must be KP/ℓ'' where K is the characteristic scale of $\mathcal{K}_{ij}$. If on the right-hand side of Eq. (3.7.2), x_i' is normalized by the mesoscale ℓ', then the ratio of two sides is

$$\frac{\langle u_i \rangle}{-\mathcal{K}_{ij}(\partial p/\partial x_j')} = \mathcal{O}\left(\frac{\ell'}{\ell''}\right) = \mathcal{O}(\epsilon). \tag{3.7.4}$$

To highlight this ordering relationship, we rewrite Darcy's law as

$$\epsilon \langle u_i \rangle = -\mathcal{K}_{ij} \frac{\partial p}{\partial x_j'}. \tag{3.7.5}$$

Now two-scale expansions are introduced:

$$\langle u_i \rangle = \langle u_i^{(0)} \rangle + \epsilon \langle u_i^{(1)} \rangle + \cdots, \tag{3.7.6}$$

$$p = p^{(0)} + \epsilon p^{(1)} + \cdots, \tag{3.7.7}$$

where all perturbation unknowns are functions of $\boldsymbol{x}'$ and $\boldsymbol{x}''$. At the first two orders we have from continuity

$$\frac{\partial \langle u_i^{(0)} \rangle}{\partial x_i'} = 0, \tag{3.7.8}$$

$$\frac{\partial \langle u_i^{(0)} \rangle}{\partial x_i''} + \frac{\partial \langle u_i^{(1)} \rangle}{\partial x_i'} = 0, \tag{3.7.9}$$

and from the mesoscale Darcy's law

$$0 = \frac{\partial p^{(0)}}{\partial x_i'}, \tag{3.7.10}$$

$$\langle u_i^{(0)} \rangle = -\mathcal{K}_{ij} \left(\frac{\partial p^{(1)}}{\partial x_i'} + \frac{\partial p^{(0)}}{\partial x_i''} \right). \tag{3.7.11}$$

Clearly,

$$p^{(0)} = p^{(0)}(\boldsymbol{x}'') \,. \tag{3.7.12}$$

Linearity suggests the following representations:

$$\langle u_i^{(0)} \rangle = -\widetilde{K}_{ik} \frac{\partial p^{(0)}}{\partial x_k''} \,, \tag{3.7.13}$$

and

$$p^{(1)} = -\widetilde{A}_k \frac{\partial p^{(0)}}{\partial x_k''} + p_o^{(1)} \,. \tag{3.7.14}$$

It follows by substituting Eq. (3.7.13) into Eq. (3.7.8) that

$$\frac{\partial \widetilde{K}_{ij}}{\partial x_i'} = 0 \,. \tag{3.7.15}$$

Because of Eq. (3.7.11), Eqs. (3.7.13) and (3.7.14) also imply the following relation between $\widetilde{K}_{ik}$ and $\widetilde{A}_k$:

$$\widetilde{K}_{ik} = \mathcal{K}_{ij} \left(\delta_{jk} - \frac{\partial \widetilde{A}_k}{\partial x_j'} \right) . \tag{3.7.16}$$

It follows from Eq. (3.7.15) that

$$\frac{\partial}{\partial x_i'} \left(\mathcal{K}_{ij} \frac{\partial \widetilde{A}_k}{\partial x_j'} \right) = \frac{\partial \mathcal{K}_{ik}}{\partial x_i'} \,, \tag{3.7.17}$$

which is the governing equation for $\widetilde{A}_k$ in the mesoscale cell. Supplemented by the requirement of Ω'-periodicity, the elliptic problem for $\widetilde{A}_k$ can be solved numerically. $\widetilde{K}_{ik}$ then follows from Eq. (3.7.16). Afterward the macroscale Darcy's law follows by Ω'-averaging (Eq. (3.7.13)),

$$\langle \langle u_i^{(0)} \rangle \rangle' = -\mathcal{K}_{ik}' \frac{\partial p^{(0)}}{\partial x_k''} \,, \tag{3.7.18}$$

where $\langle \cdot \rangle'$ denotes the mesoscale average over an Ω'-cell. The effective permeability on the macroscale is

$$\mathcal{K}_{ik}' \equiv \langle \widetilde{K}_{ik} \rangle' = \left\langle \mathcal{K}_{ij} \left(\delta_{jk} - \frac{\partial \widetilde{A}_k}{\partial x_j'} \right) \right\rangle' . \tag{3.7.19}$$

Note that $\mathcal{K}_{ij}' \neq \langle \mathcal{K}_{ij} \rangle'$.

Taking the Ω'-average of Eq. (3.7.9) and invoking periodicity we get

$$\frac{\partial \langle \langle u^{(0)} \rangle \rangle'}{\partial x_i''} = 0 \,, \tag{3.7.20}$$

which implies

$$\frac{\partial}{\partial x'_i}\left(\mathcal{K}'_{ik}\frac{\partial p^{(0)}}{\partial x''_k}\right) = 0\,. \tag{3.7.21}$$

After solving for $p^{(0)}$ from this equation and macroscale boundary conditions, we get the macrovelocity field from Eq. (3.7.18).

These results can also be obtained by a three-scale analysis in coordinates x_k, x'_k, x''_k by starting from Stokes equations at the microscale (Mei and Auriault, 1989).

3.7.2. Properties of Hydraulic Conductivity

We shall first prove that the macroscale conductivity tensor $\mathcal{K}'_{ij}$ is symmetric. Let us multiply both sides of Eq. (3.7.17) by $\widetilde{A}_n$, and then integrate the result over a Ω' cell. By partial integration, the left side gives

$$-\iiint_{\Omega'} \mathcal{K}_{ij}\frac{\partial \widetilde{A}_n}{\partial x'_i}\frac{\partial \widetilde{A}_k}{\partial x'_j}d\Omega'\,, \tag{3.7.22}$$

while the right side becomes

$$\iiint_{\Omega'} \mathcal{K}_{ik}\frac{\partial \widetilde{A}_n}{\partial x'_i}d\Omega' \tag{3.7.23}$$

and thus

$$\iiint_{\Omega'} \mathcal{K}_{ij}\left(\delta_{jk} - \frac{\partial \widetilde{A}_k}{\partial x'_j}\right)\frac{\partial \widetilde{A}_n}{\partial x'_i}d\Omega' = 0\,. \tag{3.7.24}$$

From here and from Eq. (3.7.19), it follows that

$$\mathcal{K}'_{nk} = \left\langle \mathcal{K}_{ij}\left(\delta_{jk} - \frac{\partial \widetilde{A}_k}{\partial x'_j}\right)\left(\delta_{in} - \frac{\partial \widetilde{A}_n}{\partial x'_i}\right)\right\rangle'\,. \tag{3.7.25}$$

Since $\mathcal{K}_{ij}$ is symmetric, so is $\mathcal{K}'_{nk}$.

We next show that $\mathcal{K}'_{nk}$ is positive-definite. Indeed, since $\mathcal{K}_{ij}$ is positive-definite, the following must be true for any ξ_i:

$$\left\langle K_{ij}\left(\xi_j - \frac{\partial(\widetilde{A}_k\xi_k)}{\partial x'_j}\right)\left(\xi_i - \frac{\partial(\widetilde{A}_n\xi_n)}{\partial x'_i}\right)\right\rangle' \geq 0\,. \tag{3.7.26}$$

Equality holds only if $(\xi_j - (\partial(\widetilde{A}_k\xi_k)/\partial x'_j)) = 0$ which implies $\xi_i = 0$. Hence $\mathcal{K}'_{nk}$ is positive-definite.

3.7.3. Macropermeability of a Laminated Medium

As an application, we consider a layered medium whose material properties on the ℓ' scale are known and vary periodically in depth x_1' only. For simplicity we first assume $\mathcal{K}_{ij}$ to be diagonal:

$$[\mathcal{K}_{ij}] = \begin{bmatrix} \mathcal{K}_{11}(x_1') & 0 & 0 \\ 0 & \mathcal{K}_{22}(x_1') & 0 \\ 0 & 0 & \mathcal{K}_{33}(x_1') \end{bmatrix} . \tag{3.7.27}$$

This includes as a special case of local isotropy where $\mathcal{K}_{11} = \mathcal{K}_{22} = \mathcal{K}_{33}$ at the same x_1'.

The vector component $\widetilde{A}_m$ must depend only on x_1'. From Eq. (3.7.17), we get

$$\frac{d}{dx_1'}\left(\mathcal{K}_{11}\frac{d\widetilde{A}_1}{dx_1'}\right) = \frac{dK_{11}'}{dx_1'} , \quad \frac{d}{dx_1'}\left(\mathcal{K}_{11}\frac{d\widetilde{A}_2}{dx_1'}\right) = 0 , \quad \frac{d}{dx_1'}\left(\mathcal{K}_{11}\frac{d\widetilde{A}_3}{dx_1'}\right) = 0 . \tag{3.7.28}$$

Integrating once, we get

$$\frac{d\widetilde{A}_1}{dx_1'} = 1 + \frac{C_1}{\mathcal{K}_{11}} , \quad \frac{d\widetilde{A}_2}{dx_1'} = \frac{C_2}{\mathcal{K}_{11}} , \quad \frac{d\widetilde{A}_3}{dx_1'} = \frac{C_3}{\mathcal{K}_{11}} .$$

The constants of integration are determined by invoking periodicity in x_1', i.e.,

$$\left\langle \frac{d\widetilde{A}_k}{dx_1'} \right\rangle' = 0 , \quad k = 1, 2, 3 ,$$

which gives

$$0 = 1 + C_1 \left\langle \frac{1}{\mathcal{K}_{11}} \right\rangle' , \quad C_2 = C_3 = 0 .$$

We finally obtain

$$-\frac{d\widetilde{A}_1}{dx_1'} = -1 + \frac{1/\mathcal{K}_{11}}{\langle 1/\mathcal{K}_{11}\rangle'} , \quad \frac{d\widetilde{A}_2}{dx_1'} = \frac{d\widetilde{A}_3}{dx_1'} = 0 . \tag{3.7.29}$$

Substituting Eqs. (3.7.27) and (3.7.29) into Eq. (3.7.19), we get

$$[\mathcal{K}_{ij}'] = \begin{bmatrix} \dfrac{1}{\langle 1/\mathcal{K}_{11}\rangle'} & 0 & 0 \\ 0 & \langle \mathcal{K}_{22}\rangle' & 0 \\ 0 & 0 & \langle \mathcal{K}_{33}\rangle' \end{bmatrix} . \tag{3.7.30}$$

Now consider the simplest special case in which each layer on the ℓ' scale is isotropic and has a constant hydraulic conductivity, In the nth intermediate layer, let

$$\mathcal{K}_{11} = \mathcal{K}_{22} = \mathcal{K}_{33} = \begin{cases} K_1\,, & 0 < x_1' - a_n' < \theta\ell'\,, \\ K_2\,, & \theta\ell' < x_1' - a_n' < \ell'\,, \end{cases} \tag{3.7.31}$$

with $\theta < 1$ and ℓ' being the spatial period. It follows readily that

$$\begin{aligned} \mathcal{K}_{11}' &= \frac{1}{\langle 1/\mathcal{K}_{11}\rangle'} = \frac{K_1 K_2}{(1-\theta)K_1 + \theta K_2}\,, \\ \mathcal{K}_{22}' &= \mathcal{K}_{33}' = \theta K_1 + (1-\theta)K_2\,, \\ \mathcal{K}_{ij}' &= 0\,, \quad i \neq j\,. \end{aligned} \tag{3.7.32}$$

Thus the transverse conductivity $\mathcal{K}_{11}'$ to the layers is the harmonic average, while the longitudinal conductivities $\mathcal{K}_{22}'$ and $\mathcal{K}_{33}'$ are the volume averages, of the layer conductivities. These results are well-known (Bear, 1972; Freeze and Cherry, 1979). If layer 1 is highly porous and layer 2 highly impervious, $K_1 \gg K_2$, then

$$\mathcal{K}_{11}' \cong \frac{K_2}{1-\theta}\,, \quad \mathcal{K}_{22}' = \mathcal{K}_{33}' \cong \theta\, K_1\,. \tag{3.7.33}$$

Thus $\mathcal{K}_{11}' \ll \mathcal{K}_{22}' = \mathcal{K}_{33}'$, so that the composite medium is highly anisotropic on the macroscale.

An advantage of the homogenization theory is that one can go down to the microscale after the macroscale problem is solved. This advantage may be important in some problems such as laminated materials where failure is due to excessive concentration of local stresses. To illustrate this point let us consider the flow into a point sink of flux rate M located at the origin: $x_1'' = x_2'' = x_3'' = 0$ of a horizontally layered composite which is infinite in extent on the ℓ'' scale in all directions.

From Eq. (3.7.20), the pressure $p^{(0)}$ satisfies

$$\mathcal{K}_{11}' \frac{\partial^2 p^{(0)}}{\partial x_1''^2} + \mathcal{K}_{22}' \frac{\partial^2 p^{(0)}}{\partial x_2''^2} + \mathcal{K}_{33}' \frac{\partial^2 p^{(0)}}{\partial x_3''^2} = 0\,. \tag{3.7.34}$$

The solution for a point sink is given by

$$p^{(0)} = -\frac{M}{4\pi R} \frac{1}{\sqrt{\mathcal{K}_{11}'\mathcal{K}_{22}'\mathcal{K}_{33}'}}\,, \tag{3.7.35}$$

where

$$R^2 = \frac{x_1''^2}{\mathcal{K}_{11}'} + \frac{x_2''^2}{\mathcal{K}_{22}'} + \frac{x_3''^2}{\mathcal{K}_{33}'} \tag{3.7.36}$$

with $\mathcal{K}_{ii}'$, $i = 1, 2, 3$, given by Eq. (3.7.32). The isobaric (constant-pressure) surface is an ellipsoid.

The macroscale velocity components are, from Eqs. (3.7.18) and (3.7.35), radial

$$\langle\langle u_i^{(0)}\rangle\rangle' = -\frac{M}{4\pi\sqrt{\mathcal{K}'_{11}\mathcal{K}'_{22}\mathcal{K}'_{33}}}\frac{x''_i}{R^3}, \quad i = 1, 2, 3.$$

Now we examine the velocity field in each layer, i.e., on the ℓ' scale. According to Eqs. (3.7.13), (3.7.27), and (3.7.30), we have

$$\langle u_1^{(0)}\rangle = -\widetilde{K}_{11}\frac{\partial p^{(0)}}{\partial x''_1} = -\mathcal{K}'_{11}\frac{\partial p^{(0)}}{\partial x''_1} = \langle\langle u_1^{(0)}\rangle\rangle. \tag{3.7.37}$$

However, from Eq. (3.7.29),

$$1 - \frac{dA'_1}{dx_1} = \frac{1/\mathcal{K}_{11}}{\langle 1/\mathcal{K}_{11}\rangle'},$$

hence

$$\widetilde{K}_{11} = \frac{1}{\langle 1/\mathcal{K}_{11}\rangle'} = \mathcal{K}'_{11},$$

and

$$\langle u_1^{(0)}\rangle = -\mathcal{K}'_{11}\frac{\partial p^{(0)}}{\partial x''_1} = \langle\langle u_1^{(0)}\rangle\rangle. \tag{3.7.38}$$

Thus the transverse component $\langle u_1^{(0)}\rangle = \langle\langle u_1^{(0)}\rangle\rangle'$ is the same constant in all layers, as expected from continuity. On the other hand

$$\langle u_2^{(0)}\rangle = -\widetilde{K}_{22}\frac{\partial p^{(0)}}{\partial x''_2} = -\mathcal{K}_{22}\frac{\partial p^{(0)}}{\partial x''_2} = \frac{\mathcal{K}_{22}}{\mathcal{K}'_{22}}\langle\langle u_2^{(0)}\rangle\rangle. \tag{3.7.39}$$

The component $\langle u_3^{(0)}\rangle$ is similar to $\langle u_2^{(0)}\rangle$ if the index 2 is changed to 3. While $\langle u_j^{(0)}\rangle \neq \langle\langle u_j^{(0)}\rangle\rangle$ with $j = 2, 3$, each is a different constant in alternating layers.

Consider the ratio

$$\frac{\langle u_2^{(0)}\rangle}{\langle u_1^{(0)}\rangle} = \frac{\mathcal{K}_{22}}{\mathcal{K}'_{22}}\frac{\langle\langle u_2^{(0)}\rangle\rangle'}{\langle\langle u_1^{(0)}\rangle\rangle'} = \frac{\mathcal{K}_{22}}{\mathcal{K}'_{22}}\frac{x''_2}{x''_1} \equiv J'(x'_1, x''_1)\frac{x''_2}{x''_1}. \tag{3.7.40}$$

The factor J' whose departure from unity is a measure of anisotropy on the ℓ' scale is

$$J'(x_1', x_1'') = \frac{\mathcal{K}_{22}}{\mathcal{K}_{22}'} = \begin{cases} \dfrac{K_1}{\theta K_1 + (1-\theta)K_2} & \text{Layer 1}, \\[2ex] \dfrac{K_2}{\theta K_1 + (1-\theta)K_2} & \text{Layer 2}. \end{cases} \tag{3.7.41}$$

For very small K_2/K_1 and fixed θ we have

$$J' = \begin{cases} \dfrac{1}{\theta} & \text{Layer 1}, \\[2ex] \dfrac{K_2}{\theta K_1} & \text{Layer 2}. \end{cases} \tag{3.7.42}$$

In general $1/\theta > 1$; the velocity vector in the more porous layer 1 is almost parallel to the layers. Clearly leakage from and into the less porous layers is difficult. In contrast, the flow inside is almost transverse to the layers, as sketched in Fig. 3.8. This feature is known in hydrological studies of pumping through a stratum composed of alternating clay/sand layers (Hantusch, 1960).

To solve the problem of pumping from a well in a soil consisting of a few layers the usual method in hydraulics is to seek numerical solution of the ℓ' scale equation in each layer, subject to continuity requirements at the interfaces. As the number of layers increases, the required computational task can become large, especially if the medium is deformable. As demonstrated by the example here, the present theory is advantageous for a large number of periodic layers and can readily yield analytical results.

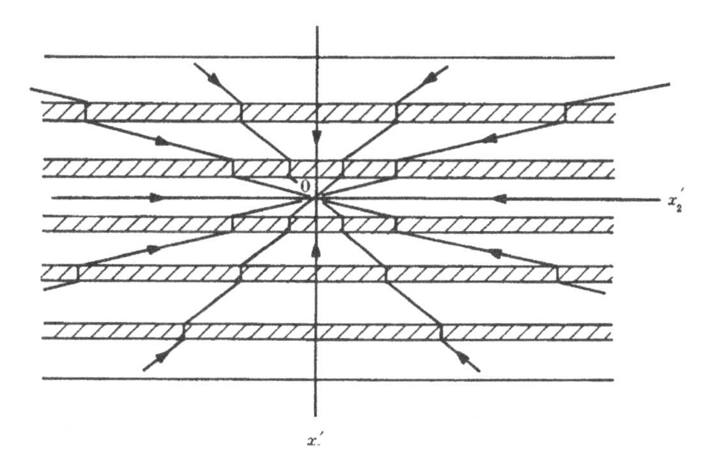

Fig. 3.8. Flow into a point sink in a layered soil saturated by water.

3.8. Brinkman's Modification of Darcy's Law

In order to bridge the two extremes of densely and sparsely packed grains, Brinkman (1947) proposed on heuristic ground the following modification of Darcy's law,

$$0 = -\frac{\partial p}{\partial x} + \mu \nabla^2 \boldsymbol{u} - \frac{\boldsymbol{u}}{K} \tag{3.8.1}$$

for the averaged pressure and flow velocity on the global scale. Theoretical justifications for its realm of validity have been offered by several authors (see reviews in Rubinstein, 1986; Nield and Bejan, 1992). Let the characteristic particle radius be R, the separation (microscale) distance be ℓ, and the global (macro)scale of pressure gradient be ℓ'. Let us first show that Brinkman's laws requires not only that $R \ll \ell \ll \ell'$, but also

$$\frac{R}{\ell'} \sim \left(\frac{\ell}{\ell'}\right)^3 \ll 1\,. \tag{3.8.2}$$

This ordering relation can be explained by examining the scale ratios so that all three terms in Eq. (3.8.1) are of comparable importance on the global scale (Levy, 1981).

Let the scales of global pressure variation and flow velocity be Δp and U, respectively. Each term in Eq. (3.8.1) represents a force per unit volume. For estimate, we use the well-known Stokes formula for the drag on a spherical particle of radius R, i.e., $\boldsymbol{F} = 6\pi\mu R\boldsymbol{u}$. The number of particles per unit volume of fluid is clearly of the order $\mathcal{O}(1/\ell^3)$. Equating the order of magnitude of viscous force in unit volume of fluid and the particle drag, we get

$$\frac{\mu U}{\ell'^2} \sim \frac{6\pi\mu R U}{\ell^3}\,, \tag{3.8.3}$$

which implies immediately Eq. (3.8.2).[3] Balancing the pressure gradient and viscous stress also gives

$$\Delta P = \mathcal{O}\left(\frac{\mu U}{\ell'}\right). \tag{3.8.4}$$

We can now give a formal derivation of Eq. (3.8.1) with a view to determining the coefficient K for a periodic array of identical particles. Let

[3] 6π is regarded as being of order unity.

us first normalize all variables as follows:

$$\boldsymbol{x} = \ell \boldsymbol{x}^\dagger\,, \quad \boldsymbol{u} = U\boldsymbol{u}^\dagger\,, \quad P = (\Delta P)p^\dagger\,, \tag{3.8.5}$$

where ΔP is given by Eq. (3.8.4). We also model the particles by a sum of concentrated forces to be represented later by delta functions

$$f_i = f_0 \sum_n \left(f_n^\dagger\right)_i, \tag{3.8.6}$$

where $f_0 = 6\pi\mu RU/\ell^3$ is the scale of the drag force per periodic cell according to Eq. (3.8.3) and $\left(f_n^\dagger\right)_i$ is the ith component of the normalized drag in the nth cell. The dimensionless Navier–Stokes equations read

$$\frac{\partial u_j^\dagger}{\partial x_j^\dagger} = 0\,, \tag{3.8.7}$$

$$\left(\frac{\rho U \ell}{\mu}\right) u_j^\dagger \frac{\partial u_i^\dagger}{\partial x_j^\dagger} = -\left(\frac{\Delta P \ell}{\mu U}\right)\frac{\partial p^\dagger}{\partial x_i^\dagger} + \frac{\partial^2 u_i^\dagger}{\partial x_j^\dagger \partial x_j^\dagger} + \frac{f_0}{\mu U/\ell^2}\sum_n \left(f_n^\dagger\right)_i. \tag{3.8.8}$$

In view of Eqs. (3.8.3) and (3.8.4) we get

$$\mathcal{O}\left(\frac{\Delta P \ell}{\mu U}\right) = \frac{\ell}{\ell'} = \epsilon\,, \quad \mathcal{O}\left(\frac{f_o}{\mu U/\ell^2}\right) = \mathcal{O}\left(\frac{R}{\ell}\right) = \epsilon^2\,. \tag{3.8.9}$$

Assume that the Reynolds number to be small so that

$$\frac{\rho U \ell}{\mu} \le \mathcal{O}(\epsilon^2)\,, \tag{3.8.10}$$

which implies

$$\frac{\rho U \ell'}{\mu} \le \mathcal{O}(\epsilon) \quad \text{and} \quad \frac{\rho(\Delta P)\ell\ell'}{\mu^2} \le \mathcal{O}(\epsilon^2)\,. \tag{3.8.11}$$

Expressing the viscous stress tensor in terms of the strain-rate tensor

$$\varepsilon_{ij}(\boldsymbol{u}) \equiv \frac{1}{2}\left(\frac{\partial u_i}{\partial x_j} + \frac{\partial u_j}{\partial x_i}\right), \tag{3.8.12}$$

we may return to dimensional equations with proper ordering symbols

$$\frac{\partial u_i}{\partial x_j} = 0\,, \tag{3.8.13}$$

$$\epsilon^2 \rho u_j \frac{\partial u_i}{\partial x_j} = -\epsilon \frac{\partial p}{\partial x_i} + \mu \frac{\partial \varepsilon_{ij}}{\partial x_j} + \epsilon^2 \sum_n f_i \delta(\boldsymbol{x} - \boldsymbol{x}_n)\,. \tag{3.8.14}$$

Introducing the slow variable $\boldsymbol{x}' = \epsilon \boldsymbol{x}$ and the multiple-scale expansions

$$p = p^{(0)} + \epsilon p^{(1)} + \epsilon^2 p^{(2)} + \cdots\,, \quad \boldsymbol{u} = \boldsymbol{u}^{(0)} + \epsilon \boldsymbol{u}^{(1)} + \epsilon^2 \boldsymbol{u}^{(2)} + \cdots\,, \tag{3.8.15}$$

where $p^{(n)}$ and $u^{(n)}$ are functions of $\boldsymbol{x}$ and $\boldsymbol{x}'$, we note first that the strain tensor becomes

$$\varepsilon_{ij}(\boldsymbol{u}) = \varepsilon_{ij}(\boldsymbol{u}^{(0)}) + \epsilon\left[\varepsilon_{ij}(\boldsymbol{u}^{(1)}) + \varepsilon'_{ij}(\boldsymbol{u}^{(0)})\right] + \epsilon^2\left[\varepsilon_{ij}(\boldsymbol{u}^{(2)}) + \varepsilon'_{ij}(\boldsymbol{u}^{(1)})\right] + \cdots, \tag{3.8.16}$$

where

$$\varepsilon_{ij}(\boldsymbol{u}) = \frac{1}{2}\left(\frac{\partial u_i}{\partial x_j} + \frac{\partial u_j}{\partial x_i}\right) \quad \text{and} \quad \varepsilon'_{ij}\boldsymbol{u} = \frac{1}{2}\left(\frac{\partial u_i}{\partial x'_j} + \frac{\partial u_j}{\partial x'_j}\right). \tag{3.8.17}$$

We obtain from Eqs. (3.8.13) and (3.8.14)

$$\frac{\partial u_j^{(0)}}{\partial x_j} = 0, \tag{3.8.18}$$

and

$$\frac{\partial \varepsilon_{ij}(\boldsymbol{u}^{(0)})}{\partial x_j} = 0 \tag{3.8.19}$$

at order $\mathcal{O}(1)$,

$$\frac{\partial u_j^{(1)}}{\partial x_j} + \frac{\partial u_j^{(0)}}{\partial x'_j} = 0, \tag{3.8.20}$$

$$0 = -\frac{\partial p^{(0)}}{\partial x_j} + \mu\frac{\partial \varepsilon_{ij}(\boldsymbol{u}^{(1)})}{\partial x_j} + \mu\frac{\partial \varepsilon'_{ij}(\boldsymbol{u}^{(0)})}{\partial x_j} + \mu\frac{\partial \varepsilon_{ij}(\boldsymbol{u}^{(0)})}{\partial x'_j} \tag{3.8.21}$$

at $\mathcal{O}(\epsilon)$ and

$$\frac{\partial u_j^{(2)}}{\partial x_j} + \frac{\partial u_j^{(1)}}{\partial x'_j} = 0, \tag{3.8.22}$$

$$\rho u_j^{(0)}\frac{\partial u_i^{(0)}}{\partial x_j} = -\frac{\partial p^{(0)}}{\partial x'_j} - \frac{\partial p^{(1)}}{\partial x_j} + \mu\frac{\partial \varepsilon_{ij}(\boldsymbol{u}^{(2)})}{\partial x_j} + \mu\frac{\partial \varepsilon_{ij}(\boldsymbol{u}^{(1)})}{\partial x'_j} + \mu\frac{\partial \varepsilon'_{ij}(\boldsymbol{u}^{(0)})}{\partial x'_j} + \sum_n f_i\delta\left(\boldsymbol{x} - \boldsymbol{x}_n\right) \tag{3.8.23}$$

at $\mathcal{O}(\epsilon^2)$. Ω-periodicity with period $\mathcal{O}(\ell)$ is assumed for all variables.

At order unity the solution is clearly $u_i^{(0)}(\boldsymbol{x}')$ which is independent of $\boldsymbol{x}$. This can be formally verified by scalar-multiplying Eq. (3.8.19) with $u_j^{(0)}$ so that

$$u_i^{(0)} \frac{\partial \varepsilon_{ij}(\boldsymbol{u}^{(0)})}{\partial x_j} = 0 = \frac{\partial}{\partial x_j}\left[u_i^{(0)} \varepsilon_{ij}(\boldsymbol{u}^{(0)})\right] - \frac{\partial u_i^{(0)}}{\partial x_j} \varepsilon_{ij}(\boldsymbol{u}^{(0)}) . \tag{3.8.24}$$

Integrating Eq. (3.8.24) over an Ω-cell and invoking periodicity, we get

$$\iiint_\Omega \frac{\partial u_i^{(0)}}{\partial x_j} \varepsilon_{ij}(\boldsymbol{u}^{(0)}) d\Omega = 0 .$$

Since $\varepsilon_{ij}(\boldsymbol{u}) = \varepsilon_{ji}(\boldsymbol{u})$,

$$\frac{\partial u_i^{(0)}}{\partial x_j} \varepsilon_{ij}(\boldsymbol{u}^{(0)}) = \frac{\partial u_i^{(0)}}{\partial x_j} \varepsilon_{ji}(\boldsymbol{u}^{(0)}) = \frac{\partial u_j^{(0)}}{\partial x_i} \varepsilon_{ij}(\boldsymbol{u}^{(0)}) .$$

Hence, the integral above may be written as

$$\frac{1}{2} \iiint_\Omega \varepsilon_{ij}(\boldsymbol{u}^{(0)}) \varepsilon_{ij}(\boldsymbol{u}^{(0)}) d\Omega = 0 .$$

This is possible only if

$$\varepsilon_{ij}(\boldsymbol{u}^{(0)}) \equiv 0 \tag{3.8.25}$$

throughout Ω, which implies that

$$u_i^{(0)} = u_i^{(0)}(\boldsymbol{x}') \tag{3.8.26}$$

on account of periodicity.

Taking Ω-average of Eq. (3.8.20), we get

$$\frac{\partial \langle u_j^{(0)} \rangle}{\partial x_j'} = \frac{\partial u_j^{(0)}}{\partial x_j'} = 0 , \tag{3.8.27}$$

hence

$$\frac{\partial u_j^{(1)}}{\partial x_j} = 0 \tag{3.8.28}$$

also. In Eq. (3.8.21) the last two terms vanish by virtue of Eq. (3.8.26). Hence

$$0 = -\frac{\partial p^{(0)}}{\partial x_i} + \mu \frac{\partial \varepsilon_{ij}(\boldsymbol{u}^{(1)})}{\partial x_j} . \tag{3.8.29}$$

Thus $u_j^{(1)}$ and $p^{(0)}$ are governed by a homogeneous Stokes problem subject to Ω-periodicity. It is easy to see that the solution is uniform in $\boldsymbol{x}$. This can be seen more formally by scalar-multiplying Eq. (3.8.29) by $u_i^{(1)}$ and integrating over a Ω-cell

$$0 = \iiint_\Omega \left\{ -\frac{\partial(u_i^{(1)} p^{(0)})}{\partial x_i} + \frac{\partial}{\partial x_j}\left[u_i^{(1)} \varepsilon_{ij}(\boldsymbol{u}^{(1)})\right] - \frac{\partial u_i^{(1)}}{\partial x_j} \varepsilon_{ij}(\boldsymbol{u}^{(1)}) \right\} d\Omega .$$

The first two terms vanish after applying Gauss' theorem and periodicity. The last term may be written as

$$0 = -\frac{1}{2} \iiint_\Omega \varepsilon_{ij}(\boldsymbol{u}^{(1)}) \varepsilon_{ij}(\boldsymbol{u}^{(1)}) d\Omega ,$$

which can be true if and only if $\boldsymbol{u}^{(1)}$ is independent of $\boldsymbol{x}$, i.e., $\boldsymbol{u}^{(1)} = \boldsymbol{u}^{(1)}(\boldsymbol{x}')$. It then follows from Eq. (3.8.29) that $p^{(0)}(\boldsymbol{x}')$ is independent of $\boldsymbol{x}$ also.

Finally, using Eq. (3.8.26) and taking Ω-average of Eq. (3.8.23), we get

$$0 = -\frac{\partial \langle p_o \rangle}{\partial x_i'} + \mu \frac{\partial \varepsilon_{ij}'(\boldsymbol{u}^{(0)})}{\partial x_j'} + f_i + \iiint_\Omega \frac{\partial}{\partial x_j}\left[-p^{(1)}\delta_{ij} + \mu \varepsilon_{ij}(\boldsymbol{u}^{(2)})\right] d\Omega ,$$

where the defining property of the delta function has been used. Invoking Gauss' theorem and periodicity we get

$$0 = -\frac{\partial p^{(0)}}{\partial x_i'} + \mu \frac{\partial' u_i^{(0)}}{\partial x_j' \partial x_j'} + f_i . \tag{3.8.30}$$

Small terms of the relative order $\mathcal{O}(C) = \mathcal{O}(R^3/\ell^3)$, where C is the particle concentration by volume, have been ignored. For particles of general shape

$$f_i = -F_{ij} u_j ,$$

where F_{ij} can in principle be obtained by solving a Stokes problem, Eq. (3.8.30) becomes

$$0 = -\frac{\partial p^{(0)}}{\partial x_i'} + \mu \frac{\partial' u_i^{(0)}}{\partial x_j' \partial x_j'} - F_{ij} u_j^{(0)} , \tag{3.8.31}$$

which is Brinkman's equation.

For a cubic array of identical spheres

$$f_i = -\frac{6\pi\mu R u_i^{(0)}}{\ell^3} ,$$

we get

$$0 = -\frac{\partial p^{(0)}}{\partial x'_j} + \mu \frac{\partial^2 u_i^{(0)}}{\partial x'_j \partial x'_j} - \frac{6\pi\mu R}{\ell^3} u_i^{(0)} . \tag{3.8.32}$$

The conductivity can be defined as

$$K = \frac{\ell^3}{6\pi\mu R} = \frac{\ell^3}{6\pi\mu R} \frac{4/3\pi R^3}{4/3\pi R^3} = \frac{2}{9} \frac{R^2}{\mu} \frac{1}{C} , \tag{3.8.33}$$

where

$$C = \frac{4\pi R^3}{3\ell^3} . \tag{3.8.34}$$

Thus K decreases with increasing particle concentration or fluid viscosity.

In summary Eq. (3.8.30) or (3.8.31) is the leading-order approximation with a relative error of $\mathcal{O}(\epsilon)$. Since from Eq. (3.8.9) or (3.8.34)

$$\epsilon = \mathcal{O}\left(\sqrt{\frac{R}{\ell}}\right) = \mathcal{O}(C^{1/6}) .$$

Brinkman equation is valid with an error of order $\mathcal{O}(C^{1/6})$.

3.9. Effects of Weak Fluid Inertia

Though Darcy' law was established mainly for low Reynolds number flows, numerous evidences have shown that it holds up to quite high Reynolds numbers up to order unity (Scheidegger, 1960; Dullien, 1979; Kovacs, 1981; Hannoura and Barrends, 1981). Many empirical attempts have also been made to extend the relation between the seepage velocity and the mean pressure gradient for high Reynolds numbers. For one-dimensional seepage flows the empirical relation usually takes the power-law form:

$$-K \frac{\partial p}{\partial x} = \langle u \rangle + b \langle u \rangle^n , \tag{3.9.1}$$

where b and n are empirical constants. For example Forcheimer (1901) found $1.6 < n < 2$; the higher value $n = 2$ has been the most popular in the literature (e.g., Scheidegger, 1960).

Theoretical confirmation of relations like Eq. (3.9.1) can in principle be obtained by strictly numerical solution of Navier–Stokes equations. Such efforts are however quite laborious as can be seen from the two-dimensional computations for steady flow in a corrugated tube by Payatakes *et al.* (1973). Fully nonlinear numerical theories for three-dimensional grains are even more demanding.

By extending the homogenization method, Mei and Auriault (1989) and Wodie and Levy (1991) have studied the case of weak inertia. In particular Mei and Auriault (1989) examined the case where the length ratio and the Reynolds number are both small. For $\ell/\ell' = \epsilon$ and $Re = \mathcal{O}(\epsilon^{1/2})$ with $\epsilon \ll 1$, they showed that the power in Eq. (3.9.1) should be $n = 3$, i.e.,

$$-K\nabla p = \langle \boldsymbol{u} \rangle + b\langle \boldsymbol{u} \rangle |\langle \boldsymbol{u} \rangle|^2 \tag{3.9.2}$$

for a porous medium which is isotropic and homogeneous on the macroscale. This theoretical result has been verified experimentally in a one-dimensional porous channel by Rasoloarijaona and Auriault (1994) for a wide range of Reynolds numbers (from 0 to 150) and also by Firdaouss *et al.* (1998) based on the measurements of Hazen (1895) and Chauveeau (1965). This cubic correction was also confirmed for smaller Reynolds numbers $Re = \mathcal{O}(\epsilon)$ by Rasoloarijaona and Auriault (1994), and under more generous conditions without requiring macroscale isotropy by Firdaouss *et al.* (1998).

We describe below some details of the theory along the lines of Mei and Auriault (1989), under the assumption that $Re = \mathcal{O}(\epsilon)$. In addition we assume that the medium is homogeneous on the macroscale, i.e., $\theta = \Omega_f/\Omega$ is a constant. The perturbation equations (3.1.13)–(3.1.17) and (3.1.20) still hold, and can be used to get at $\mathcal{O}(\epsilon)$

$$\frac{\partial u_j^{(1)}}{\partial x_j} = -\frac{\partial u_j^{(0)}}{\partial x'_j} = K_{ij}\frac{\partial^2 p^{(0)}}{\partial x'_i \partial x'_j}, \quad \boldsymbol{x} \in \Omega_f, \tag{3.9.3}$$

and

$$\begin{aligned} \frac{\partial p^{(2)}}{\partial x_i} - \mu\nabla^2 u_i^{(1)} &= -\frac{\partial p^{(1)}}{\partial x'_i} - \rho u_j^{(0)}\frac{\partial u_i^{(0)}}{\partial x_j} + 2\mu\frac{\partial^2 u_i^{(0)}}{\partial x_k \partial x'_k} \\ &= A_j\frac{\partial^2 p^{(0)}}{\partial x'_i \partial x'_j} - \frac{\partial p_o^{(1)}}{\partial x'_i} - \rho K_{gj}\frac{\partial K_{ik}}{\partial x_g}\frac{\partial p^{(0)}}{\partial x'_k}\frac{\partial p^{(0)}}{\partial x'_j} \\ &\quad - 2\mu\frac{\partial K_{ij}}{\partial x_k}\frac{\partial^2 p^{(0)}}{\partial x'_j \partial x'_k}, \quad \boldsymbol{x} \in \Omega_f, \end{aligned} \tag{3.9.4}$$

with the boundary conditions:

$$u_i^{(1)} = 0 \quad \text{on } \Gamma, \tag{3.9.5}$$

$$u_i^{(1)} \text{ is } \Omega\text{-periodic}. \tag{3.9.6}$$

Due to the assumed macroscale homogeneity, K_{ij} and A_j do not depend on $\boldsymbol{x}'$. Since the operators on the left-hand sides of Eqs. (3.9.3) and (3.9.4) are linear, we may assume formally

$$u_i^{(1)} = -L_{ijk}\frac{\partial p^{(0)}}{\partial x'_j}\frac{\partial p^{(0)}}{\partial x'_k} - M_{ijk}\frac{\partial^2 p^{(0)}}{\partial x'_j \partial x'_k} - K_{ij}\frac{\partial p_o^{(1)}}{\partial x'_j}, \qquad (3.9.7)$$

and

$$p^{(2)} = -B_{jk}\frac{\partial p^{(0)}}{\partial x'_j}\frac{\partial p^{(0)}}{\partial x'_k} - C_{jk}\frac{\partial^2 p^{(0)}}{\partial x'_j \partial x'_k} - A_j\frac{\partial p_o^{(1)}}{\partial x'_j} + p_o^{(2)} \qquad (3.9.8)$$

into Eqs. (3.9.3) and (3.9.4). The new coefficient tensors must then satisfy the following pairs of governing equations:

$$\frac{\partial L_{ijk}}{\partial x_i} = 0, \quad \boldsymbol{x} \in \Omega_f, \qquad (3.9.9)$$

$$\frac{\partial B_{jk}}{\partial x_i} - \mu\nabla^2 L_{ijk} = \rho K_{gi}\frac{\partial K_{ik}}{\partial x_g}, \quad \boldsymbol{x} \in \Omega_f \qquad (3.9.10)$$

for L_{ijk} and B_{jk}, and

$$\frac{\partial M_{ijk}}{\partial x_i} = -K_{kj}, \quad \boldsymbol{x} \in \Omega_f, \qquad (3.9.11)$$

$$\frac{\partial C_{jk}}{\partial x_i} - \mu\nabla^2 M_{ijk} = -A_j\delta_{ik} + 2\mu\frac{\partial K_{ij}}{\partial x_k}, \quad \boldsymbol{x} \in \Omega_f \qquad (3.9.12)$$

for M_{ijk} and C_{jk}. The boundary conditions (Eqs. (3.9.5) and (3.9.6)) are satisfied by imposing

$$L_{ijk} = 0, \quad M_{ijk} = -\frac{1}{S_\Gamma}\left(\iiint_{\Omega_f} K_{kj}dx\right)n_i, \quad \text{on } \Gamma, \qquad (3.9.13)$$

$$L_{ijk}, B_{ij}; \quad M_{ijk}, C_{jk} \text{ are } \Omega\text{-periodic}, \qquad (3.9.14)$$

where S_Γ is area of the wetted surface Γ in the unit cell, and n_i, $i = 1, 2, 3$, are the components of the unit normal vector to S_Γ pointing out from the pore fluid. The inhomogeneous boundary value for M_{ijk} is needed in order to counter-balance the source term in Eq. (3.9.11), hence to conserve mass in the unit cell. Indeed from Eqs. (3.9.7), (3.9.13), and Darcy's law (Eq. (3.1.33)), it follows that

$$u_i^{(1)} = \frac{\Omega}{S_\Gamma}\langle K_{kj}\rangle\frac{\partial^2 p^{(0)}}{\partial x'_j \partial x'_k}n_i = 0 \quad \text{on } \Gamma, \qquad (3.9.15)$$

and hence the boundary condition (Eq. (3.9.5)) is satisfied. To render the solutions unique we further require

$$\langle B_{jk}\rangle = \langle C_{jk}\rangle = 0. \qquad (3.9.16)$$

These inhomogeneous cell problems for L_{ijk}, M_{ijk}, B_{jk}, and C_{jk} can be solved numerically as in Sec. 4.4. Afterward the Ω-averages may be taken to give

$$\langle u_i^{(1)}\rangle = -\langle L_{ijk}\rangle \frac{\partial p^{(0)}}{\partial x'_j}\frac{\partial p^{(0)}}{\partial x'_k} - \langle M_{ijk}\rangle \frac{\partial^2 p^{(0)}}{\partial x'_j \partial x'_k} - \langle K_{ij}\rangle \frac{\partial p_o^{(1)}}{\partial x'_j}, \tag{3.9.17}$$

$$\langle p^{(2)}\rangle = \theta \bar{p}^{(2)}. \tag{3.9.18}$$

The Ω-average of Eq. (3.7.4) gives

$$\frac{\partial \langle u_j^{(1)}\rangle}{\partial x'_j} = 0, \tag{3.9.19}$$

which implies the governing equation for the first-order correction to the mean pore pressure $\bar{p}^{(1)}$:

$$\frac{\partial}{\partial x'_i}\left(\langle K_{ij}\rangle \frac{\partial \bar{p}^{(1)}}{\partial x'_j}\right) = -\frac{\partial}{\partial x'_i}\left(\langle L_{ijk}\rangle \frac{\partial p^{(0)}}{\partial x'_j}\frac{\partial p^{(0)}}{\partial x'_k} + \langle M_{ijk}\rangle \frac{\partial^2 p^{(0)}}{\partial x'_j \partial x'_k}\right). \tag{3.9.20}$$

For further discussion we assume the medium to be *isotropic* on the macroscale. It is known that any isotropic tensor of rank 3 must be proportional to the permutation tensor ϵ_{ijk} (Fung, 1977),

$$\langle M_{ijk}\rangle = M\epsilon_{ijk}, \quad \langle L_{ijk}\rangle = L\epsilon_{ijk}, \quad \langle K_{jk}\rangle = K\delta_{jk}. \tag{3.9.21}$$

The permutation tensor has the property that

$$\epsilon_{123} = \epsilon_{231} = \epsilon_{312} = 1, \quad \epsilon_{132} = \epsilon_{321} = \epsilon_{213} = -1, \tag{3.9.22}$$

while all other components vanish. From the defining Eqs. (3.9.11) and (3.9.12), $\langle M_{ijk}\rangle$ and $\langle L_{ijk}\rangle$ are symmetric in (j, k), but the permutation tensor ϵ_{ijk} is antisymmetric.

Now Eq. (3.9.21) implies that

$$\langle u_i^{(1)}\rangle = -K\frac{\partial p_o^{(1)}}{\partial x'_i}. \tag{3.9.23}$$

This can be seen by checking any of the three velocity components in Eq. (3.9.17), say $\langle u_1^{(1)}\rangle$:

$$-\langle L_{1jk}\rangle \frac{\partial p^{(0)}}{\partial x'_j}\frac{\partial p^{(0)}}{\partial x'_k} = -L\left(\epsilon_{123}\frac{\partial p^{(0)}}{\partial x'_2}\frac{\partial p^{(0)}}{\partial x'_3} + \epsilon_{132}\frac{\partial p^{(0)}}{\partial x'_3}\frac{\partial p^{(0)}}{\partial x'_2}\right) = 0,$$

$$-\langle M_{1jk}\rangle \frac{\partial^2 p^{(0)}}{\partial x'_j \partial x'_k} = -M\left(\epsilon_{123}\frac{\partial^2 p^{(0)}}{\partial x'_2 \partial x'_3} + \epsilon_{132}\frac{\partial^2 p^{(0)}}{\partial x'_3 \partial x'_2}\right) = 0,$$

and

$$-\langle K_{1j}\rangle\frac{\partial p_o^{(1)}}{\partial x_j'} = -K\frac{\partial p^{(0)}}{\partial x_i'}\,.$$

Hence Eq. (3.9.23) follows. Now Eq. (3.9.19) implies that

$$\frac{\partial^2 p_o^{(1)}}{\partial x_k'\partial x_k'} = 0\,, \tag{3.9.24}$$

which is the same as that for $p^{(0)}$. Therefore, we can absorb $p_o^{(1)}$ in $p^{(0)}$, or, equivalently, set $p_o^{(1)} = 0$. This leads to the simple result that

$$\left\langle u_i^{(1)}\right\rangle = 0\,, \quad \langle p^{(1)}\rangle = \bar{p}^{(1)} = 0\,. \tag{3.9.25}$$

Thus the nonlinear correction for convective inertia to Darcy's law is at most a third-order effect. This is indeed consistent with existing experiments all of which show that the linear law of Darcy can be good even up to $Re = \mathcal{O}(1)$ (Kovacs, 1981).

Let us now go to the next order. At $\mathcal{O}(\epsilon^2)$, the continuity equation gives, in vector form,

$$\nabla'\cdot\boldsymbol{u}^{(2)} = -\nabla\cdot\boldsymbol{u}^{(1)}\,. \tag{3.9.26}$$

Clearly the Ω-average gives

$$\nabla'\cdot\langle\boldsymbol{u}^{(1)}\rangle = 0\,. \tag{3.9.27}$$

At $\mathcal{O}(\epsilon^3)$, the momentum equation gives, in vector form,

$$\begin{aligned}\nabla p^{(3)} - \mu\nabla^2\boldsymbol{u}^{(2)} &= -\nabla' p^{(2)} - \rho\left(\boldsymbol{u}^{(0)}\cdot\nabla\boldsymbol{u}^{(1)} + \boldsymbol{u}^{(1)}\cdot\nabla\boldsymbol{u}^{(0)} + \boldsymbol{u}^{(0)}\cdot\nabla'\boldsymbol{u}^{(0)}\right)\\ &\quad + 2\mu\nabla\cdot\nabla'\boldsymbol{u}^{(1)} + \mu\nabla'^2\boldsymbol{u}^{(0)}\,.\end{aligned} \tag{3.9.28}$$

By using Eqs. (3.1.21) and (3.9.7), the right-hand side can be expressed in terms of the derivatives of $p^{(0)}$ with respect to $\boldsymbol{x}'$, for example,

$$-\nabla' p^{(2)} = -C_{jk}\frac{\partial^3 p^{(0)}}{\partial x_i'\partial x_j'\partial x_k'} - B_{jk}\frac{\partial}{\partial x_i'}\left(\frac{\partial p^{(0)}}{\partial x_j'}\frac{\partial p^{(0)}}{\partial x_k'}\right) + \frac{\partial p_o^{(2)}}{\partial x_i'}\,,$$

$$-\boldsymbol{u}^{(1)}\cdot\nabla\boldsymbol{u}^{(0)} = -M_{\ell jk}\frac{\partial K_{ig}}{\partial x_\ell}\frac{\partial^2 p^{(0)}}{\partial x_j'\partial x_k'}\frac{\partial p^{(0)}}{\partial x_g'} - L_{\ell jk}\frac{\partial K_{ig}}{\partial x_\ell}\frac{\partial p^{(0)}}{\partial x_j'}\frac{\partial p^{(0)}}{\partial x_k'}\frac{\partial p^{(0)}}{\partial x_g'}\,,$$

etc., where use is made of $p_o^{(1)} = 0$.

Again linearity suggests the following representations as linear combinations of these derivatives,

$$\begin{pmatrix} u_i^{(2)} \\ p^{(3)} \end{pmatrix} = -\begin{pmatrix} U_{ijkg}^{(I)} \\ P_{jkg}^{(I)} \end{pmatrix} \frac{\partial p^{(0)}}{\partial x'_j} \frac{\partial p^{(0)}}{\partial x'_k} \frac{\partial p^{(0)}}{\partial x'_g} - \begin{pmatrix} U_{ijkg}^{(II)} \\ P_{jkg}^{(II)} \end{pmatrix} \frac{\partial^2 p^{(0)}}{\partial x'_j \partial x'_k} \frac{\partial p^{(0)}}{\partial x'_g}$$
$$- \begin{pmatrix} U_{ijkg}^{(III)} \\ P_{jkg}^{(III)} \end{pmatrix} \frac{\partial^3 p^{(0)}}{\partial x'_j \partial x'_k \partial x'_g} - \begin{pmatrix} K_{ij} \\ A_j \end{pmatrix} \frac{\partial p_o^{(2)}}{\partial x'_j} + \begin{pmatrix} 0 \\ p_o^{(3)} \end{pmatrix} . \tag{3.9.29}$$

Here the coefficient tensors $U_{ijkg}^{(m)}$ and $P_{jkg}^{(m)}$ are unknown functions of $\boldsymbol{x}$, and are solutions to a set of inhomogeneous Stokes problems in a Ω-cell, similar to Eqs. (3.9.9) and (3.9.10), subject to the constraints of periodicity and

$$\langle P_{jkg}^{(m)} \rangle = 0 , \quad m = I, II, III . \tag{3.9.30}$$

We omit the lengthy equations here.

The Ω-average of $u_i^{(2)}$ involves the averages of the fourth-rank tensors $U_{ijkg}^{(m)}$. Again we consider only a medium which is isotropic on the macroscale. It is known that an isotropic Cartesian tensor of rank 4 can be written as

$$\langle U_{ijkg}^{(m)} \rangle = \lambda_m (\delta_{ij}\delta_{kg}) + \mu_m (\delta_{ik}\delta_{jg} + \delta_{ig}\delta_{jk}) + \nu_m (\delta_{ik}\delta_{jg} - \delta_{ig}\delta_{jk}) , \tag{3.9.31}$$

which is characterized by three coefficients λ_m, μ_m, and ν_m (Fung, 1977). It follows after some straightforward algebra that

$$\langle U_{ijkg}^{(I)} \rangle \frac{\partial p^{(0)}}{\partial x'_j} \frac{\partial p^{(0)}}{\partial x'_k} \frac{\partial p^{(0)}}{\partial x'_g} = (\lambda_I + 2\mu_I) \frac{\partial p^{(0)}}{\partial x'_i} \frac{\partial p^{(0)}}{\partial x'_k} \frac{\partial p^{(0)}}{\partial x'_k} , \tag{3.9.32}$$

$$\langle U_{ijkg}^{(II)} \rangle \frac{\partial^2 p^{(0)}}{\partial x'_j \partial x'_k} \frac{\partial p^{(0)}}{\partial x'_g} = (\lambda_{II} + \mu_{II} + \nu_{II}) \frac{\partial^2 p^{(0)}}{\partial x'_i \partial x'_k} \frac{\partial p^{(0)}}{\partial x'_k} , \tag{3.9.33}$$

$$\langle U_{ijkg}^{(III)} \rangle \frac{\partial^3 p^{(0)}}{\partial x'_j \partial x'_k \partial x'_g} = (\lambda_{III} + 2\mu_{III}) \frac{\partial^3 p^{(0)}}{\partial x'_i \partial x'_k \partial x'_k} = 0 , \tag{3.9.34}$$

where Eq. (3.1.35) has been used. In vector form Eq. (3.9.29) reads

$$\langle \boldsymbol{u}^{(2)} \rangle = -\beta \nabla' p^{(0)} (\nabla' p^{(0)})^2 - \gamma \nabla' p^{(0)} \cdot \nabla' \nabla' p^{(0)} - K \nabla' \bar{p}^{(2)} , \tag{3.9.35}$$

where

$$\beta = \lambda_I + 2\mu_I , \quad \gamma = \lambda_{II} + \mu_{II} + \nu_{II} . \tag{3.9.36}$$

Combining the first three orders we get

$$\langle \boldsymbol{u} \rangle = -K\nabla' p_o - \epsilon^2 \left(\frac{\beta}{K^3} \langle \boldsymbol{u} \rangle |\langle \boldsymbol{u} \rangle|^2 + \frac{\gamma}{K^2} \langle \boldsymbol{u} \rangle \cdot \nabla' \langle \boldsymbol{u} \rangle \right), \tag{3.9.37}$$

where $p_o = p_o^{(0)} + \epsilon^2 p_o^{(2)}$. Finally we have

$$\epsilon^2 \frac{\gamma}{K^2} \langle \boldsymbol{u} \rangle \cdot \nabla' \langle \boldsymbol{u} \rangle = -K\nabla' \overline{p} - \left(\langle \boldsymbol{u} \rangle + \epsilon^2 \frac{\beta}{K^3} \langle \boldsymbol{u} \rangle |\langle \boldsymbol{u} \rangle|^2 \right). \tag{3.9.38}$$

At $\mathcal{O}(\epsilon^3)$, the continuity equation gives

$$\nabla' \cdot \boldsymbol{u}^{(2)} = -\nabla \cdot \boldsymbol{u}^{(3)}. \tag{3.9.39}$$

By Ω-averaging we get

$$\nabla' \cdot \langle \boldsymbol{u}^{(2)} \rangle = 0. \tag{3.9.40}$$

By combining Eqs. (3.1.31), (3.9.27) and (3.9.40), it follows that

$$\nabla' \cdot \langle \boldsymbol{u} \rangle = 0. \tag{3.9.41}$$

Together with Eq. (3.9.38), we now have a complete set of effective equations for the macroscale variations of seepage velocity and pore pressure, where the coefficients β and γ can in principle be computed for any cell geometry.

In all reported experiments, the seepage flow in a channel or a tube was one dimensional and uniform, therefore $\partial\langle u \rangle/\partial x' = 0$, and Eq. (3.9.38) reduces trivially to Eq. (3.9.2). Thus the correction to Darcy's law is a term cubic in seepage velocity. Mei and Auriault (1989) considered a case where the Reynolds number defined here is of the order $\mathcal{O}(\epsilon^{3/2})$. The inertia term in Eq. (3.9.38) is also negligible.

Firdaouss *et al.* (1998) have carefully examined the experimental data by Hazen (1895) and Chauveeau (1965) and plotted them in a diagram of the normalized mean velocity x versus the deviation from Darcy's law y, where

$$x = \frac{Re}{Re_{\max}}, \quad \text{with } Re = \frac{\langle u \rangle \ell}{\nu},$$

and

$$y = \frac{1 + (K/\langle u \rangle)(\partial p_o/\partial x)}{(1 + (K/\langle u \rangle)(\partial p_o/\partial x))_{\max}}.$$

As shown in Fig. 3.9 the data points indeed fall on a quadratic curve $y = \alpha x^2$ which is equivalent to Eq. (3.9.2). Hence the cubic correction is confirmed. Similar confirmation with the data of Chauveeau (1965) can be found in Firdaouss *et al.* (1998).

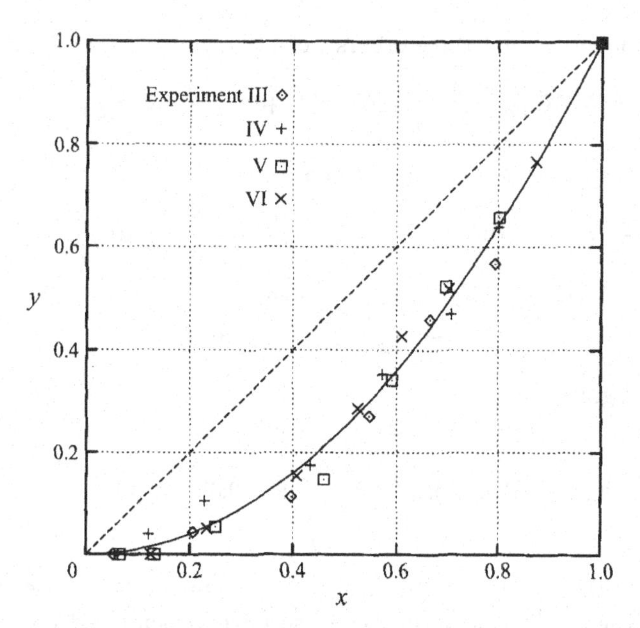

Fig. 3.9. Deviation of experimental data of Hazen (symbols) from linear Darcy's law (dashes) and cubic Darcy's law (solid). From Firdaouss *et al.* (1998), *J. Fluid Mech.*

Appendix 3A. Spatial Averaging Theorem

In deriving the effective equations a frequent operation is to interchange differentiation with respect to the macroscale variable and volume-averaging over a unit cell, hence one needs to know the relation between

$$\left\langle \frac{\partial f_i}{\partial x_i'} \right\rangle \quad \text{and} \quad \frac{\partial \langle f_i \rangle}{\partial x_i'} \,.$$

The interchange of differentiation and integration is a trivial matter if Ω_f is uniform with respect to the macroscale coordinates, i.e., when the medium is homogeneous. However, when Ω_f or, equivalently, the porosity $\theta \equiv \Omega_f/\Omega$ varies in x_i', the two operations do not commute. Greater care is needed and a theorem for spatial averaging is needed and derived by Slattery (1967), Gray and Lee (1977), and Whitaker (1986). We give the version described by Mei *et al.* (1996).

In a unit cell Ω, we let the fluid–solid interface Γ be described by the equation $S(\boldsymbol{x}, \boldsymbol{x}') = 0$, the fluid domain Ω_f by $S > 0$, and the solid domain Ω_s by $S < 0$. For example if the solid phase is composed of nontouching spheres of radius a, the surface of a particle centered at $\boldsymbol{x}_n$ is $S(\boldsymbol{x}, \boldsymbol{x}') = |\boldsymbol{x} - \boldsymbol{x}_n(\boldsymbol{x}')| - a = 0$. The cell volume of Ω is assumed to be spatially invariant but the fluid volume hence θ may vary with respect to $\boldsymbol{x}'$. An integral of any vector $f_i(\boldsymbol{x}, \boldsymbol{x}')$ over $\Omega_f(\boldsymbol{x}')$ can be replaced by an integral over Ω with the help of the Heaviside step function $H(S)$:

$$\iiint_{\Omega_f} f_i \, d\Omega = \iiint_{\Omega} f_i H(S) d\Omega \,, \tag{A.1}$$

where

$$H(S) = \begin{cases} 0, & \text{if } S < 0 \text{ (in solid)}\,, \\ 1, & \text{if } S > 0 \text{ (in fluid)}\,. \end{cases} \tag{A.2}$$

After changing the integration domain to Ω, we may now proceed as follows:

$$\begin{aligned} \frac{\partial}{\partial x_i'} \iiint_{\Omega_f} f_i \, d\Omega &= \frac{\partial}{\partial x_i'} \iiint_{\Omega} f_i H(S) d\Omega \\ &= \iiint_{\Omega} \frac{\partial f_i H}{\partial x_i'} d\Omega \\ &= \iiint_{\Omega_f^-} \frac{\partial f_i}{\partial x_i'} d\Omega + \iiint_{\Omega_\epsilon} \left[\frac{\partial f_i}{\partial x_i'} H + f_i \frac{\partial H}{\partial x_i'} \right] d\Omega \,, \end{aligned} \tag{A.3}$$

where Ω_ϵ is a thin shell enclosing the interface Γ with a thickness 2ϵ, where $\epsilon \ll 1$, and $\Omega_f^- = \Omega_f - \Omega_\epsilon$. While the first Ω_ϵ-integral in the square brackets vanishes as $\epsilon \to 0$, the second one requires further manipulations. First note that

$$\frac{\partial H(S)}{\partial x_i'} = \frac{dH(S)}{dS} \frac{\partial S}{\partial x_i'} = \delta(S) \frac{\partial S}{\partial x_i'} \,, \tag{A.4}$$

where $\delta(S)$ is the Dirac delta function in S. Hence,

$$\iiint_{\Omega_\epsilon} f_i \frac{\partial H}{\partial x_i'} d\Omega = \iiint_{\Omega_\epsilon} f_i \delta(S) \frac{\partial S}{\partial x_i'} d\Omega = \int_\Gamma \int_{-\epsilon}^{\epsilon} f_i \cdot \delta(S) \frac{\partial S}{\partial x_i'} d\zeta dA \,, \tag{A.5}$$

where ζ is the normal distance from Γ (or $S = 0$). Changing the independent variable from ζ to S by

$$dS = |\nabla S| d\zeta \,, \tag{A.6}$$

we can write the last integral as

$$\int_\Gamma dA \int_{-\epsilon}^{\epsilon} f_i \left[\delta(S) \frac{\partial S}{\partial x_i'} \right] \frac{dS}{|\nabla S|} + \int_\Gamma f_i \frac{\partial S}{\partial x_i'} \frac{dA}{|\nabla S|} \,. \tag{A.7}$$

Putting this result into Eq. (A.5) and then Eq. (A.3) and letting $\epsilon \to 0$, we obtain the averaging theorem

$$\frac{\partial}{\partial x_i'} \iiint_{\Omega_f} f_i d\Omega = \iiint_{\Omega_f} \frac{\partial f_i}{\partial x_i'} d\Omega + \iint_\Gamma f_i \frac{\partial S}{\partial x_i'} \frac{dA}{|\nabla S|} \tag{A.8}$$

or, in terms of the averaging symbols,

$$\frac{\partial \langle f_i \rangle}{\partial x_i'} = \left\langle \frac{\partial f_i}{\partial x_i'} \right\rangle + \frac{1}{\Omega} \iint_\Gamma f_i \frac{\partial S}{\partial x_i'} \frac{dA}{|\nabla S|} \,. \tag{A.9}$$

This result is equivalent to the averaging theorem of Slattery (1976) (see also, Gray and Lee, 1977; and Whitaker, 1985), now expressed specifically for two-scale problems.

As an immediate application, we let $f_i = u_i$, then the surface integral vanishes on account of the no-slip boundary condition on Γ. In this case volume-averaging and macroscale differentiation commute.

For simplicity we assume in this chapter that the porosity is uniform on the large scale, so that $\partial S/\partial x_i' = 0$; interchanges are allowable for all averages.

References

Bear, J. (1972). *Dynamics of Fluids in Porous Media*, Elsevier, Amsterdam.

Bensoussan, A., J. L. Lions and G. Papanicolaou (1978). *Asymptotic Analysis for Periodic Structures*, North Holland Publ., Amsterdam.

Brinkman, H. C. (1947). A calculation of viscous force exerted by a flowing fluid on a dense swarm of particle. *Appl. Sci. Res.* **A1**: 27–34.

Burns, G. (1985). *Solid State Physics*, Academic Press, New York.

Carman, P. C. (1937). Fluid flow through granular beds. *Trans. Inst. Chem. Eng.* **15**: 150–166.

Chauveeau, G. (1965). *Essai sur la loi de Darcy.* Thése, Universite de Toulouse.

Dagan, G. (1989). *Flow and Transport in Porous Formations*, Academic Press, New York.

Dullien, F. A. L. (1979). *Porous Media: Fluid Transport and Pore Structure*, Academic Press, New York.

Ene, H. L. and E. Sanchez-Palencia (1975). Equations et phénomène de surface pour I'écoulement dan un modèle de milieu poreux. *J. Mécanique* **4**: 73–108.

Firdaouss, M., J.-L. Guermond and P. Le Quéré (1998). Nonlinear corrections to Darcy's law at low Reynolds numbers. *J. Fluid Mech.* **343**: 331–350.

Forcheimer, P. (1901). Wasserbewegen durch Boden. *Z. Vereines Deutscher Ingneiure* **45**: 1782–1788.
Freeze, R. A. and J. A. Cherry (1979). *Groundwater*, Prentice Hall, Englewood Cliffs, New Jersey.
Fung, Y. C. (1977). *A First Course in Continuum Mechanics*, 2nd edn., Prentice-Hall, Inc., Englewood Cliffs, NJ.
Fung, Y. C. (1997). *Biomechanics: Circulation*, Springer-Verlag, New York.
Fung, Y. C. and S. S. Sobin (1969). Theory of sheet flow in lung alveoli. *J. Appl. Physiol.* **26**: 472–488.
Gelhar, L. W. (1993). *Stochastic Subsurface Hydrogeology*, Prentice-Hall, Englewood Cliffs, NJ.
Gray, W. G. and P. C. Y. Lee (1977). On the theorems for local volume averaging of multiphase systems. *Int. J. Multiphase Flow* **3**: 333–340.
Hannoura, A. A. and F. Barends (1981). Non-Darcy flow: A state of the art. *Flow and Transport in Porous Media*, eds. A. Veruijt and F. Barends, Balkema.
Hantusch, M. S. (1960). Modification of the theory of leaky aquifers. *J. Geophys. Res.* **65**: 3713–3725.
Hazen, A. (1895). *The Filtration of Public Water-Supplies*, Wiley.
Hornung, U. (1997). *Homogenization and Porous Media*, Springer, 279 pp.
Keller, J. B. (1980). Darcy's law for flow in porous media and the two space method. *Nonlinear Partial Differential Equations in Engineering and Applied Science*, eds. R. L. Sternberg, A. J. Kalinowski and J. S. Papadakis, Dekker, New York, pp. 429–443.
Kovacs, G. (1981). *Seepage Hydraulics*, Elsevier, Amsterdam.
Lee, C. K., C. C. Sun and C. C. Mei (1996). Computations of pesmeability and dispersivites of solute or heat in periods porous media. *Int. J. Heat Mass Transfer* **39**(4): 661–676.
Lee, J. S. (1969). Slow viscous flow in a lung alveoli model. *J. Biomechanics* **2**: 187–198.
Levy, T. (1981). Loi de Darcy ou loi de Brinkman? *C. R. Acad. Sci. Paris Sèrie II* **292**: 872–874.
Mei, C. C. and J.-L. Auriault (1989). Mechanics of heterogeneous porous media with several spatial scales. *Proc. R. Soc. Lond. A* **426**: 391–423.
Mei, C. C. and J.-L. Auriault (1991). The effect of weak inertia on flow through a porous medium. *J. Fluid Mech.* **222**: 647–663.
Mei, C. C., J.-L. Auriault and C. O. Ng (1996). Some applications of the homogenization theory. *Adv. Appl. Mech.* **32**: 277–348.

Nield, D. A. and A. Bejan (1992). *Convection in Porous Media*, Springer-Verlag, New York.

Payatakes, A. C., C. Tien and R. M. Turian (1973). Numerical solution of steady-state incompressible Newtonian flow through periodically constricted tubes. *AICHE J.* **19**: 67–76.

Rasoloarijaona, M. and J.-L. Auriault (1994). Nonlinear seepage flow through a rigid porous medium. *Eur. J. Mech. B/Fluids* **13**: 177–195.

Rubinstein, J. (1986). Effective equations for flow in random porous media with a large number of scales. *J. Fluid Mech.* **170**: 379–383.

Scheidegger, A. E. (1960). *The Physics of Flow Through Porous Media*, University of Toronto Press.

Slattery, J. C. (1967). Flow of viscoelastic fluids through porous media. *AIChE J.* **13**: 1066–1071.

Snyder, L. J. and W. E. Stewart (1966). Velocity and pressure profiles for Newtonian creeping flow in regular packed beds of spheres. *AIChE J.* **12**(1): 167–173.

Whitaker, S. (1986). Flow in porous media I: A theoretical derivation of Darcy's law. *Transport Porous Media* **1**: 3–25.

Wodie, J.-C. and T. Levy (1991). Correction non lineaire de la loi de Darcy. *C. R. Acad. Sci. Paris II* **312**: 157–161.

Yen R. T. and Y. C. Fung (1973). Model experiments on apparent blood viscosity and hematocrit in pulmonary alveoli. *J. Appl. Physiol.* **35**: 510–517.

Zhong, Z., Y. Dai, C. C. Mei and P. Tong (2002). A micromechanical theory of flow in pulmonary alveolar sheet. *Comput. Model. Eng. Sci.* (CMES) **3**: 77–86.

Zick, A. A. and G. M. Homsy (1980). Stokes flow through periodic arrays of spheres. *J. Fluid Mech.* **115**: 13–26.

Dispersion in Periodic Media or Flows

4

In this chapter we demonstrate for two examples the derivation of effective equations for the convective diffusion of a passive solvent in a flow with shear. This type of problem is an extension of Taylor's theory of dispersion in a straight pipe or channel described in Sec. 1.5. Two types of microscale periodicity will be illustrated: one is spatial only and the other is both spatial and temporal. In the first, we shall consider the dispersion of a passive solute in a seepage flow through a spatially periodic porous medium. Both two- and three-scale cases will be considered. In the second, homogenization is carried out for dispersion of suspended sediments in a wave-induced boundary layer at the sea bottom, where the medium is homogeneous but the fluid motion is periodic in both space and time.

4.1. Passive Solute in a Two-Scale Seepage Flow

The transport of passive or chemically reactive solvents in a fluid-saturated porous medium is of importance to a host of industrial, biomedical, and environmental problems, such as the spreading of nutrients in bones, and of contaminants in saturated soils, etc. As in the case of seepage flow in Chapter 3, one is often not directly interested in the physical processes on the small scale of grains or pores, but on the averaged picture on a much larger (macro) scale. In engineering literature there exist theoretical models that bypass the microscale details, and begin from Darcy's empirical law for the seepage flow on the scale much greater than the pore size. A convection–diffusion equation is assumed to govern the solute transport on a similarly large scale. The effective diffusion coefficients are then empirically related to the local seepage velocity (Bear, 1969; Fried and Combarnous, 1971; Dullien, 1979). Since by definition the phenomenon of dispersion depends on the flow on the microscale, a more fundamental approach is to derive the macroscale equations and the constitutive coefficients by considering the fluid mechanics on the microscale.

For fully three-dimensional models of microstructure, Brenner (1980) developed a rigorous theory for strictly periodic porous media. Starting from the Brownian theory for the probability density of suspended particles, he extended the method of moments by Aris (1956) and showed that the dispersion tensor should be calculated by first numerically solving a convection–diffusion problem in a periodic cell on the microscale.

Brenner's theory can be derived and extended rather straightforwardly by homogenization theory (Rubinstein and Mauri, 1986; Mei, 1991, 1992; Mei *et al.*, 1996; Auriault and Adler, 1992) without resorting to Brownian theory.

4.1.1. *The Solute Transport Equation and Scale Estimates*

We start from the convection–diffusion equation governing the volume concentration $C(x_i, t)$ of the solute everywhere in the pore fluid:

$$\frac{\partial C}{\partial t} + u_j \frac{\partial C}{\partial x_j} = D\nabla^2 C \quad \text{in } \Omega_f\,, \tag{4.1.1}$$

where D denotes the molecular mass diffusivity in the fluid. The velocity field of the incompressible pore fluid is assumed to be given by the theory of Chapter 3. In this analysis we shall allow the microscale geometry to vary slowly over the macroscale ($\mathcal{O}(\ell')$) and demonstrate the use of the averaging theorem in the Appendix 3A. Let the granular boundary Γ be denoted by $S(x_i) = 0$ and be impermeable, then the normal vector $\{n_i\}$ to Γ is proportional to $\{\partial S/\partial x_i\}$, so that the no-flux condition on Γ reads:

$$\frac{\partial C}{\partial x_i}\frac{\partial S}{\partial x_i} = 0 \quad \text{on } \Gamma\,. \tag{4.1.2}$$

Introducing the normalization

$$C \to (C_o)C^\dagger\,, \quad u \to Uu^\dagger\,, \quad t \to Tt^\dagger\,, \quad x_i = \ell x_i^\dagger\,, \tag{4.1.3}$$

where ΔC and T are, respectively, the scales of concentration variation and time, we get:

$$\frac{\ell^2}{TD}\frac{\partial C^\dagger}{\partial t^\dagger} + Pe\, u_j^\dagger \frac{\partial C^\dagger}{\partial x_j^\dagger} = \nabla_\dagger^2 C^\dagger\,, \tag{4.1.4}$$

where Pe is the Péclet number based on the grain size:

$$Pe = \frac{U\ell}{D} = \frac{P\ell^2}{\mu L}\frac{\ell}{D} = Re\frac{\nu}{D} = RePr\,. \tag{4.1.5}$$

For many solutes in groundwater flows the Prandtl number $Pr = \nu/D$ is large though Re is small (e.g., for salt in water $D = 10^{-5}\,\text{cm}^2/\text{s}$ while $\nu = 10^{-2}\,\text{cm}^2/\text{s}$ so that $Pr = 10^3$), therefore we shall allow $Pe = \mathcal{O}(1)$ for generality.[1] There are three time scales in this problem. The shortest is the diffusion time across the small pores $T_d = \ell^2/D$, followed by the convection time T_c across the macroscale ℓ'

$$T_c = \frac{\ell'}{U}\,, \tag{4.1.6}$$

which is much longer than T_d:

$$\frac{T_c}{T_d} = \mathcal{O}\left(\frac{\ell'}{\ell}\frac{D}{U\ell}\right) = \mathcal{O}\left(\frac{\ell'}{\ell}\right) = \mathcal{O}\left(\frac{1}{\epsilon}\right) \gg 1$$

since $\ell/\ell' \equiv \epsilon \ll 1$ and $U\ell/D = \mathcal{O}(1)$. The longest is the diffusion time across the macroscale $T_{d'}$

$$T_d' = \frac{\ell'^2}{D} \tag{4.1.7}$$

since

$$\frac{T_d'}{T_d} = \mathcal{O}\left(\frac{\ell'^2}{\ell^2}\right) = \mathcal{O}\left(\frac{1}{\epsilon^2}\right),$$

and

$$\frac{T_d'}{T_c} = \mathcal{O}\left(\frac{U\ell'}{D}\right) = \mathcal{O}\left(\frac{U\ell}{D}\frac{\ell'}{\ell}\right) = \mathcal{O}\left(\frac{Pe}{\epsilon}\right).$$

Let us ignore the very rapid diffusion across the pores and choose the convection time $T = T_c$ for normalization in Eq. (4.1.4), which then becomes,

$$Pe\left(\epsilon\frac{\partial C^\dagger}{\partial t^\dagger} + u_j^\dagger\frac{\partial C^\dagger}{\partial x_j^\dagger}\right) = \nabla^{\dagger 2} C^\dagger\,. \tag{4.1.8}$$

[1] If the pore fluid is air, as in unsaturated soils, these estimates can be greatly different.

From here on we shall work with the dimensional equation with the artifice of inserting the ordering parameter $\epsilon = \ell/\ell' \ll 1$ as follows,

$$\epsilon \frac{\partial C}{\partial t} + u_j \frac{\partial C}{\partial x_j} = D\nabla^2 C \quad \text{in } \Omega_f\,, \tag{4.1.9}$$

and examine the evolution over the two longer time scales, T_c and T'_d.

4.1.2. *Macroscale Transport Equation*

Let us introduce two time variables: t for convection and $t' = \epsilon t$ for macroscale diffusion, as well as two space variables $x_j, x'_j = \epsilon x_j$. Assuming multiple-scale expansions:

$$C = C^{(0)} + \epsilon C^{(1)} + \epsilon^2 C^{(2)} + \cdots\,, \quad u_i = u_i^{(0)} + \epsilon u_i^{(1)} + \epsilon^2 u_i^{(2)} + \cdots\,, \tag{4.1.10}$$

where $C^{(m)}$, $m = 0, 1, 2, \ldots$ depend on x_j, x'_j, t and t', it follows easily that at the orders $\mathcal{O}(\epsilon^0)$, $\mathcal{O}(\epsilon)$ and $\mathcal{O}(\epsilon^2)$:

$$\frac{\partial u_j^{(0)} C^{(0)}}{\partial x_j} = D\nabla^2 C^{(0)}\,, \tag{4.1.11}$$

$$\frac{\partial C^{(0)}}{\partial t} + \frac{\partial u_j^{(0)} C^{(1)}}{\partial x_j} + \frac{\partial u_j^{(1)} C^{(0)}}{\partial x_j} + \frac{\partial u_j^{(0)} C^{(0)}}{\partial x'_j} = D(\nabla^2 C^{(1)} + \nabla \cdot \nabla' C^{(0)} + \nabla' \cdot \nabla C^{(0)})\,, \tag{4.1.12}$$

$$\frac{\partial C^{(0)}}{\partial t'} + \frac{\partial C^{(1)}}{\partial t} + \frac{\partial u_j^{(0)} C^{(2)}}{\partial x_j} + \frac{\partial u_j^{(1)} C^{(1)}}{\partial x_j} + \frac{\partial u_j^{(0)} C^{(1)}}{\partial x'_j} + \frac{\partial u_j^{(1)} C^{(0)}}{\partial x'_j} + \frac{\partial u_j^{(2)} C^{(0)}}{\partial x_j} = D[\nabla^2 C^{(2)} + \nabla' \cdot \nabla C^{(1)} + \nabla \cdot \nabla' C^{(1)} + \nabla'^2 C^{(0)}]\,, \tag{4.1.13}$$

where

$$\nabla' \equiv \left\{ \frac{\partial}{\partial x'_j} \right\}, \quad \nabla'^2 \equiv \left\{ \frac{\partial^2}{\partial x'_j \partial x'_j} \right\}. \tag{4.1.14}$$

Since the periodic microstructure is allowed to change slowly in x'_i, S depends on x_i and x'_i. At successive orders the boundary conditions on the grain/fluid interface Γ and on the outer boundary $\partial\Omega$ of an Ω cell are

$$\frac{\partial C^{(0)}}{\partial x_i}\frac{\partial S}{\partial x_i} = 0\,, \tag{4.1.15}$$

$$\left(\frac{\partial C^{(1)}}{\partial x_i} + \frac{\partial C^{(0)}}{\partial x_i'}\right)\frac{\partial S}{\partial x_i} + \frac{\partial C^{(0)}}{\partial x_i}\frac{\partial S}{\partial x_i'} = 0\,, \tag{4.1.16}$$

and

$$\left(\frac{\partial C^{(2)}}{\partial x_i} + \frac{\partial C^{(1)}}{\partial x_i'}\right)\frac{\partial S}{\partial x_i} + \left(\frac{\partial C^{(1)}}{\partial x_i} + \frac{\partial C^{(0)}}{\partial x_i'}\right)\frac{\partial S}{\partial x_i'} = 0\,, \tag{4.1.17}$$

and

$$C^{(m)} \text{ is } \Omega\text{-periodic} \tag{4.1.18}$$

for $m = 0, 1, 2, \ldots$.

The cell problem at the order $\mathcal{O}(\epsilon^0)$ defined by Eqs. (4.1.11), (4.1.15), and (4.1.18) with $m = 0$ can be satisfied by

$$C^{(0)} = C^{(0)}(\boldsymbol{x}', t, t')\,, \tag{4.1.19}$$

which is independent of the local spatial coordinates. This can be shown systematically by multiplying Eq. (4.1.11) by $C^{(0)}$ and integrating the product over a microcell. After partial integration, the integral becomes

$$\begin{aligned} Pe \iiint_\Omega &\left\{\frac{\partial}{\partial x_j}\left(u_j^{(0)}\frac{(C^{(0)})^2}{2}\right) - \frac{(C^{(0)})^2}{2}\frac{\partial u_j^{(0)}}{\partial x_j}\right\} d\Omega \\ &= \iiint_\Omega \left\{\frac{\partial}{\partial x_j}\left(DC^{(0)}\frac{\partial C^{(0)}}{\partial x_j}\right) - D\frac{\partial C^{(0)}}{\partial x_j}\frac{\partial C^{(0)}}{\partial x_j}\right\} d\Omega\,. \end{aligned} \tag{4.1.20}$$

By invoking Gauss' theorem, periodicity and mass conservation, Eq. (4.2.33), the following is true:

$$-\iiint_\Omega D\left\{\frac{\partial C^{(0)}}{\partial x_j}\frac{\partial C^{(0)}}{\partial x_j}\right\} d\Omega = 0\,, \tag{4.1.21}$$

which implies that $C^{(0)}(\boldsymbol{x}', t, t')$ is independent of the microscale.

An immediate consequence is that Eq. (4.1.16) reduces to

$$\left(\frac{\partial C^{(1)}}{\partial x_i} + \frac{\partial C^{(0)}}{\partial x'_i}\right)\frac{\partial S}{\partial x_i} = 0\,. \qquad (4.1.22)$$

Using Eq. (4.1.19),

$$\frac{\partial u_j^{(1)}}{\partial x_j} + \frac{\partial u_j^{(0)}}{\partial x'_j} = 0\,,$$

and then taking the Ω average of Eq. (4.1.12), we obtain by Gauss' theorem and the boundary conditions,

$$\theta\frac{\partial C^{(0)}}{\partial t} + \langle u_j^{(0)}\rangle\frac{\partial C^{(0)}}{\partial x'_j} = \frac{D}{\Omega}\iint_\Gamma\left(\frac{\partial C^{(1)}}{\partial x_i} + \frac{\partial C^{(0)}}{\partial x'_i}\right)\left(-\frac{\partial S}{\partial x_i}\right)\frac{dA}{|\nabla S|}\,.$$

Clearly, the right-hand side vanishes because of the no-flux boundary condition (Eq. (4.1.22)). Hence,

$$\boxed{\theta\frac{\partial C^{(0)}}{\partial t} + \langle u_j^{(0)}\rangle\frac{\partial C^{(0)}}{\partial x'_j} = 0}\,. \qquad (4.1.23)$$

Thus, over the time range $t = \mathcal{O}(T_1)$, the solute concentration is simply convected by the seepage velocity.

Subtracting Eq. (4.1.23) from Eq. (4.1.12), we get

$$u_j^{(0)}\frac{\partial C^{(1)}}{\partial x_j} - D\nabla^2 C^{(1)} = -\tilde{u}_j^{(0)}\frac{\partial C^{(0)}}{\partial x'_j}\,, \qquad (4.1.24)$$

where

$$\widetilde{u_j^{(0)}} = u_j^{(0)} - \frac{\langle u_j^{(0)}\rangle}{\theta} \qquad (4.1.25)$$

is the deviation from the mean. To solve the linear equation (4.1.24) for $C^{(1)}$, subject to the linear boundary condition (Eq. (4.1.22)) and Ω-periodicity, we let

$$C^{(1)} = -N_\ell\frac{\partial C^{(0)}}{\partial x'_\ell}\,. \qquad (4.1.26)$$

It then follows that $N_\ell(x_i, x_i')$ must be the solution to the following canonical boundary value problem in an Ω cell:

$$u_j^{(0)}\frac{\partial N_\ell}{\partial x_j} - D\nabla^2 N_\ell = \widetilde{u}_\ell^{(0)}, \tag{4.1.27}$$

$$n_j\frac{\partial N_\ell}{\partial x_j} = n_\ell \quad \text{on } \Gamma, \tag{4.1.28}$$

$$N_\ell \text{ is } \Omega\text{-periodic}. \tag{4.1.29}$$

In addition, we require

$$\langle N_\ell \rangle = 0, \tag{4.1.30}$$

so that

$$\langle C^{(1)}(x_\ell', t, t')\rangle = 0. \tag{4.1.31}$$

Mathematically $\langle C^{(1)}(x_\ell', t, t')\rangle$ is another homogeneous solution of the steady cell problem and can be omitted by defining $C^{(0)}$ to be the Ω-average of C with an error of $O(\epsilon^2)$.

The linear cell problem defined by Eqs. (4.1.27)–(4.1.30) for N_ℓ can in principle be solved numerically for a known cell geometry. In contrast to the case of simple diffusion, the cell problem depends on the interstitial flow, which can in general vary on the global scale (i.e., in x_i'). Hence N_ℓ depends on x_i and x_i'; discretely many problems must be solved for cells separated by distances of $\mathcal{O}(\ell')$ apart. This is still less of a numerical task than multi-scale computations of the entire region of interest.

Lastly, we average Eq. (4.1.13) over the Ω-cell. By using Gauss' theorem and boundary conditions, we get

$$\begin{aligned}\theta\frac{\partial C^{(0)}}{\partial t'} &+ \frac{\partial}{\partial x_j'}\{\langle u_i^{(0)}C^{(1)}\rangle + \langle u_i^{(1)}\rangle C^{(0)}\}\\ &= \frac{D}{\Omega}\iint_\Gamma \left(\frac{\partial C^{(2)}}{\partial x_i} + \frac{\partial C^{(1)}}{\partial x_i'}\right)\left(-\frac{\partial S}{\partial x_i'}\right)\frac{dA}{|\nabla S|}\\ &\quad + D\langle \nabla'\cdot\nabla C^{(1)}\rangle + D\langle \nabla'^2 C^{(0)}\rangle.\end{aligned} \tag{4.1.32}$$

Analysis at this order requires in principle the solution of the perturbation velocity $u_j^{(1)}$ as found in Sec. 3.8. However, use may be made of the result (Eq. (3.8.25)) that $\langle u_\ell^{(1)} \rangle = 0$. Invoking the averaging theorem (A.9), Appendix 3A, Chapter 3, the last two terms on the right become

$$D\langle \nabla' \cdot \nabla C^{(1)} \rangle + D\langle \nabla'^2 C^{(0)} \rangle$$
$$= D\nabla' \cdot \langle \nabla C^{(1)} \rangle + D\nabla'^2 C^{(0)} - \frac{D}{\Omega} \iint_\Gamma (\nabla C^{(1)} + \nabla' C^{(0)}) \frac{\partial S}{\partial x_i'} \frac{dS}{|\nabla S|} . \tag{4.1.33}$$

After some algebra as described in Appendix 4A, including the use of the boundary condition (Eq. (4.1.17)), Eq. (4.1.32) is reduced to

$$\theta \frac{\partial C^{(0)}}{\partial t'} + \langle u_\ell^{(1)} \rangle \frac{\partial C^{(0)}}{\partial x_\ell'} = \frac{\partial}{\partial x_j'} \left(\mathcal{F}_{j\ell} \frac{\partial C^{(0)}}{\partial x_\ell'} \right) + \theta D \nabla'^2 C^{(0)} , \tag{4.1.34}$$

where

$$\mathcal{F}_{j\ell} = \langle \tilde{u}_j^{(0)} N_\ell \rangle - D \left\langle \frac{\partial N_\ell}{\partial x_j'} \right\rangle . \tag{4.1.35}$$

The tensor $\mathcal{F}_{j\ell}$ can be split into a symmetric and an anti-symmetric part:

$$\mathcal{F}_{j\ell} = \mathcal{D}_{j\ell} + \mathcal{E}_{j\ell} , \tag{4.1.36}$$

where the symmetric part $\mathcal{D}_{j\ell}$ is the dispersivity tensor

$$\mathcal{D}_{j\ell} = \frac{1}{2} \langle \tilde{u}_j^{(0)} N_\ell + \tilde{u}_\ell^{(0)} N_j \rangle - \frac{D}{2} \left\langle \frac{\partial N_\ell}{\partial x_j'} + \frac{\partial N_j}{\partial x_\ell} \right\rangle , \tag{4.1.37}$$

which boosts the effecive diffusivity. The anti-symmetric part

$$\mathcal{E}_{j\ell} = \frac{1}{2} \langle \tilde{u}_j^0 N_\ell - \tilde{u}_\ell^0 N_j \rangle + \frac{D}{2} \left\langle \left(\frac{\partial N_\ell}{\partial x_j'} - \frac{\partial N_j}{\partial x_\ell'} \right) \right\rangle \tag{4.1.38}$$

can be used to define a velocity,

$$\langle \widehat{u}_\ell^{(1)} \rangle = - \frac{\partial \mathcal{E}_{j\ell}}{\partial x_j'} , \tag{4.1.39}$$

which is a correction to the effective convection velocity at $\mathcal{O}(\epsilon)$. Equation (4.1.34) can then be rewritten as

$$\boxed{\theta \frac{\partial C^{(0)}}{\partial t'} + (\langle u_\ell^{(1)} \rangle + \langle \widehat{u}_\ell^{(1)} \rangle) \frac{\partial C^{(0)}}{\partial x_\ell'} = \frac{\partial}{\partial x_j'} \left(\mathcal{D}_{j\ell} \frac{\partial C^{(0)}}{\partial x_\ell'} \right) + \theta D \nabla'^2 C^{(0)}} . \tag{4.1.40}$$

We shall define the total effective diffusivity as the sum of mesoscale molecular diffusivity and flow-induced dispersivity:

$$\mathcal{D}_{ij}^T = \mathcal{D}_{ij} + nD\delta_{ij} \,. \tag{4.1.41}$$

We point out that $\mathcal{E}_{ij}$ does not have the physical effect of diffusion. Although the associated convection term $\langle \widehat{u}^{(1)} \rangle$ on the left of Eq. (4.1.40) can be formally moved to the right and expressed in the form,

$$\frac{\partial}{\partial x_j'}\left(\mathcal{E}_{j\ell}\frac{\partial C^{(0)}}{\partial x_\ell'}\right) = \frac{\partial \mathcal{E}_{j\ell}}{\partial x_j'}\frac{\partial C^{(0)}}{\partial x_\ell'} + \mathcal{E}_{j\ell}\frac{\partial^2 C^{(0)}}{\partial x_j' \partial x_\ell'} \,,$$

the diffusion-like term vanishes identically

$$\mathcal{E}_{j\ell}\frac{\partial^2 C^{(0)}}{\partial x_j' \partial x_\ell'} = \mathcal{E}_{\ell j}\frac{\partial^2 C^{(0)}}{\partial x_\ell' \partial x_j'} = -\mathcal{E}_{j\ell}\frac{\partial^2 C^{(0)}}{\partial x_j' \partial x_\ell'} = 0 \,,$$

due to the anti-symmetry, $\mathcal{E}_{j\ell} = -\mathcal{E}_{\ell j}$. Hence $\mathcal{E}_{j\ell}$ does not represent diffusivity.

Thus, over the longer time $t' = O(T_c/\epsilon)$, $C^{(0)}$ undergoes convective diffusion governed by Eq. (4.1.40). Finally we add Eq. (4.1.23) and $\epsilon \times$ (Eq. (4.1.40)) to get the convective dispersion–diffusion equation for $C^{(0)}$, valid for $t = \mathcal{O}(1/\epsilon)$ and $x' = \mathcal{O}(1)$,

$$\boxed{\theta\frac{\partial C}{\partial t} + \mathcal{U}_\ell\frac{\partial C}{\partial x_\ell'} = \epsilon\frac{\partial}{\partial x_j'}\left(\mathcal{D}_{j\ell}^T\frac{\partial C}{\partial x_\ell'}\right) = \epsilon\frac{\partial}{\partial x_j'}\left[(\mathcal{D}_{j\ell} + \theta D\delta_{j\ell})\frac{\partial C}{\partial x_\ell'}\right]} \,, \tag{4.1.42}$$

where

$$\mathcal{U}_\ell = \langle u_\ell^{(0)} \rangle + \epsilon(\langle u_\ell^{(1)} \rangle + \langle \widehat{u}_\ell^{(1)} \rangle) \tag{4.1.43}$$

is the total convection velocity.

In general the flow can be nonuniform on the macroscale, hence $u_i^{(0)}$ depends on both $\boldsymbol{x}$ and $\boldsymbol{x}'$. It follows that N_ℓ also depends on $\boldsymbol{x}$ and $\boldsymbol{x}'$ through the local Péclet number; the canonical cell problem for N_i needs to be solved numerically at discrete number of macrostations separated at distances of order $\Delta \boldsymbol{x} = \mathcal{O}(\ell')$.

From Eq. (4.1.27) and the boundary conditions (Eqs. (4.1.28) and (4.1.29)), it can be shown that

$$\langle \widetilde{u}_\ell^o N_j + \widetilde{u}_j^0 N_\ell \rangle = \left\langle u_k^{(0)} \frac{\partial N_\ell}{\partial x_k} N_j + u_k^{(0)} \frac{\partial N_j}{\partial x_k} N_\ell \right\rangle - D \langle N_j \nabla^2 N_\ell + N_\ell \nabla^2 N_j \rangle$$

$$= 2D \left\langle \frac{\partial N_j}{\partial x_k} \frac{\partial N_\ell}{\partial x_k} \right\rangle - D \left\langle \frac{\partial N_\ell}{\partial x_j'} + \frac{\partial N_j}{\partial x_\ell} \right\rangle \quad (4.1.44)$$

after using continuity and Eq. (4.1.12). Hence, Eq. (4.1.38) may be written in terms of N_ℓ only,

$$\boxed{\mathcal{D}_{j\ell} = D \left\langle \frac{\partial N_j}{\partial x_k} \frac{\partial N_\ell}{\partial x_k} \right\rangle - D \left\langle \frac{\partial N_\ell}{\partial x_j} + \frac{\partial N_j}{\partial x_\ell} \right\rangle}, \quad (4.1.45)$$

which of course still depends on the flow field in the pores since N_i is governed by Eq. (4.1.27).

The dispersivity tensor may also be written as:

$$\mathcal{D}_{j\ell} = D \left\langle \frac{\partial b_j}{\partial x_k} \frac{\partial b_\ell}{\partial x_k} \right\rangle - \theta D \delta_{jl}, \quad (4.1.46)$$

where $b_\ell = N_\ell - x_\ell$. The last form (Eq. (4.1.46)) was first derived by Brenner (1980) via Brownian motion theory.

Since for any constant vector f_j, the bilinear form

$$f_j \left\langle \frac{\partial b_j}{\partial x_k} \frac{\partial b_\ell}{\partial x_k} \right\rangle f_\ell = \left\langle \frac{\partial b_j f_j}{\partial x_k} \frac{\partial b_\ell f_\ell}{\partial x_k} \right\rangle$$

is positive definite, the total dispersivity–diffusivity tensor $\mathcal{D}_{ij} + \theta D \delta_{j\ell}$ is positive definite (Brenner, 1980). For the same reason, the part

$$\left\langle \frac{\partial N_j}{\partial x_k} \frac{\partial N_\ell}{\partial x_k} \right\rangle \quad (4.1.47)$$

must be also positive definite.

The preceding derivation has been extended to the convective diffusion of heat in a fluid saturated porous medium where the solid matrix has finite conductivity (Mei, 1991).

As always the remaining and essential task is to solve the cell problem numerically for the vector N_i. Before discussing the computations we first give a proof that the solution of the cell problem defined by Eqs. (4.1.27)–(4.1.30) is unique.

If $\widehat{N}_\ell = N_\ell^{(i)} - N_\ell^{(ii)}$ is the difference of two solutions to the same cell problem, it must satisfy the following homogeneous equations,

$$u_j^{(0)} \frac{\partial \widehat{N}_\ell}{\partial x_j} - D\nabla^2 \widehat{N}_\ell = 0 \,, \quad x_j \in \Omega_f \,, \tag{4.1.48}$$

$$n_j \frac{\partial \widehat{N}_\ell}{\partial x_j} = 0 \quad \text{on } \Gamma \,, \tag{4.1.49}$$

$$\widehat{N}_\ell \text{ is } \Omega\text{-periodic} \,. \tag{4.1.50}$$

In addition, we require

$$\langle \widehat{N}_\ell \rangle = 0 \,. \tag{4.1.51}$$

Multiplying Eq. (4.1.48) by $\widehat{N}_\ell$ and rewriting the product as

$$\frac{\partial}{\partial x_j}\left(u_j^{(0)} \frac{\widehat{N}_\ell \widehat{N}_\ell}{2} \right) - \frac{\widehat{N}_\ell \widehat{N}_\ell}{2} \frac{\partial u_j^{(0)}}{\partial x_j} - D \frac{\partial}{\partial x_j}\left(\widehat{N}_\ell \frac{\partial \widehat{N}_\ell}{\partial x_j} \right) + D \frac{\partial \widehat{N}_\ell}{\partial x_j} \frac{\partial \widehat{N}_\ell}{\partial x_j} = 0 \,,$$

we integrate the result over Ω_f, and obtain,

$$\iiint_{\Omega_f} \frac{\partial \widehat{N}_\ell}{\partial x_j} \frac{\partial \widehat{N}_\ell}{\partial x_j} d\Omega = 0 \tag{4.1.52}$$

after invoking incompressibility, Gauss' theorem, and the boundary conditions. Thus $\widehat{N}_\ell$ is at the most a constant, which must be zero because of Eq. (4.1.51). The solution for $\widehat{N}_\ell$ is therefore unique.

4.1.3. *Numerical Computation of Dispersivity*

As a prerequisite the full solution of the Stokes problem must be obtained for $u_i^{(0)}, \tilde{u}_i^{(0)}$ everywhere in the cell, by, say, the finite-element computation in Sec. 3.4. Numerical solution of the canonical problem for N_ℓ can be carried out in similar manner. It suffices to describe the variational principle for such finite-element computations (Lee *et al.*, 1995).

Let us verify that the boundary-value problem in the cell for N_ℓ ensures the stationarity of the following functional $\mathcal{F}$

$$\mathcal{F} = \iiint_{\Omega_f} \left(-\tilde{u}_m^{(0)} N_m + u_j^{(0)} N_m \frac{\partial N_m}{\partial x_j} + D \frac{\partial N_m}{\partial x_j} \frac{\partial N_m}{\partial x_j} - D \frac{\partial N_m}{\partial x_m} \right) d\Omega$$
$$+ \lambda_m \iiint_{\Omega_f} N_m d\Omega \tag{4.1.53}$$

in which λ_m is the Lagrange multiplier.

By taking the first variation of J we get

$$\delta\mathcal{F} = \iiint_{\Omega_f}\left(-\tilde{u}_m^{(0)} + u_j^{(0)}\frac{\partial N_m}{\partial x_j}\right)(\delta N_m)$$
$$+\iiint_{\Omega_f}\left[u_j^{(0)}\frac{\partial(\delta N_m)}{\partial x_j}N_m + 2D\frac{\partial N_m}{\partial x_j}\frac{\partial(\delta N_m)}{\partial x_j} - D\frac{\partial(\delta N_m)}{\partial x_m}\right]d\Omega$$
$$+\left(\iiint_{\Omega_f} N_m d\Omega\right)\delta\lambda_m + \lambda_m\left(\iiint_{\Omega_f}\delta N_m d\Omega\right). \qquad (4.1.54)$$

Using the governing equations for N_m and δN_m, the integrand of the second integral can be further rewritten as,

$$DN_m\frac{\partial^2(\delta N_m)}{\partial x_j\partial x_j} + 2D\frac{\partial N_m}{\partial x_j}\left(\frac{\partial(\delta N_m)}{\partial x_j}\right) - D\frac{\partial(\delta N_m)}{\partial x_m}$$
$$= D\frac{\partial}{\partial x_j}\left(N_m\frac{\partial(\delta N_m)}{\partial x_j}\right) + D\frac{\partial}{\partial x_j}\left(\frac{\partial N_m}{\partial x_j}(\delta N_m)\right)$$
$$- D\frac{\partial^2 N_m}{\partial x_j\partial x_j}(\delta N_m) - D\frac{\partial(\delta N_m)}{\partial x_m}. \qquad (4.1.55)$$

Note that δN_m satisfies the same conditions (Eqs. (4.1.48)–(4.1.51)), as N'_m, since $u_j^{(0)}$ is known from the Stokes problem, i.e., $\delta u_j^{(0)} \equiv 0$. Substituting Eq. (4.1.55) into Eq. (4.1.54) and making use of Gauss's theorem, we obtain

$$\delta\mathcal{F} = \iiint_{\Omega_f}\left(-\tilde{u}_m^{(0)} + u_j^{(0)}\frac{\partial N_m}{\partial x_j} - D\frac{\partial^2 N_m}{\partial x_j\partial x_j}\right)(\delta N_m)d\Omega$$
$$+\iint_{\Gamma}\left[DN_m\frac{\partial(\delta N_m)}{\partial x_j}\right]n_j dS$$
$$+\iint_{\Gamma}\left[\left(D\frac{\partial N_m}{\partial x_j}n_j - Dn_m\right)(\delta N_m)\right]dS$$
$$+\iint_{\Gamma_f}\left\{D\left[N_m\frac{\partial(\delta N_m)}{\partial x_j} + \frac{\partial N_m}{\partial x_j}(\delta N_m)\right]n_j - D(\delta N_m)n_m\right\}dS$$
$$+\left(\iiint_{\Omega_f} N_m d\Omega\right)\delta\lambda_m + \lambda_m\left(\iiint_{\Omega_f}\delta N_m d\Omega\right), \qquad (4.1.56)$$

where Γ_f is the fluid portion on the outer surface of the microcell. The surface integrals on the interface Γ vanish because of the boundary conditions. The integral over Γ_f is also zero due to the Ω-periodicity conditions and the fact that n_j is of opposite sign on opposite side of Γ_f. Finally, if the zero average condition (Eq. (4.1.30)) is imposed, the last two integrals in Eq. (4.1.56) also vanish. Thus, the governing equations (4.1.27) and (4.1.28) guarantee the stationarity of $\mathcal{F}$, i.e.,

$$\delta\mathcal{F} = 0\,. \tag{4.1.57}$$

The inverse can also be shown by the standard argument of contradiction. This proves the equivalence of $\delta\mathcal{F} = 0$ to the boundary-value problem (Eqs. (4.1.27)–(4.1.30)).

Prior to the numerical task it is convenient to first nondimensionalize the cell boundary-value problem according to the scaling of Eq. (4.1.3) and define the normalized $N_m^\dagger$ by $N_m = \ell N_m^\dagger$ where ℓ is the size of the cell. The governing equation becomes, with the daggers omitted,

$$Pe\left(\widetilde{u}_j^{(0)} - u_j^{(0)}\frac{\partial N_m}{\partial x_j}\right) = \frac{\partial^2 N_m}{\partial x_j \partial x_j}\,. \tag{4.1.58}$$

Otherwise N_m obeys the boundary conditions (Eqs. (4.1.28)–(4.1.30)). Since the dimensionless N_m depends on the Péclet number Pe as a parameter, so does the normalized dispersivity tensor $\mathcal{D}_{ij}/D$ and the total effective diffusivity $\mathcal{D}_{ij}^T$.

Using finite elements to approximate the vector N_m in Ω_f, extremization of J leads to a matrix equation for the nodal coefficients. For a cubically packed array of Wigner–Seitz grains, the Ω cell is a unit cube containing just one grain. Using the flow field $u_i^{(0)}$ found from the Stokes' problem in Chapter 3, Lee *et al.* (1995) have computed the dispersivity tensor. In their computations, the mean flow is directed along the x axis. $\mathcal{D}_{ij}$ is diagonal with two independent components which are the longitudinal $\mathcal{D}_L$ and transverse $\mathcal{D}_T$ dispersivities: $\mathcal{D}_{11} = \mathcal{D}_L$ and $\mathcal{D}_{22} = \mathcal{D}_{33} = \mathcal{D}_T$. For any other flow direction in the xy-plane, there are four nonzero independent dispersivity coefficients: $\mathcal{D}_{11}, \mathcal{D}_{22}, \mathcal{D}_{33}$, and $\mathcal{D}_{12} = \mathcal{D}_{21}$ by symmetry and $\mathcal{D}_{13} = \mathcal{D}_{23} = 0$.

Computed values of the total longitudinal and transverse diffusivity coefficients $\mathcal{D}_L^T/D$ and $\mathcal{D}_T^T/D$ are shown in Figs. 4.1(a) and 4.1(b), respectively, for a large range of Pe and for two porosities $\theta = 0.38, 0.5$. The Péclet numbers are defined in terms of the mean flow velocity averaged over Ω_f only, i.e., $\langle u\rangle\ell/\theta D = Pe/\theta$. In Fig. 4.1(a), the longitudinal dispersivity for simple cubic packing of uniform spheres is also compared with the measured data by Gunn and Pryce (1969) and the calculations by Salles *et al.* (1993) that $\theta = 0.48, 0.74$, and 0.82. The results for $\theta = 0.48$ by Koch *et al.* (1989) based on an approximate analysis for dilute concentration are also

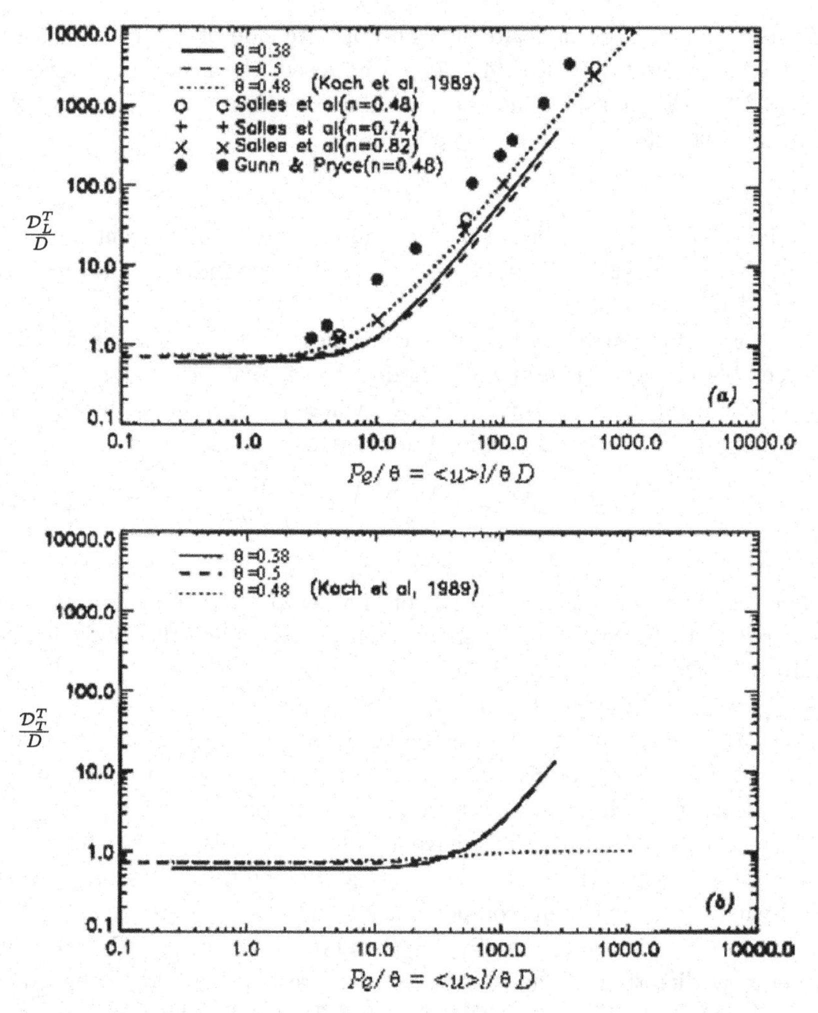

Fig. 4.1. Normalized dispersivities for a cubic array of Wigner–Seitz grains. The flow is in the x_1 direction. The ordinate is normalized in units of D, the molecular diffusivity. Numerical solution by the present scheme is compared with other computations for different grain geometries. (a): Total longitudinal diffusivity $\mathcal{D}_L^T/D = \mathcal{D}_{11}^T/D = (\mathcal{D}_{11} + \theta D)/D$. (b): Total transverse diffusivity $\mathcal{D}_T^T/D = \mathcal{D}_{22}^T/D = (\mathcal{D}_{22} + \theta D)/D$. From Lee, Sun and Mei (1995), *Int. J. Heat Mass Transfer.*

included. All are in qualitative agreement for $\mathcal{D}_L^T = \mathcal{D}_L + \theta D$. In Figs. 4.2 and 4.3, we further compare the computed results with a number of experimental data for natural grains of nonuniform shapes and sizes reported by Ebach and White (1958), Edwards and Richardson (1968), Harleman and

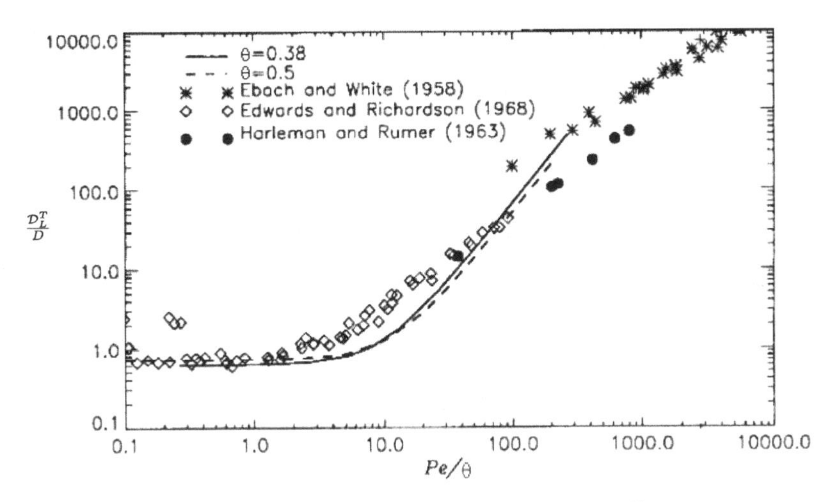

Fig. 4.2. Comparison of computed total longitudinal diffusivity $\mathcal{D}_L^T/D = (\mathcal{D}_{11}+nD)/D$ for a cubic array of Wigner–Seitz grains with experiments for natural grains. From Lee, Sun and Mei (1995), *Int. J. Heat Mass Transfer.*

Rumer (1963), Blackwell (1962), and Hiby (1962). As can be seen, the measured dispersion coefficients are close to being linear in the Péclet number. This linear dependence is likely the result of granular irregularity and is predicted by Saffman (1960), de Jong (1958), and Haring and Greenhorn (1970) for random networks of capillaries and by Koch and Brady (1985) for randomly packed spheres.

In general the local flow direction can be arbitrary. One must first decompose the local seepage flow velocity into three mutually orthogonal components, and obtain first the dispersion tensor $\mathcal{D}_{ij}$ for each component. The results are then summed to get the final local dispersivity tensor at each point. With these one can solve the averaged concentration equation, subject to proper boundary conditions on the macroscale.

4.2. Macrodispersion in a Three-Scale Porous Medium

Many natural and man-made materials are structured on a hierarchy of scales. One example of environmental importance is transport of the pollutants in groundwater through inhomogeneous soils where highly impervious lenses or strata are embedded in highly porous sand or rocks. Theoretical models sufficient for laboratory samples can be inadequate for physical predictions in the field. In the literature on hydrology the prevailing approach

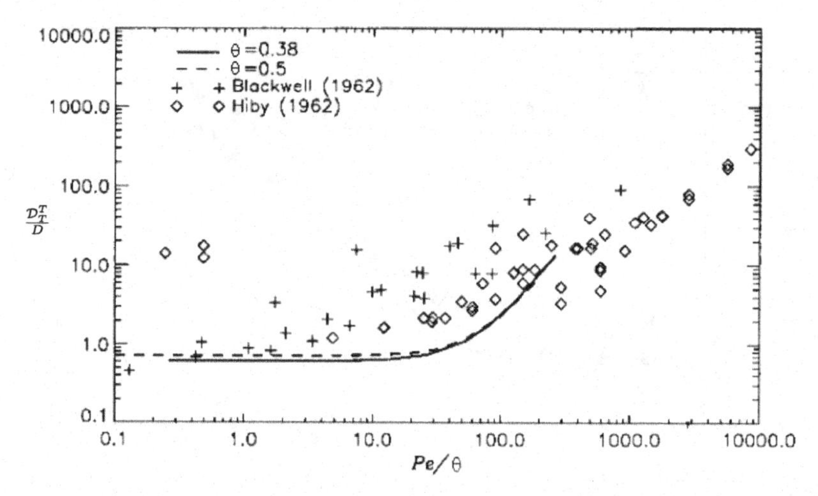

Fig. 4.3. Comparision of computed transverse dispersivity $\mathcal{D}_T^T/D = \mathcal{D}_{33}/D$ for a cubic array of Wigner–Seitz grains with experiments for natural grains. From Lee, Sun and Mei (1995), *Int. J. Heat Mass Transfer*.

begins from a continuum model of porous media where the permeability and the diffusivity defined for scales much larger than pores and grains are described by certain empirical and statistical rules. Statistical analysis is then applied to derive the averaged behavior on the larger scale. Such treatments of macroscale dispersion of passive solute or heat can be found in Dagan (1989) and Gelhar (1993).

Despite the prevalence of randomness in natural materials, it is instructive to extend the theory of homogenization to some manufactured porous media which are characterized by three distinct scales and periodic on the micro- and mesoscales. This section[2] is devoted to the derivation of the macroscale transport equation for a passive solute in seepage flow through a fluid-saturated porous medium. In principle one can carry out a three-scale analysis. However, as in the case of seepage flow with three scales, the analysis can be broken up into two steps: from micro- to mesoscale and then from meso- to macroscale. We shall take the latter approach, and show the differences from and the similarities to the two-scale problem. The treatment can be extended to the transport of heat or mass through a multiscaled rigid or deformable material.

[2]Materials in this section are adapted from the doctoral thesis by Lee (1994) who treated heat transport in a deformable porous medium.

4.2.1. *From Micro- to Mesoscale*

Starting from the microscale $\mathcal{O}(\ell)$, the solute concentration is governed by the convective diffusion equation:

$$\frac{\partial C}{\partial t} + u_j \frac{\partial C}{\partial x_j} = \frac{\partial}{\partial x_j}\left(D\frac{\partial C}{\partial x_j}\right), \quad \boldsymbol{x} \in \Omega_f . \tag{4.2.1}$$

We are interested in the effect of convection on the mesoscale $\mathcal{O}(\ell')$ described by $\boldsymbol{x}' = \epsilon \boldsymbol{x} = \mathcal{O}(1)$ where $\epsilon = \ell/\ell' \ll 1$, and will assume that

$$Pe' = \frac{U\ell'}{D} = \mathcal{O}(1) . \tag{4.2.2}$$

Note that

$$Pe = \frac{U\ell}{D} = \mathcal{O}(\epsilon) , \tag{4.2.3}$$

which is comparable to that in Sec. 2.12 but much smaller than the case treated in the last section. The effective equation on the mesoscale can be derived as a special case of Sec. 2.12. In particular, by taking $C_w = 0$ and $\theta_g = \theta =$ porosity, we get from Eq. (2.12.20) the mesoscale equation for $C^{(0)}$:

$$\theta \frac{\partial C^{(0)}}{\partial t} + \langle u_i^{(0)} \rangle \frac{\partial C^{(0)}}{\partial x_i'} = \frac{\partial}{\partial x_i'}\left(\mathcal{D}_{i\ell}\frac{\partial C^{(0)}}{\partial x_\ell'}\right) . \tag{4.2.4}$$

The effective diffusivity is

$$\mathcal{D}_{i\ell} \equiv -D\left\langle \frac{\partial N_\ell}{\partial x_i}\right\rangle + \theta D \delta_{i\ell} , \tag{4.2.5}$$

where N_ℓ is found after solving the Ω-cell problem defined in Sec. 2.12. $\mathcal{D}_{ij}$ is a symmetric tensor. We stress that because the microscale Péclet number is small, the microscale cell problem is not the one defined in Eqs. (4.1.27)–(4.1.30). Dispersion is unimportant on the mesoscale.

Let us first estimate the relative magnitudes of various terms in the mesoscale equations.

Following the same reasoning leading to Eq. (4.1.8), we normalize the distances by ℓ' and time by $T_c' = \ell'/U$. After normalization the term $\partial C^{(0)}/\partial t$ is found to be of the order $\epsilon = \ell'/\ell''$ smaller than other terms in the convection–diffusion equation.

The seepage flow is assumed to be steady, and is governed by

$$\langle u_i^{(0)} \rangle = -\mathcal{K}_{ij} \frac{\partial p^{(0)}}{\partial x'_j}, \tag{4.2.6}$$

$$\frac{\partial \langle u_i^{(0)} \rangle}{\partial x'_i} = 0, \tag{4.2.7}$$

on the mesoscale. The conductivity $\mathcal{K}_{ij}(\boldsymbol{x}')$ and the effective diffusivity $\mathcal{D}_{ij}(\boldsymbol{x}')$ can vary periodically on the mesoscale. Since the flow is driven by the macroscale pressure gradient $\Delta P/\ell''$, from Darcy's law (Eq. (4.2.6)) the velocity scale is implied

$$U = K_0 \frac{\Delta P}{\ell''}. \tag{4.2.8}$$

It can be seen from Eq. (4.2.6) that

$$\frac{\langle u_i^{(0)} \rangle}{-\mathcal{K}_{ij}(\partial p^{(0)}/\partial x'_j)} = \mathcal{O}(\epsilon).$$

Based on these estimates we rewrite Eqs. (4.2.7), (4.2.6), and (4.2.4) by inserting order symbol,

$$\frac{\partial \bar{u}_j}{\partial x'_j} = 0, \tag{4.2.9}$$

$$\epsilon \bar{u}_i = -\mathcal{K}_{ij} \frac{\partial \bar{p}}{\partial x'_j}, \tag{4.2.10}$$

$$\epsilon n \frac{\partial \bar{C}}{\partial t} + \bar{u}_i \frac{\partial \bar{C}}{\partial x'_i} = \frac{\partial}{\partial x'_i}\left(\mathcal{D}_{i\ell} \frac{\partial \bar{C}}{\partial x'_\ell}\right), \tag{4.2.11}$$

where the following changes of notations have been made for simplicity,

$$\langle u_i^{(0)} \rangle \to \bar{u}_i, \quad p^{(0)} \to \bar{p}, \quad C^{(0)} \to \bar{C}. \tag{4.2.12}$$

We shall be interested in the transport over the macroscale. The mesoscale equations above will be used as the starting basis for homogenization analysis.

4.2.2. *Mass Transport Equation on the Macroscale*

As usual we substitute $\bar{C} = \bar{C}^{(0)}(\boldsymbol{x}', \boldsymbol{x}''; t, t') + \epsilon \bar{C}^{(1)}(\boldsymbol{x}', \boldsymbol{x}''; t, t') + \cdots$ in Eq. (4.2.11), where $t' = \epsilon t$. For consistency we also expand formally the seepage velocity $\bar{u}_i$ in powers of ϵ but will work out the specifics later. The transport process is quasi-steady at the leading order, i.e.,

$$\bar{u}_j^{(0)} \frac{\partial \bar{C}^{(0)}}{\partial x'_j} = \frac{\partial}{\partial x'_i} \left(\mathcal{D}_{ij} \frac{\partial \bar{C}^{(0)}}{\partial x'_j} \right) . \tag{4.2.13}$$

In addition $\bar{C}^{(0)}$ must be Ω'-periodic as $\bar{u}_j^{(0)}$. Clearly $\bar{C} = \bar{C}^{(0)}(\boldsymbol{x}'', t, t')$ is independent of $\boldsymbol{x}'$. Using this result the perturbation equation at order $\mathcal{O}(\epsilon)$ reduces to,

$$\theta \frac{\partial \bar{C}^{(0)}}{\partial t} + \bar{u}_j^{(0)} \left(\frac{\partial \bar{C}^{(1)}}{\partial x'_j} + \frac{\partial \bar{C}^{(0)}}{\partial x''_j} \right) = \frac{\partial}{\partial x'_i} \left(\mathcal{D}_{ij} \frac{\partial \bar{C}^{(1)}}{\partial x'_j} \right) + \frac{\partial}{\partial x'_i} \left(\mathcal{D}_{ij} \frac{\partial \bar{C}^{(0)}}{\partial x''_j} \right) , \tag{4.2.14}$$

whose Ω'-average is

$$\langle \theta \rangle' \frac{\partial \bar{C}^{(0)}}{\partial t} + \langle \bar{u}_j^{(0)} \rangle' \frac{\partial \bar{C}^{(0)}}{\partial x'_j} = 0 , \tag{4.2.15}$$

which can also be written as

$$\theta \frac{\partial \bar{C}^{(0)}}{\partial t} + \frac{\theta}{\langle \theta \rangle'} \langle \bar{u}_j^{(0)} \rangle' \frac{\partial \bar{C}^{(0)}}{\partial x'_j} = 0 , \tag{4.2.16}$$

where $\langle \cdot \rangle'$ denotes the average over an Ω'-cell. The difference of Eqs. (4.2.14) and (4.2.16) is a steady diffusion equation for $C^{(1)}$,

$$\bar{u}_j^{(0)} \frac{\partial \bar{C}^{(1)}}{\partial x'_j} + \tilde{\bar{u}}_j^{(0)} \frac{\partial \bar{C}^{(0)}}{\partial x''_j} = \frac{\partial}{\partial x_i} \left(\mathcal{D}_{ij} \frac{\partial \bar{C}^{(1)}}{\partial x'_j} \right) + \frac{\partial}{\partial x'_i} \left(\mathcal{D}_{ij} \frac{\partial \bar{C}^{(0)}}{\partial x''_j} \right) , \tag{4.2.17}$$

where

$$\tilde{\bar{u}}_j^{(0)} = u_j^{(0)} - \frac{\theta}{\langle \theta \rangle'} \langle u_j^{(0)} \rangle' \tag{4.2.18}$$

is the deviation from the mean. Let the solution for $C^{(1)}$ be represented formally by

$$C^{(1)} = B_k \frac{\partial C^{(0)}}{\partial x''_k} , \tag{4.2.19}$$

then B_k is governed by the inhomogeneous equation

$$\boxed{\bar{u}_j^{(0)} \frac{\partial B_k}{\partial x'_j} = \frac{\partial}{\partial x'_i} \left(\mathcal{D}_{ij} \frac{\partial B_k}{\partial x'_j} \right) - \tilde{\bar{u}}_k^{(0)} + \frac{\partial \mathcal{D}_{kj}}{\partial x'_j}} \tag{4.2.20}$$

in an Ω'-cell, subject further to Ω'-periodicity and

$$\langle B_k \rangle' = 0\,. \tag{4.2.21}$$

This cell problem can be solved numerically in general, and analytically for a layered stratum.

At the next order $\mathcal{O}(\epsilon^2)$, the transport law (Eq. (4.2.11)), gives

$$\theta \frac{\partial \bar{C}^{(0)}}{\partial t'} + \theta \frac{\partial \bar{C}^{(1)}}{\partial t} + \bar{u}_j^{(1)} \frac{\partial \bar{C}^{(1)}}{\partial x_j'} + \bar{u}_j^{(1)} \frac{\partial \bar{C}^{(0)}}{\partial x_j''} + \bar{u}_j^{(0)} \frac{\partial \bar{C}^{(1)}}{\partial x_j'} + \bar{u}_j^{(0)} \frac{\partial \bar{C}^{(2)}}{\partial x_j'}$$

$$= \frac{\partial}{\partial x_i'}\left(\mathcal{D}_{ij} \frac{\partial \bar{C}^{(2)}}{\partial x_j'}\right) + \frac{\partial}{\partial x_i'}\left(\mathcal{D}_{ij} \frac{\partial \bar{C}^{(1)}}{\partial x_j''}\right) + \frac{\partial}{\partial x_i''}\left(\mathcal{D}_{ij} \frac{\partial \bar{C}^{(1)}}{\partial x_j'}\right)$$

$$+ \frac{\partial}{\partial x_i''}\left(\mathcal{D}_{ij} \frac{\partial \bar{C}^{(0)}}{\partial x_j''}\right).$$

With the help of partial integration and Gauss' law, the mesoscale cell average of the preceding equation reduces to

$$\langle\theta\rangle' \frac{\partial \bar{C}^{(0)}}{\partial t'} + \left\langle \theta \frac{\partial \bar{C}^{(1)}}{\partial t} \right\rangle' + \langle \bar{u}_j^{(1)} \rangle' \frac{\partial \bar{C}^{(0)}}{\partial x_j''} + \frac{\partial}{\partial x_j''} \langle \bar{u}_j^{(0)} \bar{C}^{(1)} \rangle'$$

$$= \frac{\partial}{\partial x_i''}\left\langle \mathcal{D}_{ij} \frac{\partial \bar{C}^{(1)}}{\partial x_j'} \right\rangle' + \frac{\partial}{\partial x_i''}\left(\langle \mathcal{D}_{ij} \rangle' \frac{\partial \bar{C}^{(0)}}{\partial x_j''} \right). \tag{4.2.22}$$

Periodicity and continuity are also used. Replacing $\bar{C}^{(1)}$ by Eq. (4.2.19) and using Eq. (4.2.15), the second term above can be written as

$$\left\langle \theta \frac{\partial \bar{C}^{(1)}}{\partial t} \right\rangle' = \left\langle \theta B_i \frac{\partial}{\partial t} \frac{\partial \bar{C}^{(0)}}{\partial x_i''} \right\rangle' = \left\langle \theta B_i \frac{\partial}{\partial x_i''} \frac{\partial \bar{C}^{(0)}}{\partial t} \right\rangle'$$

$$= \frac{\partial}{\partial x_i''}\left\langle \theta B_i \frac{\partial \bar{C}^{(0)}}{\partial t} \right\rangle' - \frac{\partial \bar{C}^{(0)}}{\partial t} \frac{\partial \langle \theta B_i \rangle'}{\partial x_i''}$$

$$= -\frac{\partial}{\partial x_i''}\left(\frac{\langle \theta B_i \rangle'}{\langle\theta\rangle'} \langle \bar{u}_k^{(0)} \rangle' \frac{\partial \bar{C}^{(0)}}{\partial x_k''} \right) + \frac{\partial \langle \theta B_i \rangle'}{\partial x_i''} \frac{\langle \bar{u}_k^{(0)} \rangle'}{\langle\theta\rangle'} \frac{\partial \bar{C}^{(0)}}{\partial x_k''},$$

we finally get after using Eqs. (4.2.15), (4.2.21), and some algebra

$$\langle\theta\rangle' \frac{\partial \bar{C}^{(0)}}{\partial t'} + \left(\langle \bar{u}_i^{(1)} \rangle' + \frac{\partial \langle \theta B_k \rangle}{\partial x_k'} \frac{\langle \bar{u}_i^{(0)} \rangle'}{\langle\theta\rangle'} \right) \frac{\partial \bar{C}^{(0)}}{\partial x_i''}$$

$$= \frac{\partial}{\partial x_i''}\left\{ \left(\langle \mathcal{D}_{ij} \rangle' \delta_{jk} - \langle \mathcal{U}_i^{(0)} B_k \rangle' + \left\langle D_{ij} \frac{\partial B_k}{\partial x_j'} \right\rangle' \right) \frac{\partial \bar{C}^{(0)}}{\partial x_k''} \right\}, \tag{4.2.23}$$

where

$$\widehat{\bar{u}}_i^{(0)} = \bar{u}_i^{(0)} - \langle \bar{u}_i^{(0)} \rangle', \tag{4.2.24}$$

which is different from $\widetilde{\bar{u}}_j^{(0)}$ defined in Eq. (4.2.18). Use has been made of $\langle B_k \rangle' = 0$, so that $\langle \bar{u}_i^{(0)} B_k \rangle' = \langle \widehat{\bar{u}}_i^{(0)} B_k \rangle'$. Equation (4.2.23) can be simplified. Adding Eq. (4.2.15) and $\epsilon \times$(Eq. (4.2.23)) and removing the artifice of multiple time scales we have

$$\begin{aligned} \langle \theta \rangle' \frac{\partial \bar{C}^{(0)}}{\partial t} &+ \left[\langle \bar{u}_i^{(0)} \rangle' + \epsilon \left(\langle \bar{u}_i^{(1)} \rangle' + \frac{\partial \langle \theta B_k \rangle}{\partial x_k'} \frac{\langle \bar{u}_i^{(0)} \rangle'}{\langle \theta \rangle'} \right) \right] \frac{\partial \bar{C}^{(0)}}{\partial x_i''} \\ &= \epsilon \frac{\partial}{\partial x_i''} \left\{ \left(\langle \mathcal{D}_{ij} \rangle' \delta_{jk} - \langle \widetilde{\bar{u}}_i^{(0)} B_k \rangle' + \frac{\langle \theta B_k \rangle'}{\langle \theta \rangle'} \langle \bar{u}_i^{(0)} \rangle' \right. \right. \\ &\quad \left. \left. + \left\langle \mathcal{D}_{ij} \frac{\partial B_k}{\partial x_j'} \right\rangle' \right) \frac{\partial \bar{C}^{(0)}}{\partial x_k''} \right\}, \end{aligned} \tag{4.2.25}$$

which governs the convective diffusion of the solvent on the macroscales of time and space.

In Eq. (4.2.25) the tensor

$$\mathcal{F}_{ik}'' = \langle \mathcal{D}_{ij} \rangle' \delta_{jk} - \langle \widehat{\bar{u}}_i^{(0)} B_k \rangle' + \frac{\langle \theta B_k \rangle'}{\langle \theta \rangle'} \langle \bar{u}_i^{(0)} \rangle' + \left\langle \mathcal{D}_{ij} \frac{\partial B_k}{\partial x_j'} \right\rangle' \tag{4.2.26}$$

is not symmetric, i.e., $\mathcal{F}_{ik}'' \neq \mathcal{F}_{ki}''$. However, it can be rewritten as

$$\mathcal{F}_{ik}' = \langle \mathcal{D}_{ij} \rangle' \delta_{jk} + \mathcal{D}_{ik}' + \mathcal{E}_{ik}', \tag{4.2.27}$$

where

$$\begin{aligned} \begin{pmatrix} \mathcal{D}_{ik} \\ \mathcal{E}_{ik}' \end{pmatrix} &= -\frac{1}{2} \left(\langle \widehat{\bar{u}}_i^{(0)} B_k \rangle' \pm \langle \widehat{\bar{u}}_k^{(0)} B_i \rangle' \right) \\ &\quad + \frac{1}{2\langle \theta \rangle'} \left(\langle \theta B_k \rangle' \langle \bar{u}_i^{(0)} \rangle' \pm \langle \theta B_i \rangle' \langle \bar{u}_k^{(0)} \rangle' \right) \\ &\quad + \frac{1}{2} \left\langle \mathcal{D}_{i\ell} \frac{\partial B_k}{\partial x_\ell'} \pm \mathcal{D}_{k\ell} \frac{\partial B_i}{\partial x_\ell'} \right\rangle'. \end{aligned} \tag{4.2.28}$$

$\mathcal{D}_{ik}'$ will be called the tensor of macrodispersivity, which can be expressed in the alternate form:

$$\boxed{\mathcal{D}_{ik}' = \left\langle \mathcal{D}_{j\ell} \frac{\partial B_i}{\partial x_j'} \frac{\partial B_k}{\partial x_\ell'} \right\rangle' + \left\langle \mathcal{D}_{i\ell} \frac{\partial B_k}{\partial x_\ell'} + \mathcal{D}_{k\ell} \frac{\partial B_i}{\partial x_\ell'} \right\rangle'} \tag{4.2.29}$$

as described in Appendix 4B. This result is formally similar to Eq. (4.1.45), but the associated cell problem is different.

Equation (4.2.25) takes the final form:

$$\boxed{\langle n\rangle'\frac{\partial \bar{C}^{(0)}}{\partial t}+\mathcal{U}_i'\frac{\partial \bar{C}^{(0)}}{\partial x_i''}=\epsilon\frac{\partial}{\partial x_i''}\left\{\left(\langle\mathcal{D}_{ij}\rangle'\delta_{jk}+\mathcal{D}_{ik}'\right)\frac{\partial \bar{C}^{(0)}}{\partial x_k''}\right\}}, \tag{4.2.30}$$

where the effective velocity of convection is

$$\mathcal{U}_i'=\langle\bar{u}_i^{(0)}\rangle+\epsilon\left(\langle\bar{u}_i^{(1)}\rangle'+\frac{\partial\langle\theta B_k\rangle'}{\partial x_k'}\frac{\langle\bar{u}_i^{(0)}\rangle'}{\langle\theta\rangle'}+\frac{\partial\mathcal{E}_{ik}''}{\partial x_k'}\right). \tag{4.2.31}$$

We must now find the second-order correction $\langle\bar{u}_i^{(1)}\rangle'$.

4.2.3. *Second-Order Seepage Velocity*

Let us assume the multiple-scale expansions

$$\bar{u}=\bar{u}^{(0)}+\epsilon\bar{u}^{(1)}+\cdots;\quad \bar{p}=\bar{p}^{(0)}+\epsilon\bar{p}^{(1)}+\cdots; \tag{4.2.32}$$

where $(\bar{u}^{(m)},\bar{p}^{(m)})$ are functions of $(\boldsymbol{x}',\boldsymbol{x}'')$. It follows from mass conservation (Eq. (4.2.9)) that,

$$\frac{\partial\bar{u}_i^{(0)}}{\partial x_i'}=0, \tag{4.2.33}$$

$$\frac{\partial\bar{u}_i^{(1)}}{\partial x_i'}+\frac{\partial\bar{u}_i^{(0)}}{\partial x_i''}=0. \tag{4.2.34}$$

From momentum conservation (Eq. (4.2.10)) we get

$$-K_{ij}\frac{\partial\bar{p}^{(0)}}{\partial x_j'}=0, \tag{4.2.35}$$

$$\bar{u}_i^{(0)}=-K_{ij}\frac{\partial\bar{p}^{(1)}}{\partial x_j'}-K_{ij}\frac{\partial\bar{p}^{(0)}}{\partial x_j''}, \tag{4.2.36}$$

$$\bar{u}_i^{(1)}=-K_{ij}\frac{\partial\bar{p}^{(2)}}{\partial x_j'}-K_{ij}\frac{\partial\bar{p}^{(1)}}{\partial x_j''}. \tag{4.2.37}$$

It follows from Eq. (4.2.35) that $\bar{p}^{(0)}=\bar{p}^{(0)}(\boldsymbol{x}'')$. Let

$$\bar{p}^{(1)}=-\bar{A}_k\frac{\partial\bar{p}^{(0)}}{\partial x_k''}, \tag{4.2.38}$$

then from Eqs. (4.2.36) and (4.2.33) we find $\bar{A}_k(\boldsymbol{x}')$ to be governed by

$$\frac{\partial}{\partial x_i'}\left(\mathcal{K}_{ij}\frac{\partial \bar{A}_k}{\partial x_j'}\right) = \frac{\partial \mathcal{K}_{ij}\delta_{jk}}{\partial x_i'}, \tag{4.2.39}$$

and be Ω'-periodic in a mesoscale cell Ω'. After $\bar{A}_k$ is solved we get

$$\bar{u}_i^{(0)} = -\mathcal{K}_{ij}\left(\delta_{jk} - \frac{\partial \bar{A}_k}{\partial x_j}\right)\frac{\partial \bar{p}^{(0)}}{\partial x_k''}. \tag{4.2.40}$$

Upon averaging over the mesocell, the leading-order macroscale velocity components are,

$$\langle \bar{u}_i^{(0)}\rangle' = -\mathcal{K}_{ik}'\frac{\partial \bar{p}^{(0)}}{\partial x_k''} \equiv -\left[\langle \mathcal{K}_{ij}\rangle'\delta_{jk} - \left\langle \mathcal{K}_{ij}\frac{\partial \bar{A}_k}{\partial x_j'}\right\rangle'\right]\frac{\partial \bar{p}^{(0)}}{\partial x_k''}. \tag{4.2.41}$$

Next the second-order velocity $\bar{u}_i^{(1)}$. Using Eqs. (4.2.38) and (4.2.40), Eqs. (4.2.34) and (4.2.37) become

$$\frac{\partial \bar{u}_i^{(1)}}{\partial x_i'} = \frac{\partial}{\partial x_i''}\left[\mathcal{K}_{ij}\left(\delta_{jk} - \frac{\partial \bar{A}_k}{\partial x_j'}\right)\frac{\partial \bar{p}^{(0)}}{\partial x_k''}\right], \tag{4.2.42}$$

and

$$\bar{u}_i^{(1)} = -\mathcal{K}_{ij}\frac{\partial \bar{p}^{(2)}}{\partial x_j'} + \mathcal{K}_{ij}\frac{\partial}{\partial x_j''}\left(\bar{A}_k\frac{\partial \bar{p}^{(0)}}{\partial x_k''}\right). \tag{4.2.43}$$

The two can be combined to give

$$\begin{aligned}\frac{\partial}{\partial x_i'}\left(\mathcal{K}_{ij}\frac{\partial \bar{p}^{(2)}}{\partial x_j'}\right) &= \left\{\frac{\partial}{\partial x_i'}\left(\mathcal{K}_{ij}\frac{\partial \bar{A}_k}{\partial x_j''}\right) + \frac{\partial}{\partial x_i''}\left(-\mathcal{K}_{ij}\delta_{jk} + \frac{\partial \bar{A}_k}{\partial x_j'}\right)\right\}\frac{\partial \bar{p}^{(0)}}{\partial x_k''} \\ &\quad + \left\{\frac{\partial(\mathcal{K}_{ik}\bar{A}_k)}{\partial x_i'} + \mathcal{K}_{ij}\left(-\delta_{jk} + \frac{\partial \bar{A}_k}{\partial x_j'}\right)\right\}\frac{\partial^2 \bar{p}^{(0)}}{\partial x_i''\partial x_k''}.\end{aligned} \tag{4.2.44}$$

This inhomogeneous equation for $p^{(2)}$ can be solved in a mesocell by assuming

$$\bar{p}^{(2)} = -E_k\frac{\partial \bar{p}^{(0)}}{\partial x_k''} - F_{ik}\frac{\partial^2 \bar{p}^{(0)}}{\partial x_i''\partial x_k''}. \tag{4.2.45}$$

Then in an Ω'-cell E_k and F_{ik} must satisfy

$$\frac{\partial}{\partial x_i'}\left(\mathcal{K}_{ij}\frac{\partial E_k}{\partial x_j'}\right) = \frac{\partial}{\partial x_i'}\left(\mathcal{K}_{ij}\frac{\partial \bar{A}_k}{\partial x_j''}\right) + \frac{\partial}{\partial x_i''}\left[\mathcal{K}_{ij}\left(-\delta_{jk} + \frac{\partial \bar{A}_k}{\partial x_j'}\right)\right], \tag{4.2.46}$$

and

$$\frac{\partial}{\partial x'_i}\left(\mathcal{K}_{ij}\frac{\partial F_{ik}}{\partial x'_j}\right) = \frac{\partial(K_{ik}\bar{A}_k)}{\partial x'_i} + \mathcal{K}_{ij}\left(-\delta_{jk} + \frac{\partial \bar{A}_k}{\partial x'_j}\right) \tag{4.2.47}$$

subject to the condition of Ω'-periodicity and uniqueness. Afterwards, the average velocity correction can be found by Ω'-averaging Eq. (4.2.37):

$$\langle \bar{u}_i^{(1)}\rangle' = \left\langle \mathcal{K}_{ij}\left(\frac{\partial \bar{A}_k}{\partial x'_j} + \frac{\partial E_k}{\partial x'_j}\right)\right\rangle' \frac{\partial \bar{p}^{(0)}}{\partial x''_k} + \left\langle \mathcal{K}_{ij}\bar{A}_k + \mathcal{K}_{i\ell}\frac{\partial F_{jk}}{\partial x'_\ell}\right\rangle' \frac{\partial^2 p^{(0)}}{\partial x''_j \partial x''_k}, \tag{4.2.48}$$

which is needed in Eq. (4.2.31).

We have thus completed the derivation of the macroscale equations. In principle, one can use the theory in Sec. 3.7 to solve for the macroscale velocity field first, then the meso scale velocity. The latter information is then used to solve the cell problem for the vector B_k governed by Eqs. (4.2.20), (4.2.21), and Ω'-periodicity. In the special case of horizontally layered media, the mathematics involves only linear ordinary differential equations which can be handled analytically or numerically.

The theory here has been discussed in a broader context by Lee (1994) for the diffusion of heat in a seepage flow through a porous medium. Elastic deformation of solid matrix is also allowed. Explicit results for the cell problems have been solved and applied to thermal consolidation and dispersion stimulated by possible applications to geothermal reservoir and the burial of nuclear wastes.

4.3. Dispersion and Transport in a Wave Boundary Layer Above the Seabed

Erosion and transport of sediment particles near the seabed by wind-induced waves or tides are important factors responsible for shoreline formation and evolution. As human population density is often high in coastal regions, sewage or dredged silt discharged in the neighboring sea can have serious impact on the environment. In addition, planktonic larvae and nutrients can adhere to and be carried by the moving particles; their spreading affects the growth and death of marine life (Denny, 1988).

In engineering literature, it is often thought that wind generates both waves and a drift current in water. It is the waves which dislodge and suspend the particles from the seabed, but it is the drift current that spreads them around. However, it is known in the theory of water-wave dynamics that waves also generate a current, called the *mass transport*, in the bottom boundary layer where viscosity (or eddy viscosity) is important, due to the effects of Reynolds stresses (see, e.g., Longuet-Higgins, 1953; Hunt and Johns, 1963; Mei, 1989, or Mei *et al.*, 2005). Thus wind drift is not the only mechanism for redistributing suspended particles. In addition, the strong shear in the boundary layer should augment the effective diffusivity by Taylor dispersion. It is therefore of interest for coastal oceanography to study how mass transport and dispersion together contribute to the spreading of suspended particles.

In this section we shall apply the homogenization analyses to the spreading of sediments suspended in the wave boundary layer just above the seabed. Our attention is restricted to very fine particles of negligible inertia, so that the horizontal particle velocity is essentially equal to that of the surrounding fluid. In the vertical direction, gravity and the density difference between particles and water can lead to free fall even in calm water, hence the fall velocity will not be ignored. We also assume low particle concentration as to cause negligible change in the flow field. Waves are assumed to be periodic in time and of small amplitude. Detailed analysis has been discussed more extensively for nonerodible bed by Mei and Chian (1994), extended to erodible bed by Mei *et al.* (1997) and to tidal boundary layers by Mei *et al.* (1998).

4.3.1. *Depth-Integrated Transport Equation in the Boundary Layer*

In the theory of low Reynolds number flows of viscous fluids, it is known that the free-fall velocity of a spherical particle of small diameter d is

$$w_o = \frac{gd^2(\rho_s - \rho)}{18\nu\rho}, \tag{4.3.1}$$

where ν is the molecular viscosity of the fluid, ρ_s the particle density and ρ the fluid density. Hence particles can be characterized by either its diameter or its fall velocity. Let C be the volume concentration of small particles of a specific size. The concentration C of a very dilute cloud of small but heavy particles in a moving fluid is assumed to be governed by

$$\frac{\partial C}{\partial t} + \frac{\partial u_i C}{\partial x_i} + \frac{\partial}{\partial z}[(-w_o + w)C] = D\left(\frac{\partial^2 C}{\partial x_i \partial x_i} + \frac{\partial^2 C}{\partial z^2}\right), \tag{4.3.2}$$

where $i = 1, 2$ with $(x_1, x_2) \equiv (x, y)$ representing the horizontal coordinates and $(u_1, u_2) \equiv (u, v)$ the horizontal components of the fluid velocity. D is the eddy diffusivity of mass. The crudest model of constant eddy viscosity is adopted here.

At the sea bottom the boundary condition for C is the least certain when the bed surface is erodible. In principle, the net upward flux is the difference between the rates of erosion and deposition, both of which are related to the difference between the local fluid shear stress and a critical (threshold) stress. Empirical information on these is unfortunately limited, especially for wavy flows, as discussed in Mei *et al.* (1997). From data on steady flows, the threshold stresses are often very low for many cohesive materials and the erosion rate is far greater than the deposition rate. We shall take the simplest model that accounts only for erosion,

$$-D\frac{\partial C}{\partial z} - w_o C = \mathsf{E}|\tau_b|\,, \quad z = 0\,, \tag{4.3.3}$$

where $|\tau_b|$ is the magnitude of the bottom shear and E an empirical factor. We also assume that C vanishes outside the boundary layer which is much thinner than the water depth due to the relatively high frequency of gravity waves.[3] There are three characteristic length scales in the vertical direction, defined by

$$\delta_s = \frac{D}{w_o}\,, \quad \delta = \sqrt{\frac{2\nu_e}{\omega}} \quad \text{and} \quad \delta_c = \sqrt{\frac{2D}{\omega}}\,. \tag{4.3.4}$$

In order of magnitude, the first is the thickness of a steady concentration layer due to the balance of downward sedimentation by gravity and vertical diffusion, where D denotes the (eddy) diffusivity due to turbulence. The second and third are, respectively, the scales of the momentum and mass boundary-layer thicknesses in an oscillatory flow. For generality, all three vertical scales are assumed to be comparable, i.e.,

$$\mathcal{O}(\delta_s) = \mathcal{O}(\delta) = \mathcal{O}(\delta_c)\,, \tag{4.3.5}$$

which implies that the Schmidt number is of order unity,

$$Sc = \frac{\nu_e}{D} \sim \left(\frac{\delta}{\delta_c}\right)^2 = \mathcal{O}(1)\,. \tag{4.3.6}$$

In the horizontal direction the characteristic length scale is the surface wavelength $(2\pi/k)$, which can be comparable to the water depth h, i.e.,

[3]This is not the case in long-period tides.

$kh = \mathcal{O}(1)$. It will be assumed that the boundary-layer thickness is far smaller than the horizontal length scale or the sea depth, so that

$$\beta = k\delta = \mathcal{O}\left(\frac{\delta}{h}\right) \ll 1\,. \tag{4.3.7}$$

Let us also assume that the wave slope is very gentle,

$$\epsilon = kA \ll 1\,, \tag{4.3.8}$$

where A denotes the typical wave amplitude. In most natural flows of interest β is much smaller than ϵ, and we let $\beta = \mathcal{O}(\epsilon)$ for generality. A consequence of this assumption is that the horizontal diffusion is retained in the final diffusion equation, as will be shown.

Introducing the following normalization

$$\begin{aligned} x_i^\dagger &= kx_i\,, \quad z^\dagger = \frac{z}{\delta_c}\,, \quad t^\dagger = \omega t\,, \\ u_i^\dagger &= \frac{u_i}{\omega A}\,, \quad w^\dagger = \frac{w}{k\delta_c\omega A}\,, \quad C^\dagger = \frac{C}{C_0}\,, \end{aligned} \tag{4.3.9}$$

the diffusion equation (4.3.2) is rescaled to become

$$\frac{\partial C^\dagger}{\partial t^\dagger} + \epsilon\frac{\partial u_i^\dagger C^\dagger}{\partial x_i^\dagger} + \frac{\partial}{\partial z^\dagger}[(-Pe + \epsilon w^\dagger)C^\dagger] = \beta^2\frac{\partial^2 C^\dagger}{\partial x_i^\dagger \partial x_i^\dagger} + \frac{\partial^2 C^\dagger}{\partial z^{\dagger 2}}\,, \tag{4.3.10}$$

where

$$Pe = \frac{w_o\delta}{D} \tag{4.3.11}$$

is the Péclet number based on the fall velocity of a particle. Physically the two sharply contrasting length scales ($1/k$ and δ) imply two distinct time scales, one for vertical diffusion across the boundary layer, $\mathcal{O}(\omega^{-1}) = \mathcal{O}(\delta^2/D)$, which is the same as a wave period, and the other one for horizontal diffusion across a wavelength, $\mathcal{O}(1/k^2D)$, which is much longer by a factor of $\mathcal{O}(1/k^2\delta^2) = \mathcal{O}(1/\beta^2)$.

The scale C_0 of concentration should be directly related to the rate of erosion. Anticipating the C is dominated by its period average, we balance the net horizontal flux rate between two vertical sections at dx apart and the period-averaged rate of erosion from the bed surface between the two sections, and get

$$\delta_c\overline{u}\frac{\partial C}{\partial x}dx = \mathsf{E}\overline{|\tau_o|}dx\,, \tag{4.3.12}$$

where $\overline{F}$ denotes the period average of F. For small amplitude waves in shallow water the leading-order fluid velocity is $\mathcal{O}(U_b)$ where $U_b = A\sqrt{gh}/h$

and the mean streaming velocity (mass transport) is kA times smaller, i.e., $\overline{u} \sim kA(A/h)\sqrt{gh}$. Using the following estimate as the scale for the bed shear stress,

$$\tau_o \sim D\frac{\sqrt{2}\rho U_b}{\delta_c} = \frac{\sqrt{2}\rho DA\sqrt{g}}{h\delta_c}, \tag{4.3.13}$$

we get the scale of C:

$$\mathcal{O}(C) = \frac{2\sqrt{2}\rho \mathsf{E} D}{k^2\delta_c^2 A}. \tag{4.3.14}$$

With this and the small empirical value of $\mathsf{E} \sim 5.5 \times 10^{-9}$ s/m according to Sheng and Cook (1990), Mei *et al.* (1997) estimated for typical data from Lake Okeechobee, Florida that $C \sim 10^{-5}\,\text{kg/m}^3$. With these estimated scales the normalized bottom flux condition becomes

$$-Pe\,C^\dagger - \frac{\partial C^\dagger}{\partial z^\dagger} = \frac{k\delta_c^2 U_b^2}{\omega D}\frac{|\tau_b|}{\tau_o}, \quad z^\dagger = 0. \tag{4.3.15}$$

The last coefficient is of the small order

$$\frac{k\delta_c^2 U_b^2}{\omega D} = \mathcal{O}(k^2A^2) = \mathcal{O}(\epsilon^2). \tag{4.3.16}$$

Having identified the orders we return to the dimensional form (Eq. (4.3.2)) by preserving the ordering parameter ϵ as follows:

$$\frac{\partial C}{\partial t} + \epsilon\frac{\partial u_i C}{\partial x_i} + \frac{\partial}{\partial z}[(-w_o + \epsilon w)C] = D\left(\epsilon^2\frac{\partial^2 C}{\partial x_i \partial x_i} + \frac{\partial^2 C}{\partial z^2}\right). \tag{4.3.17}$$

$$-w_o C - D\frac{\partial C}{\partial z} = \epsilon^2 \mathsf{E}|\tau_b|, \quad z = 0. \tag{4.3.18}$$

Since diffusion is confined inside the thin bottom boundary layer, we require the boundary condition

$$C \to 0, \quad z \gg \delta_c. \tag{4.3.19}$$

Let us take $\beta = \mathcal{O}(\epsilon)$ and introduce multiple-scale coordinates for time: t and $t' = \epsilon^2 t$. The velocity and concentration are expanded as follows:

$$u_i = u_i^{(1)} + \epsilon u_i^{(2)} + \mathcal{O}(\epsilon^2), \quad w = w^{(1)} + \epsilon w^{(2)} + \mathcal{O}(\epsilon^3), \tag{4.3.20}$$

$$C = C^{(0)} + \epsilon C^{(1)} + \epsilon^2 C^{(2)} + \mathcal{O}(\epsilon^3), \tag{4.3.21}$$

where $u_i^{(n)}$ and $w^{(n)}$ are functions of x_i, z and t and $C^{(n)} = C^{(n)}(x_i, z, t, t')$. At the leading order $C^{(0)}(x_i, z, t')$ is expected to represent the period average and depends only on t'. The diffusion equation at $\mathcal{O}(1)$ is a quasi-steady homogeneous differential equation in z

$$-w_o \frac{\partial C^{(0)}}{\partial z} = D \frac{\partial^2 C^{(0)}}{\partial z^2} \quad 0 < z < \infty . \tag{4.3.22}$$

The boundary conditions are homogeneous

$$w_o C^{(0)} + D \frac{\partial C^{(0)}}{\partial z} = 0 , \quad z = 0 ; \tag{4.3.23}$$

$$C^{(0)} = 0 , \quad z \to \infty . \tag{4.3.24}$$

Thus, a nontrivial solution exists

$$C^{(0)} = \widehat{C}(x_i, t') F(z) , \tag{4.3.25}$$

where $\widehat{C}(x_i, t)$ is the unknown concentration at the seabed and $F(z)$ represents the vertical profile,

$$F(z) = \exp\left(-\frac{w_o z}{D}\right) = \exp\left(-\frac{z}{\delta_s}\right) . \tag{4.3.26}$$

At $\mathcal{O}(\epsilon), C^{(1)}$ represents the concentration fluctuations due to the oscillating velocity field and satisfies

$$\frac{\partial C^{(1)}}{\partial t} - w_o \frac{\partial C^{(1)}}{\partial z} - D \frac{\partial^2 C^{(1)}}{\partial z^2} = -\frac{\partial (u_i^{(1)} C^{(0)})}{\partial x_i} - \frac{\partial (w^{(1)} C^{(0)})}{\partial z} . \tag{4.3.27}$$

The homogeneous boundary conditions (Eqs. (4.3.22) and (4.3.23)) apply also to $C^{(1)}$. At $O(\epsilon^2)$, $C^{(2)}$ satisfies

$$\frac{\partial C^{(2)}}{\partial t} - w_o \frac{\partial C^{(2)}}{\partial z} - D \frac{\partial^2 C^{(2)}}{\partial z^2} = -\frac{\partial C^{(0)}}{\partial t'} - \frac{\partial u_i^{(1)} C^{(1)}}{\partial x_i} - \frac{\partial w^{(1)} C^{(1)}}{\partial z}$$
$$- \frac{\partial u_i^{(2)} C^{(0)}}{\partial x_i} - \frac{\partial w^{(2)} C^{(0)}}{\partial z} + D \frac{\partial^2 C^{(0)}}{\partial x_j \partial x_j} \tag{4.3.28}$$

and the boundary conditions

$$w_o C^{(2)} + D \frac{\partial C^{(2)}}{\partial z} = \mathsf{E} |\tau_b| , \quad z = 0 ; \tag{4.3.29}$$

$$C^{(2)} = 0 , \quad z \to \infty . \tag{4.3.30}$$

Our interest is in the slow diffusion at the leading order, and attention will be focused on the governing equation for the factor $\widehat{C}\,(x_i, t')$, defined by Eq. (4.3.25).

Let us assume that the velocity field at the leading-order $\left(u_i^{(1)}, w^{(1)}\right)$ is simple harmonic in time with the frequency ω. All the forcing terms on the right-hand side of Eq. (4.3.27) are likewise simple harmonic. Let the time average with respect to a wave period of $C^{(1)}$ be denoted by $\bar{C}^{(1)}$. Then $\bar{C}^{(1)}$ satisfies the homogeneous equation (4.3.22) also and the homogeneous boundary conditions (Eqs. (4.3.23) and (4.3.24)). Without loss of generality we shall take $\bar{C}^{(1)} = 0$ so that $C^{(1)}$, which corresponds to the departure from the zeroth-order mean $C^{(0)}$, consists only of first harmonic fluctuations:

$$C^{(1)} = \mathcal{R}e\left(C_{11}e^{-i\omega t}\right), \tag{4.3.31}$$

where $C_{11} = C_{11}(x_i, z, t')$ and $\mathcal{R}e$ denotes the real part of its argument. Averaging Eqs. (4.3.28) to (4.3.30) over the wave period, we find that the time average $\bar{C}^{(2)}$ of $C^{(2)}$ satisfies an inhomogeneous differential equation similar to Eq. (4.3.22), and the boundary conditions (Eqs. (4.3.23) and (4.3.24)). Integrating the time-average of Eq. (4.3.28) across the boundary layer and using the time-averaged boundary condition (Eq. (4.3.29)) we get

$$\frac{\partial}{\partial t'}\langle C^{(0)}\rangle + \frac{\partial}{\partial x_i}\langle \bar{u}_i^{(2)} C^{(0)}\rangle = -\frac{\partial}{\partial x_i}\overline{\langle u_i^{(1)} C^{(1)}\rangle} + D\frac{\partial^2 \langle C^{(0)}\rangle}{\partial x_i \partial x_i} + \mathsf{E}\overline{|\tau_b|}, \tag{4.3.32}$$

where $\langle f\rangle$ represents depth integration of f from $z = 0$ to $z \gg 1$. Again, this is just the solvability condition for the inhomogeneous boundary value problem of $C^{(1)}$. It follows by using Eq. (4.3.26) that

$$\boxed{\frac{\partial \widehat{C}}{\partial t'} + \frac{1}{\delta_s}\frac{\partial}{\partial x_i}\left[\langle \bar{u}_i^{(2)} F\rangle \widehat{C}\right] = -\frac{1}{\delta_s}\frac{\partial}{\partial x_i}\overline{\langle u_i^{(1)} C^{(1)}\rangle} + D\frac{\partial^2 \widehat{C}}{\partial x_i \partial x_i} + \frac{\mathsf{E}\overline{|\tau_b|}}{\delta_s}}\,. \tag{4.3.33}$$

This gives the effective convection–diffusion equation for $\widehat{C}$, where erosion from the bed becomes a source of particles. The first term on the right-hand side of Eq. (4.3.33) is the correlation of velocity and concentration fluctuations. The second term represents horizontal diffusion due directly to turbulence. We now fill in some details.

4.3.2. *Effective Convection Velocity*

The velocity field in the wave boundary layer is known in water wave theory (Mei, 1989; Mei *et al.*, 2005). For small amplitudes, the linearized inviscid theory can be used to compute the first-order flow inside the oscillatory boundary layer. At the second order in wave steepness, convective inertia in the boundary layer induces a mean shear stress which drives a second-order mean current (Eulerian streaming). For a horizontal seabed and constant eddy viscosity, both the first order oscillation $u^{(1)}$ and the second order mean drift $\bar{u}^{(2)}$ can be found from Hunt and Johns (1963) and Mei (1989). We cite below the explicit results for horizontal seabed and constant eddy viscosity ν_e.

Let the inviscid wave motion just above the boundary layer be simple harmonic in time to the leading order

$$U_i = \mathcal{R}e\left(U_{oi}e^{-i\omega t}\right), \tag{4.3.34}$$

where $U_{oi} = U_{oi}(x_j) = (U_0, V_0)$ is the amplitude of U_i. At the first order the boundary-layer velocity field is

$$u_i^{(1)} = \mathcal{R}e\left[U_{oi}F_1(z)e^{-i\omega t}\right], \quad w^{(1)} = \mathcal{R}e\left[\frac{1}{\alpha}\frac{\partial U_{oi}}{\partial x_i}F_w(z)e^{-i\omega t}\right] \tag{4.3.35}$$

with

$$F_1(z) = 1 - e^{-\alpha z}, \quad F_w(z) = 1 - e^{-\alpha z} - \alpha z, \tag{4.3.36}$$

with $\alpha = (1-i)/\delta$. The erosion rate can be computed immediately from the local stress:

$$\begin{aligned}\left[\tau_{b,x}, \tau_{b,y}\right]_{z=0} &= \rho\nu_e\left[\frac{\partial u^{(1)}}{\partial z}, \frac{\partial v^{(1)}}{\partial z}\right]_{z=0} \\ &= \frac{\sqrt{2}\rho\nu_e}{\delta_u}\mathcal{R}e\left[U_0(x,y), V_0(x,y)\right]e^{-i(\omega t+\pi/4)}.\end{aligned} \tag{4.3.37}$$

Using asterisks to denote complex conjugates, we then get at the second order the time-averaged horizontal components of Eulerian streaming velocity

$$\overline{u}^{(2)}(x_i, z) = -\frac{1}{\omega}\mathcal{R}e\left[F_2U_o\frac{\partial U_o^*}{\partial x} + F_3V_o\frac{\partial U_o^*}{\partial y} + F_4U_o\frac{\partial V_o^*}{\partial y}\right], \tag{4.3.38}$$

$$\overline{v}^{(2)}(x_i, z) = -\frac{1}{\omega}\mathcal{R}e\left[F_2V_o\frac{\partial V_o^*}{\partial y} + F_3U_o\frac{\partial V_o^*}{\partial x} + F_4V_o\frac{\partial U_o^*}{\partial x}\right], \tag{4.3.39}$$

with

$$F_2 = -\frac{1-3i}{2}e^{(-1+i)\xi} - \frac{1+i}{4}e^{-2\xi} + \frac{1+i}{2}\xi e^{(-1+i)\xi} + \frac{3}{4}(1-i)\,, \tag{4.3.40}$$

$$F_3 = \frac{1}{2}ie^{(-1+i)\xi} - \frac{i}{2}e^{-(1+i)\xi} - \frac{1}{4}e^{-2\xi} + \frac{1}{4}\,, \tag{4.3.41}$$

$$F_4 = -\frac{1-2i}{2}e^{(-1+i)\xi} + \frac{1+i}{2}\xi e^{-(1-i)\xi} - \frac{i}{4}e^{-2\xi} + \frac{1}{4}(2-3i)\,, \tag{4.3.42}$$

where $\xi = z/\delta$ is the normalized boundary-layer coordinate. By evaluating the depth integrals, the weighted average of the Eulerian-induced streaming velocity $\langle \bar{u}_i^{(2)} F\rangle/\delta_s$ can be found to give the effective convection velocity components,

$$\mathcal{U} = \frac{1}{\delta_s}\langle \bar{u}^{(2)} F\rangle = \frac{1}{\omega}\mathcal{R}e\left(H_1 U_o \frac{\partial U_o^*}{\partial x} + H_2 V_o \frac{\partial U_o^*}{\partial y} + H_3 U_o \frac{\partial V_o^*}{\partial y}\right), \tag{4.3.43}$$

$$\mathcal{V} = \frac{1}{\delta_s}\langle \bar{v}^{(2)} F\rangle = \frac{1}{\omega}\mathcal{R}e\left(H_1 V_o \frac{\partial V_o^*}{\partial y} + H_2 U_o \frac{\partial V_o^*}{\partial x} + H_3 V_o \frac{\partial U_o^*}{\partial x}\right). \tag{4.3.44}$$

The complex coefficients H_1, H_3, and the real coefficient of H_2 are functions of Pe and Sc. We omit the lengthy formulas (see Appendix in Mei and Chian, 1994) and only display their sample values in Fig. 4.4.

4.3.3. Correlation Coefficients $\overline{\langle u_i^{(1)} C^{(1)}\rangle}$ and Dispersivity Tensor

At the first order, the concentration fluctuation, $C^{(1)}$, formally given by Eq. (4.3.31), must satisfy Eq. (4.3.27). By substituting the preceding result of Eulerian streaming in terms of the oscillatory velocity components U_0 and V_0 just above the boundary layer and Eq. (4.3.26) into the right-hand side of Eq. (4.3.27), then extracting the coefficient of the first harmonic part, we get

$$D\frac{d^2 C_{11}}{dz^2} + w_0 \frac{dC_{11}}{dz} + i\omega C_{11} = DG(x_i, z) \tag{4.3.45}$$

subject to the boundary conditions

$$w_0 C_{11} + D\frac{dC_{11}}{dz} = 0\,, \quad z = 0\,; \quad C_{11} \to 0\,, \quad z \to \infty\,. \tag{4.3.46}$$

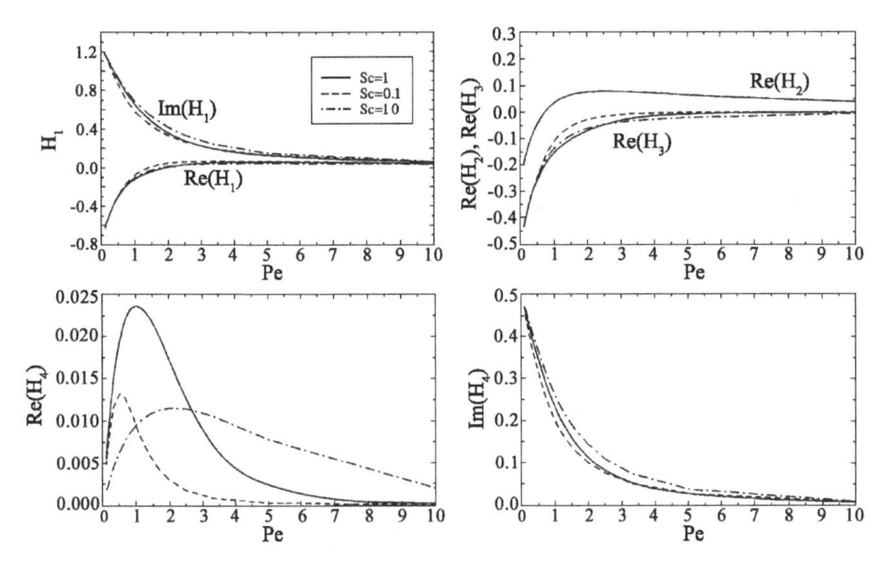

Fig. 4.4. Coefficients H_1, H_2, H_3, H_4 as functions of Pe, for Schmidt numbers $Sc =$ 0.1, 1.0, and 10.0. Top Left: $\mathcal{R}e(H_1)$ and $\mathcal{I}m(H_1) = \mathcal{I}m(H_3)$. Top Right: $\mathcal{R}e(H_2)$ and $\mathcal{R}e(H_3)$. $\mathcal{I}m(H_2) = 0$. Bottom Left: $\mathcal{R}e(H_4)$. Bottom Right: $\mathcal{I}m(H_4)$. Dashes: $Sc = 0.1$; solid: $Sc = 1$; dash-dots: $Sc = 10$. From Mei and Chian (1994), *J. Phys. Oceanogr.*

The forcing term is

$$G(x_i, z) = \frac{1}{D}\left[\left(U_{oi}\frac{\partial \widehat{C}}{\partial x_i}\right)e^{-z/d}F_1(z) - \frac{w_0\widehat{C}}{\alpha D}\left(\frac{\partial U_{oi}}{\partial x_i}\right)\exp\left(-\frac{w_0 z}{D}\right)F_w(z)\right], \tag{4.3.47}$$

where $F_1(z)$ and $F_w(z)$ are given by Eq. (4.3.36). Since G involves both $\partial\widehat{C}/\partial x_i$ and $\widehat{C}$ linearly, C_{11} must be likewise. Leaving the lengthy but straightforward derivation of C_{11} to Mei and Chian (1994) and Mei *et al.* (1997), we cite the resulting effective diffusion equation for $\widehat{C}$:

$$\frac{\partial \widehat{C}}{\partial t'} + \frac{\partial}{\partial x_i}(\mathcal{U}_i\widehat{C}) = D\frac{\partial^2 \widehat{C}}{\partial x_j \partial x_j} + \frac{\partial}{\partial x_i}\left(\mathcal{F}_{ij}\frac{\partial \widehat{C}}{\partial x_j}\right) + \frac{\mathsf{E}|\bar{\tau}_b|}{\delta_s}, \tag{4.3.48}$$

where $\mathcal{F}_{ij}$ arises from the correlation tensor $\langle u_i^{(1)}C^{(1)}\rangle$,

$$\begin{aligned} \mathcal{F}_{xx} &= \mathcal{R}e\left[\frac{H_4}{\omega}|U_o|^2\right], \qquad \mathcal{F}_{xy} = \mathcal{R}e\left[\frac{H_4}{\omega}(U_o^* V_o)\right], \\ \mathcal{F}_{yx} &= \mathcal{R}e\left[\frac{H_4}{\omega}(U_o V_o^*)\right], \quad \mathcal{F}_{yy} = \mathcal{R}e\left[\frac{H_4}{\omega}|V_o|^2\right]. \end{aligned} \tag{4.3.49}$$

The formulas deduced here are quite general for any small amplitude wave field as long as the first-order inviscid velocities (U_o, V_o) in the tangential direction are known at the upper edge of the boundary layer. The real and imaginary parts of the complex coefficient H_4 depend also on Pe and Sc, and are plotted in Fig. 4.4.

In Eq. (4.3.48) the quantity

$$\mathcal{U}_i\widehat{C} - (\mathcal{F}_{ij} + D\delta_{ij})\frac{\partial \widehat{C}}{\partial x_i} \tag{4.3.50}$$

represents the rate-of-flux vector. On the solid coast line, we expect $\mathcal{U}_i n_i$ and the normal flux of particles to vanish, hence

$$n_i \frac{\partial \widehat{C}}{\partial x_i} = 0\,. \tag{4.3.51}$$

With a prescribed initial data for $\widehat{C}(x, y, 0)$, the two-dimensional convective diffusion problem can be solved.

Because H_4 is complex, the tensor $\mathcal{F}_{ij}$ is in general not symmetric, i.e., $\mathcal{F}_{ij} \neq \mathcal{F}_{ji}$; but one may rewrite $\mathcal{F}_{ij}$ as the sum of a symmetric and an anti-symmetric part

$$\mathcal{F}_{ij} = \mathcal{D}_{ij} + \mathcal{E}_{ij}\,, \tag{4.3.52}$$

where

$$\mathcal{D}_{ij} = \frac{1}{2}(\mathcal{F}_{ij} + \mathcal{F}_{ji}) = \mathcal{D}_{ji} \quad \text{and} \quad \mathcal{E}_{ij} = \frac{1}{2}(\mathcal{F}_{ij} - \mathcal{F}_{ji}) = -\mathcal{E}_{ji}\,, \tag{4.3.53}$$

where from Eqs. (4.3.49) and (4.3.53)

$$\mathcal{D}_{ij} = \frac{\mathcal{R}e(H_4)}{\omega}\mathcal{R}e(U_{oi}^* U_{oj})\,, \tag{4.3.54}$$

and

$$\mathcal{E}_{ij} = \frac{\mathcal{I}m(H_4)}{\omega}\mathcal{I}m(U_{oi}^* U_{oj})(\delta_{ij} - 1)\,. \tag{4.3.55}$$

Again the anti-symmetric $\mathcal{E}_{ij}$ does not represent diffusivity and the symmetric part $\mathcal{D}_{ij}$ is the dispersion tensor. By virtue of the anti-symmetry of $\mathcal{E}_{ij}$, the following is true

$$\frac{\partial^2 \mathcal{E}_{ij}}{\partial x_i \partial x_j} = \frac{\partial^2 \mathcal{E}_{ji}}{\partial x_j \partial x_i} = -\frac{\partial^2 \mathcal{E}_{ij}}{\partial x_i \partial x_j} = 0\,,$$

hence Eq. (4.3.48) can be rewritten as

$$\begin{aligned}\frac{\partial \widehat{C}}{\partial t'} + \frac{\partial \mathcal{U}_i \widehat{C}}{\partial x_i} - \frac{\partial \mathcal{E}_{ij}}{\partial x_i}\frac{\partial \widehat{C}}{\partial x_j} &= \frac{\partial \widehat{C}}{\partial t'} + \frac{\partial}{\partial x_i}\left[\left(\mathcal{U}_i + \frac{\partial \mathcal{E}_{ij}}{\partial x_j}\right)\widehat{C}\right] \\ &= \frac{\partial}{\partial x_i}\left[(\mathcal{D}_{ij} + D\delta_{ij})\frac{\partial \widehat{C}}{\partial x_j}\right] + \frac{\mathsf{E}|\bar{\tau}_b|}{\delta_s}\end{aligned} . \quad (4.3.56)$$

4.3.4. Dispersion Under a Standing Wave in a Lake

As a specific example we examine the transport in the bottom boundary layer beneath a simple standing wave in a rectangular lake (Mei *et al.*, 1997). Let the lake be a rectangular basin of width a, length b, and constant depth h. The free-surface displacement is

$$\zeta = A_o \cos\frac{m\pi x}{a}\cos\frac{n\pi y}{b}\mathcal{R}e(e^{-i\omega t}), \quad (4.3.57)$$

where A_o is constant amplitude. Maximum surface displacement occurs along the coasts, $x = \pm a/m$ or $y = \pm b/n$, and nondisplacements along the nodal lines x or y axis. For long waves in a shallow basin ($h \ll a, b$), the frequency is related to the basin dimensions by

$$\omega = k\sqrt{gh} = \pi\sqrt{gh\left(\frac{m^2}{a^2} + \frac{n^2}{b^2}\right)} . \quad (4.3.58)$$

The amplitudes of the free-stream velocities are then given by

$$\begin{aligned}U_o &= \frac{imU_b\sqrt{a^2+b^2}}{a}\sin\frac{m\pi x}{a}\cos\frac{n\pi y}{b}, \\ V_o &= \frac{inU_b\sqrt{a^2+b^2}}{b}\cos\frac{m\pi x}{a}\sin\frac{n\pi y}{b},\end{aligned} \quad (4.3.59)$$

where U_b is defined as

$$U_b = \frac{\pi g A_o}{\omega}\frac{1}{\sqrt{a^2+b^2}} . \quad (4.3.60)$$

By introducing the normalization:

$$x^\dagger = \frac{\pi x}{a}, \quad y^\dagger = \frac{\pi y}{b}, \quad U_o^\dagger = \frac{U_o}{U_b}, \quad V_o^\dagger = \frac{V_o}{U_b}, \quad (4.3.61)$$

we get in dimensionless form

$$U_o^\dagger = im\sqrt{1+s^2}\sin mx^\dagger \cos ny^\dagger\,, \quad V_o^\dagger = in\sqrt{1+s^{-2}}\cos mx^\dagger \sin ny^\dagger\,, \tag{4.3.62}$$

where s denotes the aspect ratio,

$$s \equiv \frac{b}{a}\,. \tag{4.3.63}$$

Since the products $U_{oi}U_{oj}^*$ are real for all (i,j) the dispersion tensor is symmetric according to Eq. (4.3.52),

$$[\mathcal{D}_{xx}, \mathcal{D}_{xy}, \mathcal{D}_{yy}] = \frac{1}{\omega}\mathcal{R}e(H_4)[|U_0|^2, U_0V_0^*, |V_0|^2]\,.$$

The asymmetric tensor $\mathcal{E}_{ij}$ vanishes identically.

Let us introduce the following additional normalization

$$t^\dagger = t'\frac{\omega a^2}{\pi^2 U_b^2}\,, \quad (\mathcal{U}^\dagger, \mathcal{V}^\dagger) = \frac{(\mathcal{U}, \mathcal{V})}{\pi U_b^2/\omega a}\,, \quad \mathcal{D}_{ij}^\dagger = \frac{U_b^2}{\omega}\mathcal{D}_{ij}\,, \quad \widehat{C}^\dagger = \widehat{C}\frac{E\tau_0}{\delta_s}\,, \tag{4.3.64}$$

where τ_0 is the scale of $|\bar{\tau}_b|$. The dimensionless (Eq. (4.3.65)) reads

$$\boxed{\frac{\partial \widehat{C}^\dagger}{\partial t^\dagger} + \frac{\partial \mathcal{U}_i^\dagger \widehat{C}^\dagger}{\partial x_i^\dagger} = \frac{\partial}{\partial x_i^\dagger}\left[(\mathcal{D}^\dagger{}_{ij} + D^\dagger \delta_{ij})\frac{\partial \widehat{C}^\dagger}{\partial x_j^\dagger}\right] + \frac{|\bar{\tau}_b|}{\tau_0}}\,, \tag{4.3.65}$$

where the last term $|\bar{\tau}_b|/\tau_0$ is the dimensionless erosion rate. Specifically, the dimensionless convection velocity components are,

$$\begin{aligned}\mathcal{U}^\dagger &= \frac{1}{2}\mathcal{R}e[(m^3(1+s^2)H_1 + mn^2(1+s^{-2})H_3)\cos^2 ny^\dagger \\ &\quad - mn^2(1+s^{-2})H_2 \sin^2 ny^\dagger]\sin 2mx^\dagger\,, \end{aligned}\tag{4.3.66}$$

$$\begin{aligned}\mathcal{V}^\dagger &= \frac{1}{2s}\mathcal{R}e[(n^3(1+s^{-2})H_1 + m^2n(1+s^2)H_3)\cos^2 mx^\dagger \\ &\quad - m^2n(1+s^2)H_2 \sin^2 mx^\dagger]\sin 2ny^\dagger\,. \end{aligned}\tag{4.3.67}$$

Note that along the banks $x^\dagger = (0,\pi)$ the normal convection velocity $\mathcal{U}^\dagger$ vanishes. Similarly, $\mathcal{V}^\dagger = 0$ along $y^\dagger = (0,\pi)$. The dispersivity tensor components are

$$\mathcal{D}^\dagger_{xx} = \mathcal{R}e(H_4)m^2(1+s^2)\sin^2 mx^\dagger \cos^2 ny^\dagger\,, \tag{4.3.68}$$

$$\mathcal{D}^\dagger_{yy} = \mathcal{R}e(H_4)n^2(1+s^{-2})\cos^2 mx^\dagger \sin^2 ny^\dagger\,, \tag{4.3.69}$$

$$\mathcal{D}^\dagger_{xy} = \mathcal{D}^\dagger_{yx} = \frac{1}{4}\mathcal{R}e(H_4)mn\sqrt{1+s^{-2}}\sqrt{1+s^2}\sin 2mx^\dagger \sin 2ny^\dagger\,. \tag{4.3.70}$$

The total dimensionless effective diffusivity tensor is $\mathcal{K}^\dagger_{ij} \equiv D^\dagger\delta_{ij} + \mathcal{D}^\dagger_{ij}$. Consequently $\mathcal{F}_{ij} = \mathcal{D}_{ij}$ and $\mathcal{E}_{ij} = 0$.

Accurate predictions of the erosion rate are not yet available. Using the crude model of constant eddy viscosity, the classical Stokes theory can be used to determine the flow in the oscillatory boundary layer, and consequently the shear stress at the bottom. With the first-order velocity scaled by Eq. (4.3.60), the physical bottom shear stress components are

$$\begin{pmatrix}\tau_{bx}\\ \tau_{by}\end{pmatrix} = \frac{\sqrt{2}\rho D}{\delta}U_b\sqrt{a^2+b^2}\begin{pmatrix}\frac{m}{a}\sin\frac{m\pi x}{a}\cos\frac{n\pi y}{b}\\ \frac{n}{b}\cos\frac{m\pi x}{a}\sin\frac{n\pi y}{b}\end{pmatrix}\sin\left(\omega t+\frac{\pi}{4}\right). \tag{4.3.71}$$

Its period average can be used to define

$$\tau_o = \frac{2\sqrt{2}\rho D U_b}{\pi\delta}\,, \tag{4.3.72}$$

and the dimensionless averaged erosion rate is

$$\frac{\overline{|\tau_b|}}{\tau_o} = [m^2(1+s^2)\sin^2 mx^\dagger \cos^2 ny^\dagger + n^2(1+s^{-2})\cos^2 mx^\dagger \sin^2 ny^\dagger]^{1/2}\,. \tag{4.3.73}$$

We show first in Fig. 4.6 the convection velocity field for a square lake and for the standing wave mode with $n = m = 1$. The period-averaged flow in the bottom boundary layer converges toward the four corners, where there will be upwelling. The total diffusivity tensor components for $Sc = Pe = 1$ are shown in Fig. 4.6, with $D' = 0.001$. If the coastal strip but not the center portion of the bottom is erodible, the dimensionless erosion rate is also shown in Fig. 4.5. The time evolution of the concentration of suspended sediments is shown in Fig. 4.7 showing gradual accumulation near the corners.

The fate of a particle cloud initially released at some point over a non-erodible bed has been studied by Mei and Chian (1994). Extensions to a

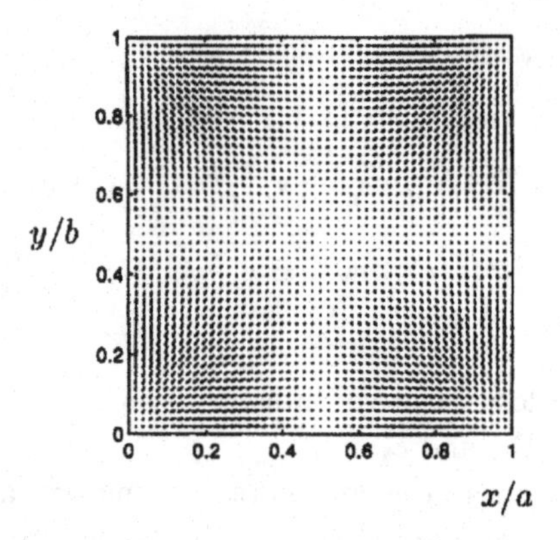

Fig. 4.5. Convection velocity vectors showing divergence from the center and convergence towards the corners. From Mei *et al.* (1997), *J. Geophys. Res.*

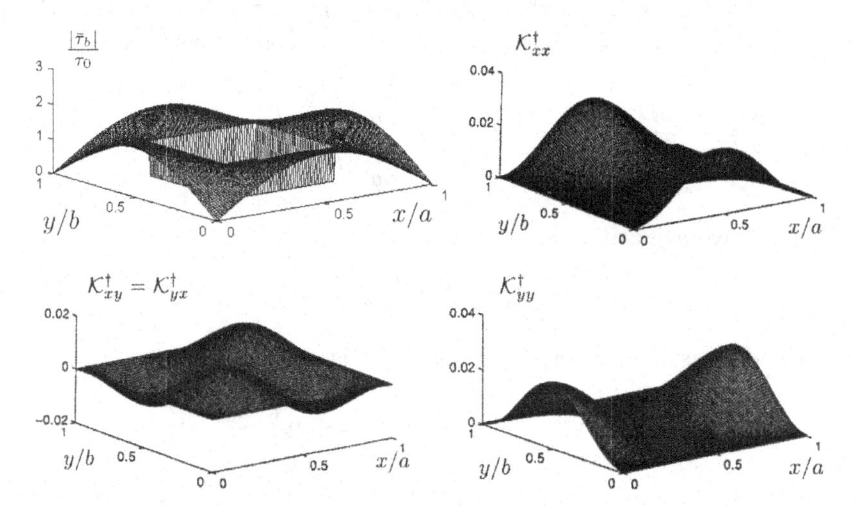

Fig. 4.6. Top Left: Normalized erosion rate on the lake bed where only a belt along the coast is erodible. Top Right: Total diffusivity $\mathcal{K}^{\dagger}_{xx}$. Bottom Left: Total diffusivity $\mathcal{K}^{\dagger}_{xy} = \mathcal{K}^{\dagger}_{yx}$. Bottom Right: Total diffusivity $\mathcal{K}^{\dagger}_{yy}$. From Mei *et al.* (1997), *J. Geophys. Res.*

tidal boundary layer where earth rotation plays a role are treated by Mei *et al.* (1998).

In conclusion, the ideas of homogenization has been shown to apply to problems with microstructures periodic in both space and time. It is therefore not surprising that the technique can be effective in wave propagation problems. Such is indeed the case in the slow evolution of a long wave envelope of a nearly sinusoidal wave train (see for example, Mei *et al.* (2005)). A few additional examples will be illustrated in the next chapter.

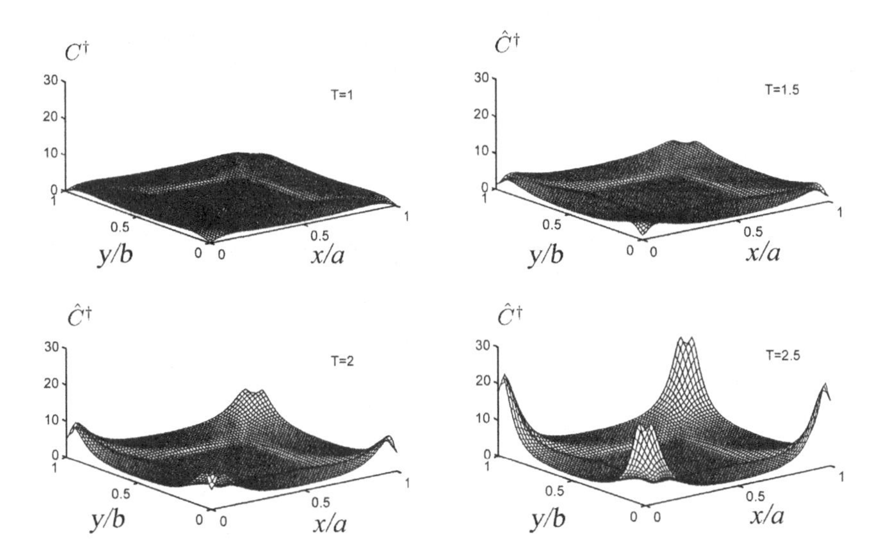

Fig. 4.7. Evolution of concentration of suspended sediments over the lake bottom. Only the coastal belt is erodible. From Mei *et al.* (1997), *J. Geophys. Res.*

Appendix 4A. Derivation of Convection–Dispersion Equation

Invoking the spatial averaging theorem, the right-hand side of Eq. (4.1.32) can be written as:

$$D\left\langle\frac{\partial}{\partial x_i'}\frac{\partial C^{(1)}}{\partial x_i}\right\rangle + D\left\langle\frac{\partial}{\partial x_i'}\frac{\partial C^{(0)}}{\partial x_i'}\right\rangle - \frac{D}{\Omega}\int_\Gamma\left(\frac{\partial C^{(2)}}{\partial x_i} + \frac{\partial C^{(1)}}{\partial x_i'}\right)\frac{\partial S}{\partial x_i}\frac{dS}{|\nabla S|}$$

$$= D\frac{\partial}{\partial x_i'}\left\langle\frac{\partial C^{(1)}}{\partial x_i}\right\rangle + D\frac{\partial}{\partial x_i'}\left\langle\frac{\partial C^{(0)}}{\partial x_i'}\right\rangle$$

$$-\frac{D}{\Omega}\int_\Gamma\left(\frac{\partial C^{(1)}}{\partial x_i} + \frac{\partial C^{(0)}}{\partial x_i'}\right)\frac{\partial S}{\partial x_i'}\frac{dS}{|\nabla S|}$$

$$-\frac{D}{\Omega}\int_\Gamma\left(\frac{\partial C^{(2)}}{\partial x_i} + \frac{\partial C^{(1)}}{\partial x_i'}\right)\frac{\partial S}{\partial x_i}\frac{dS}{|\nabla S|}$$

$$= D\frac{\partial}{\partial x_i'}\left\langle\frac{\partial C^{(1)}}{\partial x_i}\right\rangle + D\frac{\partial}{\partial x_i'}\left\langle\frac{\partial C^{(0)}}{\partial x_i'}\right\rangle$$

$$= D\frac{\partial}{\partial x_i'}\left\langle\frac{\partial C^{(1)}}{\partial x_i}\right\rangle + \theta D\frac{\partial^2 C^{(0)}}{\partial x_i'\partial x_i'}\,. \qquad \text{(A.1)}$$

The two surface integrals cancel by virtue of the boundary condition (Eq. (4.1.17)). We now substitute Eq. (4.1.26) for $C^{(1)}$ and get

$$
\begin{aligned}
D\frac{\partial}{\partial x_i'}\left\langle\frac{\partial C^{(1)}}{\partial x_i}\right\rangle &= -D\frac{\partial}{\partial x_i'}\left(\left\langle\frac{\partial N_\ell}{\partial x_i}\right\rangle\frac{\partial C^{(0)}}{\partial x_\ell'}\right)\\
&= -\frac{D}{2}\left\{\frac{\partial}{\partial x_\ell'}\left(\left\langle\frac{\partial N_i}{\partial x_\ell}\right\rangle\frac{\partial C^{(0)}}{\partial x_i'}\right)+\frac{\partial}{\partial x_i'}\left(\left\langle\frac{\partial N_\ell}{\partial x_i}\right\rangle\frac{\partial C^{(0)}}{\partial x_\ell'}\right)\right\}\\
&= -\frac{D}{2}\left\{\frac{\partial}{\partial x_\ell'}\left(\left\langle\frac{\partial N_i}{\partial x_\ell}\right\rangle\frac{\partial C^{(0)}}{\partial x_i'}\right)-\frac{\partial}{\partial x_i'}\left(\left\langle\frac{\partial N_i}{\partial x_\ell}\right\rangle\frac{\partial C^{(0)}}{\partial x_\ell'}\right)\right\}\\
&\quad -\frac{D}{2}\frac{\partial}{\partial x_i'}\left\{\left(\left\langle\frac{\partial N_\ell}{\partial x_i}\right\rangle+\left\langle\frac{\partial N_i}{\partial x_\ell}\right\rangle\right)\frac{\partial C^{(0)}}{\partial x_\ell'}\right\}\\
&= -\frac{D}{2}\left[\frac{\partial}{\partial x_\ell'}\left\langle\frac{\partial N_i}{\partial x_\ell}\right\rangle-\frac{\partial}{\partial x_\ell'}\left\langle\frac{\partial N_\ell}{\partial x_i}\right\rangle\right]\frac{\partial C^{(0)}}{\partial x_i'}\\
&\quad -\frac{D}{2}\frac{\partial}{\partial x_i'}\left\{\left(\left\langle\frac{\partial N_\ell}{\partial x_i}\right\rangle+\left\langle\frac{\partial N_i}{\partial x_\ell}\right\rangle\right)\frac{\partial C^{(0)}}{\partial x_\ell'}\right\},
\end{aligned}
$$

where interchanges of summation indices have been made.

Let us turn to the left-hand side of Eq. (4.1.32). It is easily seen by using the spatial averaging theorem and the first-order continuity equation

$$\frac{\partial u_i^{(1)}}{\partial x_i}+\frac{\partial u_i^{(0)}}{\partial x_i'}=0 \tag{A.2}$$

that

$$\left\langle\frac{\partial u_i^{(1)}C^{(0)}}{\partial x_i'}\right\rangle=\frac{\partial}{\partial x_i'}\langle u_i^{(1)}C^{(0)}\rangle=\langle u_i^{(1)}\rangle\frac{\partial C^{(0)}}{\partial x_i'}=0\,. \tag{A.3}$$

Use is made of Eq. (3.8.25) that $\langle u_i^{(1)}\rangle=0$. On the other hand,

$$
\begin{aligned}
\left\langle\frac{\partial u_i^{(0)}C^{(1)}}{\partial x_i'}\right\rangle &= \frac{\partial}{\partial x_i'}\langle u_i^{(0)}C^{(1)}\rangle = -\frac{\partial}{\partial x_i'}\left(\langle u_i^{(0)}N_\ell\rangle\frac{\partial C^{(0)}}{\partial x_\ell'}\right)\\
&= -\frac{\partial}{\partial x_i'}\left(\left\langle\widetilde{u}_i^{(0)}N_\ell\right\rangle\frac{\partial C^{(0)}}{\partial x_\ell'}\right),
\end{aligned} \tag{A.4}
$$

after using Eqs. (4.1.25) and (4.1.26).

By interchanging the summation indices we get

$$
\begin{aligned}
-\frac{\partial}{\partial x_i'}&\left(\langle\widetilde{u}_i^{(0)}N_\ell\rangle\frac{\partial C^{(0)}}{\partial x_\ell'}\right)\\
&=-\frac{1}{2}\frac{\partial}{\partial x_i'}\left(\langle\widetilde{u}_i^{(0)}N_\ell\rangle\frac{\partial C^{(0)}}{\partial x_\ell'}\right)-\frac{1}{2}\frac{\partial}{\partial x_\ell'}\left(\langle\widetilde{u}_\ell^{(0)}N_i\rangle\frac{\partial C^{(0)}}{\partial x_i'}\right)\\
&=-\frac{\partial}{\partial x_i'}\left(\langle\widetilde{u}_i^{(0)}N_\ell\rangle\frac{\partial C^{(0)}}{\partial x_\ell'}\right)-\frac{\partial}{\partial x_i'}\left(\langle\widetilde{u}_\ell^{(0)}N_i\rangle\frac{\partial C^{(0)}}{\partial x_\ell'}\right)\\
&\quad+\frac{\partial}{\partial x_i'}\left(\langle\widetilde{u}_\ell^{(0)}N_i\rangle\frac{\partial C^{(0)}}{\partial x_\ell'}\right)-\frac{\partial}{\partial x_\ell'}\left(\langle\widetilde{u}_\ell^{(0)}N_i\rangle\frac{\partial C^{(0)}}{\partial x_i'}\right)\\
&=-\frac{\partial}{\partial x_i'}\left[\left(\langle\widetilde{u}_i^{(0)}N_\ell\rangle+\langle\widetilde{u}_\ell^{(0)}N_i\rangle\right)\frac{\partial C^{(0)}}{\partial x_\ell'}\right]\\
&\quad+\left(\frac{\partial}{\partial x_i'}\langle\widetilde{u}_\ell^{(0)}N_i\rangle\right)\frac{\partial C^{(0)}}{\partial x_\ell'}-\left(\frac{\partial}{\partial x_\ell'}\langle\widetilde{u}_\ell^{(0)}N_i\rangle\right)\frac{\partial C^{(0)}}{\partial x_i'}\\
&=-\frac{\partial}{\partial x_i'}\left[\left(\langle\widetilde{u}_i^{(0)}N_\ell\rangle+\langle\widetilde{u}_\ell^{(0)}N_i\rangle\right)\frac{\partial C^{(0)}}{\partial x_\ell'}\right]\\
&\quad+\left(\frac{\partial}{\partial x_i'}\langle\widetilde{u}_\ell^{(0)}N_i\rangle-\frac{\partial}{\partial x_i'}\langle\widetilde{u}_i^{(0)}N_\ell\rangle\right)\frac{\partial C^{(0)}}{\partial x_\ell'}\,. \qquad \text{(A.5)}
\end{aligned}
$$

Making use of Eqs. (A.4), (A.5), and (A.3) in Eq. (4.1.32), one gets Eq. (4.1.40).

Appendix 4B. An Alternate Form of Macrodispersion Tensor

Since

$$
\begin{aligned}
\widetilde{u}_i^{(0)}&=\bar{u}_i^{(0)}-\frac{\theta}{\langle\theta\rangle'}\langle\bar{u}_i^{(0)}\rangle'=\bar{u}_i^{(0)}-\langle\bar{u}_i^{(0)}\rangle'-\left(1-\frac{\theta}{\langle\theta\rangle'}\right)\langle\bar{u}_i^{(0)}\rangle'\\
&\equiv\widehat{u}_i^{(0)}+\left(1-\frac{\theta}{\langle\theta\rangle'}\right)\langle\bar{u}_i^{(0)}\rangle', \qquad \text{(B.1)}
\end{aligned}
$$

let us first find $\langle\widetilde{u}_i^{(0)}B_j\rangle'$.

Multiplying Eq. (4.2.20) by B_j, we get

$$
\begin{aligned}
\widetilde{u}_i^{(0)}B_j+\bar{u}_\ell^{(0)}B_j\frac{\partial B_i}{\partial x_\ell}&=\frac{\partial B_j\mathcal{D}_{ki}}{\partial x_k}+\frac{\partial}{\partial x_k'}\left(B_j\mathcal{D}_{kn}\frac{\partial B_i}{\partial x_n'}\right)\\
&\quad-\mathcal{D}_{ki}\frac{\partial B_j}{\partial x_k'}-\mathcal{D}_{kn}\frac{\partial B_i}{\partial x_k'}\frac{\partial B_j}{\partial x_n'}\,. \qquad \text{(B.2)}
\end{aligned}
$$

After interchanging the indices i and j,

$$\widetilde{u}_j^{(0)} B_i + \bar{u}_\ell^{(0)} B_i \frac{\partial B_j}{\partial x_\ell} = \frac{\partial B_i \mathcal{D}_{kj}}{\partial x_k} + \frac{\partial}{\partial x'_k}\left(B_i \mathcal{D}_{kn} \frac{\partial B_j}{\partial x'_n}\right) - \mathcal{D}_{kj}\frac{\partial B_i}{\partial x'_k} - \mathcal{D}_{kn}\frac{\partial B_j}{\partial x'_k}\frac{\partial B_i}{\partial x'_n}, \tag{B.3}$$

we sum Eqs. (B.2) and (B.3) and taking the Ω'-average. It follows after applying Gauss' theorem and periodicity that,

$$\langle \widetilde{u}_i^{(0)} B_j \rangle' + \langle \widetilde{u}_j^{(0)} B_i \rangle' = -\left\langle \mathcal{D}_{ik}\frac{\partial B_j}{\partial x'_k} + \mathcal{D}_{jk}\frac{\partial B_i}{\partial x'_k}\right\rangle' - \left\langle \mathcal{D}_{kn}\left(\frac{\partial B_j}{\partial x'_k}\frac{\partial B_i}{\partial x'_n} + \frac{\partial B_i}{\partial x'_k}\frac{\partial B_j}{\partial x'_n}\right)\right\rangle'. \tag{B.4}$$

Substituting Eq. (B.1) into Eq. (B.4), we get, after invoking $\langle B_j \rangle' = 0$,

$$\langle \widehat{u}_i^{(0)} B_j \rangle' + \langle \widehat{u}_j^{(0)} B_i \rangle' = \frac{1}{\langle \theta \rangle'}\left[\langle \bar{u}_i^{(0)} \rangle' \langle \theta B_j \rangle' + \langle \bar{u}_j^{(0)} \rangle \langle \theta B_i \rangle'\right] - \left\langle \mathcal{D}_{ik}\frac{\partial B_j}{\partial x'_k} + \mathcal{D}_{jk}\frac{\partial B_i}{\partial x'_k}\right\rangle' - 2\left\langle \mathcal{D}_{kn}\left(\frac{\partial B_j}{\partial x'_k}\frac{\partial B_i}{\partial x'_n} + \frac{\partial B_i}{\partial x'_k}\frac{\partial B_j}{\partial x'_n}\right)\right\rangle'. \tag{B.5}$$

The expression of dispersion tensor (Eq. (4.2.29)) follows by inserting Eq. (B.5) in Eq. (4.2.28).

References

Aris, R. (1956). On the dispersion of solute in a fluid flowing in a tube. *Proc. R. Soc. Lond. A* **235**: 67–77.

Bear, J. (1969). *Hydrodynamic Dispersions in Flow Through Porous Media*, ed. R. de Wiest, Academic Press, pp. 109–199.

Blackwell, R. J. (1962). Laboratory studies of microscopic dispersion phenomena. *Soc. Pet. Eng. J.* **2**: 1–8.

Brenner, H. (1980). Dispersion resulting from flow through spatially periodic porous media. *Phil. Trans. R. Soc. Lond. A* **297**: 81–133.

Dagan, G. (1987). Theory of solute transport by groundwater. *Ann. Rev. Fluid Mech.* **19**: 183–215.

Dagan, G. (1989). *Flow and Transport in Porous Formations*, Springer-Verlag, Berlin, 465 pp.

de Jong, D. J. (1958). Longitudinal and transverse diffusion in granular deposits. *Trans. Amer. Geophys. Union* **39**: 67–74.

Denney, M. W. (1988). *Biology and the Mechanics of the Wave-Swept Environment*, Princeton University Press, N.J., 529 pp.

Dullien, F. A. L. (1979). *Porous Media: Fluid Transport and Pore Structure*, Academic Press, New York.

Ebach, E. A. and R. R. White (1958). Mixing of fluids flowing through beds of packed solids. *AIChE J.* **4**: 161–169.

Edwards, E. R. G. and J. F. Richardson (1968). Gas dispersion in packed beds. *Chem. Eng. Sci.* **23**: 109–123.

Fried, J. J. and M. A. Combarnous (1971). Dispersion in porous media. *Advances in Hydroscience*, Vol. 7, ed. V. T. Chow, pp. 169–282.

Gelhar, L. W. (1993). *Stochastic Subsurface Hydrology*, Prentice Hall, 390 pp.

Gunn D. J. and C. Pryce (1969). Dispersion in packed beds. *Chem. Eng. Res. Design* **47**: T341–T350.

Haring, R. E. and R. A. Greenhorn (1970). A statistical model of a porous medium with nonuniform pores. *AIChE J.* **16**: 477–483.

Harleman, D. F. H. and R. R. Rumer (1963). Longitudinal and lateral dispersion in an isotropic porous medium. *J. Fluid Mech.* **16**: 385–394.

Hiby, J. W. (1962). Longitudinal and transverse mixing during single phase flow through granular beds. *Proc. Symp. Interactions Between Fluids and Particles*, London, pp. 312–325.

Hunt, J. N. and B. Johns (1963). Current induced by tides and gravity waves. *Tellus.* **15**: 343–351.

Koch, D. L. and J. L. Brady (1985). Dispersion in fixed beds. *J. Fluid Mech.* **154**: 399–427.

Koch, D. L., R. G. Cox, H. Brenner and J. F. Brady (1989). The effect of order on dispersion in porous media. *J. Fluid Mech.* **200**: 173–188.

Lee, C. K. (1994). Thermal consolidation and dispersion in inhomogeneous deformable media. PhD. thesis, Department of Civil and Environmental Engineering, Mass. Inst. Tech.

Lee, C. K., C. C. Sun and C. C. Mei (1995). Computation of permeability and dispersivity of solute or heat in periodic porous media. *Int. J. Heat Mass Transfer* **39**: 661–676.

Longuet-Higgins, M. S. (1953). Mass transport in water waves. *Phil. Trans. R. Soc.* **345**: 535–581.

Mei, C. C. (1989). *Applied Dynamics of Ocean Surface Waves*, World Scientific, Singapore.

Mei, C. C. (1991). Dispersion of heat in periodic porous media by homogenization method. *Mathematical Approaches in Hydrodynamics*, ed. T. Miloh, Soc. Ind. Appl. Math., 425–435.

Mei, C. C. (1992). Method of homogenization applied to dispersion in porous media. *Transport in Porous Media* **9**: 261–274.

Mei, C. C., J.-L. Auriault and C. O. Ng (1996). Some applications of homogenization theory. *Adv. Appl. Mech.* **32**: 278–348.

Mei, C. C. and C. M. Chian (1994). Dispersion of small suspended particles in a wave boundary layer. *J. Amer. Meteorol. Soc.* **24**(12): 2479–2495.

Mei, C. C., C. Chian and F. Ye (1998). Transport and resuspension of fine particles in a tidal boundary layer near a small peninsula. *J. Phys. Oceanogr.* 2313–2331.

Mei, C. C., S. J. Fan and K. R. Jin (1997). Resuspension and transport of fine sediments by waves. *J. Geophys. Res.* **102**: 15807–15821.

Mei, C. C., M. Stiassnie and D. K.-P. Yue (2005). *Theory and Application of Ocean Surface Waves*, Vol. II, World Scientific, Singapore.

Rubinstein, J. and R. Mauri (1986). Dispersion and convection in periodic porous media. *SIAM. J. Appl. Math.* **46**(6): 1018–1023.

Saffman, P. G. (1960). Dispersion due to molecular diffusion and macroscopic mixing in flow through a network of capillaries. *J. Fluid Mech.* **7**: 194–208.

Salles, J., J.-F. Thovert, R. Delannay, L. Prevors, J.-L. Auriault and P. M. Adler (1993). Taylor dispersion in porous media, determination of the dispersion tensor. *Phys. Fluids* **5**: 2348–2376.

Sheng, Y. P. and V. Cook (1990). Resuspension and vertical mixing of fine sediments in shallow water. Part B. University of Florida, Coastal Eng. Oceanogr. Eng. Lab. Report C90-020.

Heterogeneous Elastic Materials

5

In this chapter we derive and study the effective properties of heterogeneous elastic media formed by periodic variations of either material properties (laminates or fiber-reinforced composites) or geometric construction (cellular or honeycomb microstructures). A variety of theoretical treatments of elastic composites have been described in monographs by Oleinik *et al.* (1992), Nemat-Nasser and Hori (1999), Cherkaev (2000), Milton (2002), Torquato (2002), Cristescu *et al.* (2004), while Sanchez-Palencia (1980) and Bakhvalov and Panasenko (1989) have employed the method of homogenization exclusively. We select a few topics here to demonstrate the use of homogenization theory for deriving macroscale equations, and to discuss how variational bounds based on the cell boundary-value problems can estimate the magnitudes of the effective coefficients.

We begin with a brief overview of the derivation of the effective properties of three-dimensional elastic composites and present some computed results for the special case of fiber-reinforced composites. Next we examine a case where the overall characteristic length in one direction is much smaller than those in the other two, as in honeycomb panels. Typically these panels have a two-dimensional periodic cellular core structure formed by thin-walled cells, sandwiched between two parallel plates. They are widely used due to their high strength-to-weight ratio. The thickness of these periodic plates is often of the same order of magnitude as the cellular period. By homogenization theory one can show that the effective equation governing the macroscale behavior is similar to that for a Kirchoff plate (Ciarlet and Destuynder, 1979; Caillerie, 1984; Kohn and Vogelius, 1984; Parton and Kudryavtsev, 1993; Gilbert and Hackl, 1995; Cioranescu and Saint Jean Paulin, 1999; Lewinsky and Telega, 2000). We shall describe the derivation of a two-dimensional effective equation by accounting for the three-dimensional microstructure. Details of numerical solution of the cell problems for calculating the effective moduli depend on specific microscale geometries, and can be found in, e.g., Shi and Tong (1995) and Xu and Qiao (2002).

A substantial portion of this chapter is devoted to the use of variational principles to characterize the effective elastic moduli defined by the cell problems in the homogenization analysis, and to derive upper and lower bounds for these moduli. In particular we shall derive the celebrated Voigt, Reuss, and Hashin–Shtrikman bounds (Hashin and Shtrikman, 1963).

Often elastic composites exhibit partial bonding between constituents. For such composites two different models have been used: an *interphase model* and an *interface model*. In the former one considers a third phase between constituents. The latter is the one we chose to exemplify in this chapter; it supposes that the displacement field has jumps across the interface, proportional to the normal stress. The homogenization method has been used by Lenè and Leguillon (1982) for the case of interfaces with tangential slip of the displacement. Here we follow the work of Lipton and Vernescu (1995) and Lipton and Vernescu (1996) and derive the effective properties, variational principles characterizing the effective moduli, and bounds that have as a particular case the mean bounds derived by Hashin (1992). In the case of particulate composites the bounds are used to highlight the size effects for partially bonded elastic composites and to find the critical size of particles that are *cloaked* (i.e., made invisible) by the matrix.

There is a strong connection between the analysis of materials with microstructure and the important area of structural optimization in engineering. Starting with the early work of Lurie *et al.* (1980), Lurie and Cherkaev (1984, 1987), Murat and Tartar (1985), Kohn and Strang (1986) the link between homogenization and structural optimization was put forth and is the subject of a vast literature. For many aspects not treated in this chapter, the reader is referred to Cherkaev (2000) and to the references cited therein.

5.1. Effective Equations on the Macroscale

Let us consider a three-dimensional composite which is spatially periodic, with cells of linear dimension $\mathcal{O}(\ell)$ in all directions. We denote by Ω_α the part of Ω occupied by material $\alpha, \alpha = 1, 2$. The total volume occupied by the composite has the characteristic length ℓ' which is much greater than the cell size so that $\ell/\ell' = \mathcal{O}(\epsilon) \ll 1$.

Let $\boldsymbol{v}(\boldsymbol{x})$ denote the displacement vector, σ_{ij} the components of the stress tensor, $\varepsilon_{ij}(\boldsymbol{v})$ the components of the strain tensor, and $C_{ijk\ell}$ the components of the fourth-order elasticity (or stiffness) tensor. In the framework of linear elasticity Hooke's law is given by

$$\sigma_{ij} = C_{ijk\ell}\varepsilon_{k\ell}(\boldsymbol{v}), \tag{5.1.1}$$

$$\varepsilon_{k\ell}(\boldsymbol{v}) = \frac{1}{2}\left(\frac{\partial v_k}{\partial x_\ell} + \frac{\partial v_\ell}{\partial x_k}\right) \tag{5.1.2}$$

with different elastic constants in the two materials:

$$C_{ijk\ell} = \begin{cases} (C_1)_{ijk\ell} & \text{in } \Omega_1 \\ (C_2)_{ijk\ell} & \text{in } \Omega_2 \end{cases} \tag{5.1.3}$$

satisfying symmetry and positivity properties

$$C_{ijk\ell} = C_{jik\ell} = C_{k\ell ij}, \tag{5.1.4}$$

$$C_{ijk\ell}\xi_{ij}\xi_{k\ell} \geq \alpha\xi_{ij}\xi_{ij}, \tag{5.1.5}$$

where $\alpha > 0$ and (ξ_{ij}) is any constant, symmetric second-order tensor. The elasticity problem for the composite is described by the Cauchy equation

$$\frac{\partial(\sigma_\alpha)_{ij}}{\partial x_j} = 0 \quad \text{in } \Omega_\alpha, \quad \alpha = 1, 2, \tag{5.1.6}$$

together with the continuity conditions for the displacement and stress on the interface Γ between the two materials

$$(v_1)_i = (v_2)_i \quad \text{on } \Gamma, \tag{5.1.7}$$

$$(\sigma_1)_{ij}n_j = (\sigma_2)_{ij}n_j \quad \text{on } \Gamma. \tag{5.1.8}$$

Here $(v_\alpha)_i$ and $(\sigma_\alpha)_{ij}$, $\alpha = 1, 2$ represent, respectively, the displacement and the stress of material α.

We introduce the fast and slow coordinates x_i and $x_i' = \epsilon x_i$ and the multiple scale expansions for $\boldsymbol{v}$:

$$\boldsymbol{v}_\alpha = \boldsymbol{v}_\alpha^{(0)}(x_i, x_i') + \epsilon\boldsymbol{v}_\alpha^{(1)}(x_i, x_i') + \epsilon^2\boldsymbol{v}_\alpha^{(2)}(x_i, x_i') + \cdots, \quad \text{in } \Omega_\alpha, \tag{5.1.9}$$

with $\boldsymbol{v}^{(i)}$ Ω-periodic in the variable x_i. This yields an expansion for the stress tensor of the form

$$\boldsymbol{\sigma}_\alpha = \boldsymbol{\sigma}_\alpha^{(0)}(x_i, x_i') + \epsilon\boldsymbol{\sigma}_\alpha^{(1)}(x_i, x_i') + \epsilon^2\boldsymbol{\sigma}_\alpha^{(2)}(x_i, x_i') + \cdots, \quad \text{in } \Omega_\alpha, \tag{5.1.10}$$

where

$$(\sigma_\alpha^{(0)})_{ij} = C_{ijk\ell}\varepsilon_{k\ell}(\boldsymbol{v}_\alpha^{(0)})\,, \tag{5.1.11}$$

$$(\sigma_\alpha^{(1)})_{ij} = C_{ijk\ell}\big(\varepsilon_{k\ell}(\boldsymbol{v}_\alpha^{(1)}) + \varepsilon'_{k\ell}(\boldsymbol{v}_\alpha^{(0)})\big)\,, \tag{5.1.12}$$

$$\left(\sigma_\alpha^{(2)}\right)_{ij} = C_{ijk\ell}\big(\varepsilon_{k\ell}(\boldsymbol{v}_\alpha^{(2)}) + \varepsilon'_{k\ell}(\boldsymbol{v}_\alpha^{(1)})\big)\,, \tag{5.1.13}$$

and where $(\varepsilon'_{k\ell})$ denotes the strain tensor with respect to the macroscopic variables

$$\varepsilon'_{k\ell}(\boldsymbol{v}) = \frac{1}{2}\left(\frac{\partial v_k}{\partial x'_\ell} + \frac{\partial v_\ell}{\partial x'_k}\right). \tag{5.1.14}$$

By substituting Eq. (5.1.10) into Eqs. (5.1.6)–(5.1.8) we get at order $\mathcal{O}(\epsilon^0)$:

$$\frac{\partial}{\partial x_j}\big(C_{ijk\ell}\varepsilon_{k\ell}(\boldsymbol{v}_\alpha^{(0)})\big) = 0 \quad \text{in } \Omega_\alpha\,, \tag{5.1.15}$$

$$\boldsymbol{v}_1^{(0)} = \boldsymbol{v}_2^{(0)} \quad \text{on } \Gamma\,, \tag{5.1.16}$$

$$C_{ijk\ell}\varepsilon_{k\ell}(\boldsymbol{v}_1^{(0)})n_j = C_{ijk\ell}\varepsilon_{k\ell}(\boldsymbol{v}_2^{(0)})n_j \quad \text{on } \Gamma\,, \tag{5.1.17}$$

$$\boldsymbol{v}^{(0)} \text{ is } \Omega\text{-periodic}. \tag{5.1.18}$$

The above problem has only solutions constant in $\boldsymbol{x}$, i.e., $\boldsymbol{v}^{(0)}$ depends only on the macroscopic variable:

$$\boldsymbol{v}^{(0)} = \boldsymbol{v}^{(0)}(\boldsymbol{x}')\,, \tag{5.1.19}$$

which implies that $\sigma_{ij}^{(0)} = \varepsilon_{ij}(\boldsymbol{v}^{(0)}) \equiv 0$.

At the next order we have

$$\frac{\partial}{\partial x_j}\big(C_{ijk\ell}\varepsilon_{k\ell}(\boldsymbol{v}_\alpha^{(1)})\big) = -\frac{\partial}{\partial x_j}\big(C_{ijk\ell}\varepsilon'_{k\ell}(\boldsymbol{v}_\alpha^{(0)})\big) \quad \text{in } \Omega_\alpha\,, \tag{5.1.20}$$

$$\boldsymbol{v}_1^{(1)} = \boldsymbol{v}_2^{(1)} \quad \text{on } \Gamma\,, \tag{5.1.21}$$

$$C_{ijk\ell}\varepsilon_{k\ell}(\boldsymbol{v}_1^{(1)})n_j = C_{ijk\ell}\varepsilon_{k\ell}(\boldsymbol{v}_2^{(1)})n_j \quad \text{on } \Gamma\,, \tag{5.1.22}$$

$$\boldsymbol{v}_\alpha^{(1)} \text{ is } \Omega\text{-periodic}. \tag{5.1.23}$$

The above problem has the solution of the form

$$\boldsymbol{v}^{(1)} = \boldsymbol{\chi}^{k\ell}\varepsilon'_{k\ell}(\boldsymbol{v}^{(0)}) + \text{const.}\,, \quad \text{or} \quad v_i^{(1)} = \chi_i^{k\ell}\varepsilon'_{k\ell}(\boldsymbol{v}^{(0)}) + \text{const.}\,, \tag{5.1.24}$$

where $\{\boldsymbol{\chi}^{k\ell}\}_i = \chi_i^{k\ell}$ denotes the components of a tensor of rank 3, and is governed by the cell problem

$$\frac{\partial}{\partial x_j}\left(C_{ijmn}\varepsilon_{mn}\left(\boldsymbol{\chi}_\alpha^{k\ell}\right)\right) = -\frac{\partial C_{ijk\ell}}{\partial x_j} \quad \text{in } \Omega_\alpha\,, \tag{5.1.25}$$

$$\boldsymbol{\chi}_1^{k\ell} = \boldsymbol{\chi}_2^{k\ell} \quad \text{on } \Gamma\,, \tag{5.1.26}$$

$$C_{ijmn}\varepsilon_{mn}\left(\boldsymbol{\chi}_1^{k\ell}\right)n_j = C_{ijmn}\varepsilon_{mn}\left(\boldsymbol{\chi}_2^{k\ell}\right)n_j \quad \text{on } \Gamma\,, \tag{5.1.27}$$

$$\boldsymbol{\chi}^{k\ell} \text{ is } \Omega\text{-periodic}\,. \tag{5.1.28}$$

In addition $\boldsymbol{\chi}^{k\ell}$ has zero average over Ω:

$$\left\langle \boldsymbol{\chi}_\alpha^{k\ell} \right\rangle = 0\,, \tag{5.1.29}$$

where the angle brackets denote the volume average over the cell Ω

$$\langle \boldsymbol{\chi}_\alpha \rangle = \frac{1}{\Omega}\left(\iiint_{\Omega_1} \boldsymbol{\chi}_1^{k\ell} d\Omega + \iiint_{\Omega_2} \boldsymbol{\chi}_2^{k\ell} d\Omega\right). \tag{5.1.30}$$

At the order $\mathcal{O}(\epsilon^2)$ we have

$$\frac{\partial \sigma_{ij}^{(1)}}{\partial x_j'} + \frac{\partial \sigma_{ij}^{(2)}}{\partial x_j} = 0\,. \tag{5.1.31}$$

By averaging over Ω, the macroscale equation of elastostatics is

$$\boxed{\frac{\partial}{\partial x_j'}\langle \sigma_{ij}^{(1)} \rangle = 0}\,. \tag{5.1.32}$$

The macroscopic Hooke's law follows from the average of Eq. (5.1.12) using Eq. (5.1.24)

$$\boxed{\langle \sigma_{ij}^{(1)} \rangle = \mathcal{C}_{ijk\ell}\varepsilon_{k\ell}'(\boldsymbol{v}^{(0)})}\,, \tag{5.1.33}$$

where the homogenized elastic coefficients are given by

$$\boxed{\mathcal{C}_{ijk\ell} = \left\langle C_{ijmn}\left(\delta_{mk}\delta_{n\ell} + \varepsilon_{mn}\left(\boldsymbol{\chi}^{k\ell}\right)\right)\right\rangle}\,. \tag{5.1.34}$$

Thus the macroscopic behavior of the elastic composite is similar to a homogeneous elastic material having the elasticity coefficients given by Eq. (5.1.34) in terms of the solutions $\boldsymbol{\chi}^{k\ell}$ of the cell problems (Eqs. (5.1.25)–(5.1.29)).

5.2. The Effective Elastic Coefficients

In order for the effective coefficients given by Eq. (5.1.34) to represent the coefficients of a linear elastic material, they need to satisfy the symmetry and positivity properties similar to Eqs. (5.1.4) and (5.1.5), namely

$$\mathcal{C}_{ijk\ell} = \mathcal{C}_{jik\ell} = \mathcal{C}_{k\ell ij}\,, \tag{5.2.1}$$

and

$$\mathcal{C}_{ijk\ell}\xi_{ij}\xi_{k\ell} \geq \alpha\xi_{ij}\xi_{ij}\,, \tag{5.2.2}$$

for any (ξ_{ij}) constant, symmetric second-order matrix, and some positive constant α.

The first equality in Eq. (5.2.1) is a direct consequence of Eq. (5.1.34) and the symmetry of $C_{ijk\ell}$. In order to establish the second part of Eqs. (5.2.1) and (5.2.2) we need to express the effective coefficients in a more convenient way. Multiplying the cell problem (Eq. (5.1.25)) by χ_i^{pq} in each domain Ω_1 and Ω_2 and by applying Gauss' formula we obtain

$$\begin{aligned} -\iiint_\Omega C_{ijmn}\varepsilon_{mn}(\boldsymbol{\chi}^{k\ell})\varepsilon_{ij}(\boldsymbol{\chi}^{pq})d\Omega + \iint_\Gamma [C_{ijmn}\varepsilon_{mn}(\boldsymbol{\chi}^{k\ell})n_j\chi_i^{pq}]d\Gamma \\ = \iiint_\Omega C_{ijk\ell}\varepsilon_{ij}(\boldsymbol{\chi}^{pq})d\Omega\,, \end{aligned} \tag{5.2.3}$$

where we have used the periodicity to cancel the integrals on opposite faces of $\partial\Omega$. The jump term, on the interface, represented by the square brackets, cancels due to the continuity conditions (Eqs. (5.1.26) and (5.1.27)). We can rewrite the remaining two terms as

$$\langle C_{ijmn}(\delta_{mk}\delta_{n\ell} + \varepsilon_{mn}(\boldsymbol{\chi}^{k\ell}))\varepsilon_{ij}(\boldsymbol{\chi}^{pq})\rangle = 0 \tag{5.2.4}$$

or equivalently, after renaming indices

$$\langle C_{pqmn}(\delta_{mk}\delta_{n\ell} + \varepsilon_{mn}(\boldsymbol{\chi}^{k\ell}))\varepsilon_{pq}(\boldsymbol{\chi}^{ij})\rangle = 0\,. \tag{5.2.5}$$

By adding Eq. (5.2.5) to the definition of the effective coefficients (Eq. (5.1.34)) we get a new expression for the effective coefficients

$$\mathcal{C}_{ijk\ell} = \left\langle C_{pqmn}\left(\delta_{pi}\delta_{qj} + \varepsilon_{pq}(\boldsymbol{\chi}^{ij})\right)\left(\delta_{mk}\delta_{n\ell} + \varepsilon_{mn}(\boldsymbol{\chi}^{k\ell})\right)\right\rangle\,. \tag{5.2.6}$$

From here the symmetry $\mathcal{C}_{ijk\ell} = \mathcal{C}_{k\ell ij}$ is immediate by using the symmetry of $C_{ijk\ell}$.

Equation (5.2.6) also implies the positivity of $C_{ijk\ell}$. Indeed if (ξ_{ij}) is a symmetric matrix, we use Eq. (5.1.5) to get

$$C_{ijk\ell}\xi_{ij}\xi_{k\ell} = \langle C_{pqmn}(\xi_{pq} + \varepsilon_{pq}(\chi^{ij}\xi_{ij}))(\xi_{mn} + \varepsilon_{mn}(\chi^{k\ell}\xi_{k\ell}))\rangle \geq 0\,. \quad (5.2.7)$$

In order to show Eq. (5.2.2) we need to prove that if $C_{ijk\ell}\xi_{ij}\xi_{k\ell} = 0$ then $\xi_{ij} = 0$. Indeed if Eq. (5.2.7) is zero, then from the positive definiteness of C_{pqmn} we get

$$\xi_{pq} + \varepsilon_{pq}(\chi^{ij}\xi_{ij}) = 0\,. \quad (5.2.8)$$

Integrating Eq. (5.2.8) on a cell Ω and using the conditions (Eqs. (5.1.26)–(5.1.28)) we obtain $\xi_{pq} = 0$.

5.3. Application to Fiber-Reinforced Composite

Fiber-reinforced composites have been developed for use in various structures such as aircrafts and sports equipments. To predict the effective strength of a composite, earlier theories were based on the approximation that the composite property is the volume-weighted mean or harmonic mean of the properties of the components. These and other approximations can be found in Tsai (1987) and Vinson and Sierakowski (1990). We summarize the results of Tong and Mei (1992) who employed the method of homogenization.

Consider a periodic array of long fibers imbedded in and perfectly bonded to an elastic matrix as shown in Fig. 5.1.

We choose the Ω-cell to be a hexagon with a circular fiber at the center. Both the fiber and the matrix are assumed to be homogeneous and

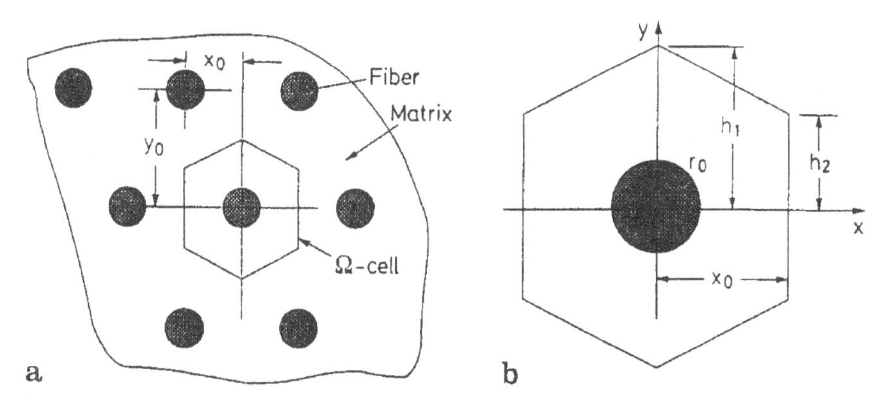

Fig. 5.1. Fiber-reinforced composite.

isotropic. The dimensions of the hexagon can be expressed in terms of the fiber spacing (x_0, y_0) by

$$h = \frac{1}{2}\sqrt{x_0^2 + y_0^2}\,, \quad h_1 = \frac{1}{2}\left(y_0 + \frac{x_0^2}{y_0}\right), \quad h_2 = \frac{1}{2}\left(y_0 - \frac{x_0^2}{y_0}\right). \tag{5.3.1}$$

A material with this microscale geometry is monoclinically symmetric with respect to the plane $z = 0$. The volume fraction of the fibers is

$$\theta_f = \frac{\pi r_0^2}{2(h_1 + h_2)x_0} = \frac{\pi r_0^2}{2x_0 y_0}\,. \tag{5.3.2}$$

Computations have been made for a hexagon with $y_0/x_0 = \sqrt{2}$, $r_0/x_0 = 0.7044$ (i.e., $h_1/x_0 = 1.155$, $h_2/x_0 = 0.5744$). The volume fraction of fibers is $\theta_f = 0.45$. Discretizing the Ω-cell into a mesh of finite elements, the cell boundary-value problem is solved by a standard method in elastostatics. For glass fibers (f) in an epoxy matrix (m), the Young's moduli are $E_f = 81.2\,\text{GPa}$, $E_m = 3.4\,\text{GPa}$, the fiber Poisson ratio is $\nu_f = 0.2$. Figure 5.2 shows the computed transverse Young's moduli E_{xx}, E_{yy}, the shear moduli E_{yz}, E_{yz}, and Poisson couplings[1] ν_{xy}, ν_{yz}, ν_{yz}, as functions of matrix Poisson ratio ν_m. For comparison the predictions by the mixture rule are simply

$$E_{zz} = \theta_f E_f + \theta_m E_m\,, \quad \frac{1}{E_{xx}} = \frac{1}{E_{yy}} = \frac{\theta_f}{E_f} + \frac{\theta_m}{E_m}\,,$$
$$\frac{1}{E_{yz}} = \frac{1}{E_{zx}} = \frac{\theta_f}{G_f} + \frac{\theta_m}{G_m}\,. \tag{5.3.3}$$

For the glass–epoxy composite $E_{xx} = E_{yy} = 5.98\,\text{GPa} = \text{constant}$. The shear moduli $E_{yz} = E_{xz}$ are much lower, as shown in Fig. 5.2.

Tong and Mei (1992) have extended the homogenization theory further to a composite of three scales, composed of periodic laminates each of which is fiber-reinforced composite.

5.4. Elastic Panels with Periodic Microstructure

Elastic panels with periodic microstructure are present in many engineering applications. The cellular or honeycomb structures are one example which

---

[1]The Poisson's coupling ratio ν_{ij}^* is defined for the transverse strain in direction j when the stress is applied in direction i only, with all other stress components vanishing, i.e., $\nu_{ij}^* = -\varepsilon_{jj}/\varepsilon_{ii}$ for $\sigma_{ii} = \sigma$. Here it is further redefined as $\nu_{ij} \equiv \nu_{ij}^*/E_{ii}$ and $\nu_{ji} \equiv \nu_{ji}^*/E_{jj}$ so that $\nu_{ij} = \nu_{ji}$ in order that the compliance matrix is symmetric (Cristescu *et al.*, 2004).

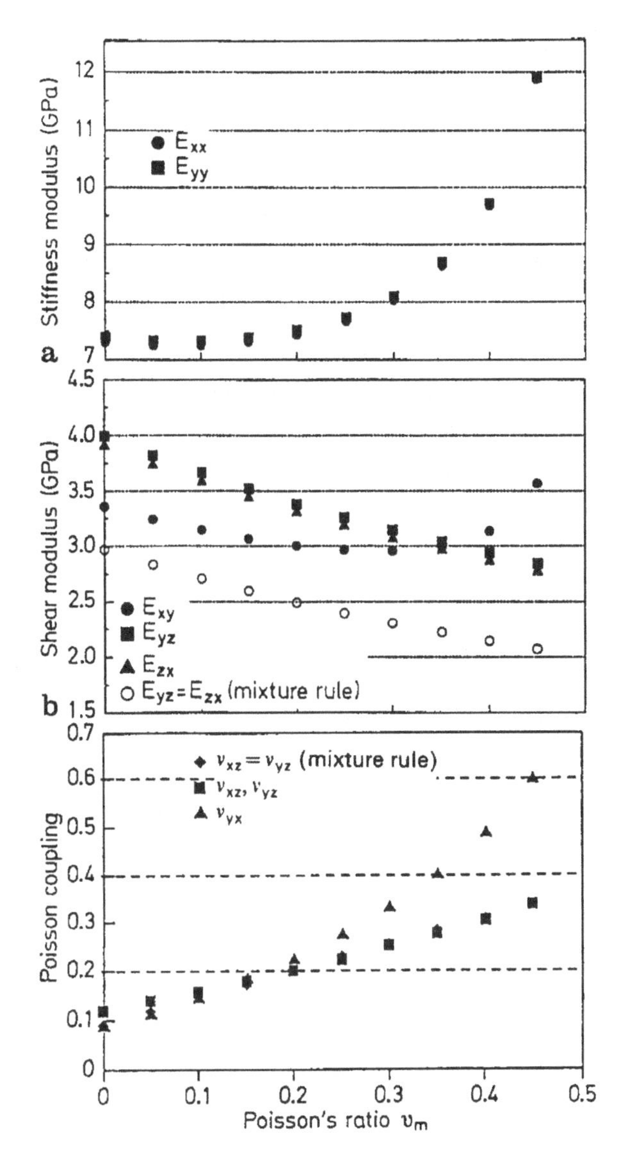

Fig. 5.2. Effective elastic moduli for a glass–epoxy composite. Comparison with predictions by the mixture rule. From Tong and Mei (1992), *Comp. Math. Eng. Sci.*

are effective and economical due to their high strength-to-weight ratio. In design applications these structures are analyzed by macroscale equations and boundary conditions, hence deriving the effective constitutive coefficients is of great importance. A general treatise on cellular structures has been given by Gibson and Ashley (1997). Homogenization theory was first used to derive rigorously the effective equations for periodic plates by Caillerie (1984) and Kohn and Vogelius (1984). Numerous results have since been obtained and published in several mathematical monographs (Gilbert and Hackl, 1995; Parton and Kudryavtsev, 1993; Cioranescu and Saint Jean Paulin, 1999; Lewinsky and Telega, 2000).

For honeycombs of infinite height, Shi and Tong (1995) have obtained the effective properties by homogenization and solved the two-dimensional cell problems analytically, treating the cell walls as thin plates. Honeycomb panels usually consist of two-dimensional periodic cells of finite height, sandwiched between two parallel plates, as shown in Fig. 5.3. The derivation of two-dimensional effective equation is not trivial. The corresponding cell problems are three-dimensional and can be solved by approximate analytical and discrete numerical methods for the effective coefficients as in Xu and Qiao (2002). A comprehensive review of the large body of work in this area can be found in Höhe and Becker (2002).

We shall only demonstrate the derivation of the linearized effective equations for infinitesimal deformation here. Consider here a plane panel, composed of a large number of periodic cells Ω, with characteristic dimension ℓ which is much smaller than the dimension ℓ' of the panel, $\ell/\ell' \ll 1$. The thickness of the panel is assumed to have also the same characteristic length ℓ. Denoting by $\epsilon = \ell/\ell'$, we shall derive the effective equations for the structure when ϵ tends to zero. The boundary of the side walls of the cell Ω will be denoted by Γ and the top and bottom walls by $S^{\pm}$, i.e., $x_3 = \pm h/2$ where $h = \mathcal{O}(\ell)$.

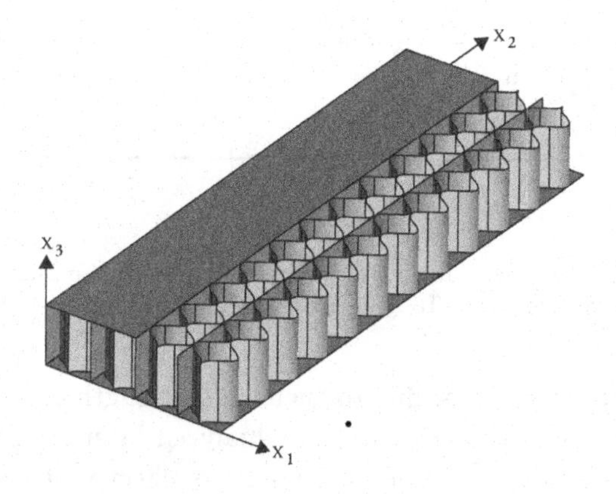

Fig. 5.3. A honeycomb panel.

5.4.1. *Order Estimates*

As a homogeneous plate should be a limiting case, let us deduce some order relations inherent in Kirchoff's theory (see Fung, 1965). For convenience, we define the transverse coordinate x_3 to be vertical and the in-plane coordinates (x_1, x_2) to be horizontal. The thickness of the plate is $h = \mathcal{O}(\ell)$ while the width and length of the plate are $\mathcal{O}(\ell')$ with $\ell/\ell' = \epsilon \ll 1$. For linearized approximation, the vertical deformation v_3 is assumed to be much smaller than the plate thickness h. Distinguishing the in-plane from the transverse components, Cauchy's equation of static equilibrium reads

$$\frac{\partial \sigma_{\alpha\beta}}{\partial x_\beta} + \frac{\partial \sigma_{\alpha 3}}{\partial x_3} = 0\,, \tag{5.4.1}$$

$$\frac{\partial \sigma_{3\alpha}}{\partial x_\alpha} + \frac{\partial \sigma_{33}}{\partial x_3} = 0 \tag{5.4.2}$$

at any interior point, where the Greek letters $\alpha, \beta = 1, 2$ signify the in-plane components only. Since $h = \mathcal{O}(\ell) = \mathcal{O}(\epsilon\ell')$, the contrast of length scales dictates that

$$\sigma_{33} = \mathcal{O}\left(\frac{\ell}{\ell'}\right)\sigma_{\alpha 3} = \mathcal{O}\left(\frac{\ell^2}{\ell'^2}\right)\sigma_{\alpha\beta}\,. \tag{5.4.3}$$

By Hooke's law the smallness of $\sigma_{\alpha 3}$ implies

$$v_\alpha = \mathcal{O}\left(\frac{\ell}{\ell'}v_3\right). \tag{5.4.4}$$

Hence,

$$\sigma_{\alpha\beta} = \mathcal{O}\left(E\frac{v_\alpha}{\ell'}\right) = \mathcal{O}\left(E\frac{\ell}{\ell'}\frac{v_\alpha}{\ell'}\right) = \mathcal{O}\left(E\frac{\ell^2}{\ell'^2}\frac{v_3}{\ell'}\right), \tag{5.4.5}$$

where E is the characteristic Young's modulus. Equation (5.4.3) then implies

$$\sigma_{\alpha 3} = \mathcal{O}\left(E\frac{\ell^3}{\ell'^3}\frac{v_3}{\ell}\right), \quad \sigma_{33} = \mathcal{O}\left(E\frac{\ell^4}{\ell'^4}\frac{v_3}{\ell}\right). \tag{5.4.6}$$

Let there be external shear stresses $\sigma_{\alpha 3}^{\pm}$ and normal stress $\sigma_{33}^{\pm}$ applied on the top and bottom surfaces $S^{\pm}$ of the panel. For continuity we must require that

$$\sigma_{\alpha 3}^{\pm} = \mathcal{O}\left(E\frac{\ell^3}{\ell'^3}\frac{v_3}{\ell}\right), \quad \text{and } \sigma_{33}^{\pm} = \mathcal{O}\left(E\frac{\ell^4}{\ell'^4}\frac{v_3}{\ell}\right), \quad \text{on } S^{\pm} : x_3 = \pm\frac{h}{2}\,. \tag{5.4.7}$$

Accordingly the surface boundary conditions on $S^{\pm}$ can be expressed as:

$$\sigma_{\alpha 3}^{\pm} = \epsilon^3 t_{\alpha}^{\pm}, \quad \sigma_{33}^{\pm} = \epsilon^4 q^{\pm}; \quad \text{on } S^{\pm} : x_3 = \pm\frac{h}{2}, \tag{5.4.8}$$

where the applied stresses t_{α}, q are of the order $\mathcal{O}(Ev_3/\ell)$.

5.4.2. *Two-Scale Analysis and Effective Equations*

Since the panel thickness is comparable to the cell size, we only need two-dimensional macroscopic coordinates $\boldsymbol{x}' = (x_1', x_2')$ for in-plane variations

$$x_1' = \epsilon x_1, \quad x_2' = \epsilon x_2. \tag{5.4.9}$$

We assume that the material properties can be inhomogeneous but periodic on the microscale. Let the following two-scale asymptotic expansion for the displacement be introduced,

$$\begin{aligned} \boldsymbol{v} = \boldsymbol{v}^{(0)}(x_1, x_2, x_3, x_1', x_2') + \epsilon \boldsymbol{v}^{(1)}(x_1, x_2, x_3, x_1', x_2') \\ + \epsilon^2 \boldsymbol{v}^{(2)}(x_1, x_2, x_3, x_1', x_2') + \cdots, \end{aligned} \tag{5.4.10}$$

with $\boldsymbol{v}^{(i)}$ periodic in (x_1, x_2). This yields an expansion for the strain tensor

$$\varepsilon_{\alpha\beta}(\boldsymbol{v}) = \varepsilon_{\alpha\beta}(\boldsymbol{v}^{(0)}) + \epsilon(\varepsilon_{\alpha\beta}'(\boldsymbol{v}^{(0)}) + \varepsilon_{\alpha\beta}(\boldsymbol{v}^{(1)})) + \cdots, \tag{5.4.11}$$

$$\varepsilon_{3\beta}(\boldsymbol{v}) = \varepsilon_{3\beta}(\boldsymbol{v}^{(0)}) + \epsilon\left(\frac{1}{2}\frac{\partial u_3^{(0)}}{\partial x_\beta'} + \varepsilon_{3\beta}(\boldsymbol{v}^{(1)})\right) + \cdots, \tag{5.4.12}$$

$$\varepsilon_{33}(\boldsymbol{v}) = \frac{\partial u_3^{(0)}}{\partial x_3} + \epsilon\frac{\partial u_3^{(1)}}{\partial x_3} + \cdots. \tag{5.4.13}$$

Recall that the Greek subscripts $\alpha, \beta, \ldots$ take only the values $1, 2$, while the Roman subscripts take values $1, 2, 3$. Formally this yields an expansion for the stress tensor of the form

$$\begin{aligned} \boldsymbol{\sigma} = \boldsymbol{\sigma}^{(0)}(x_i, x_\alpha') + \epsilon\boldsymbol{\sigma}^{(1)}(x_i, x_\alpha') + \epsilon^2\boldsymbol{\sigma}^{(2)}(x_i, x_\alpha') + \cdots, \\ \text{in } \Omega, \ i = 1, 3, \ \alpha = 1, 2, \end{aligned} \tag{5.4.14}$$

where

$$\sigma_{ij}^{(0)} = C_{ijk\ell}\frac{\partial v_\ell^{(0)}}{\partial x_k}, \tag{5.4.15}$$

$$\sigma_{ij}^{(1)} = C_{ijk\ell}\frac{\partial v_\ell^{(1)}}{\partial x_k} + C_{ij\alpha\beta}\varepsilon_{\alpha\beta}'(\boldsymbol{v}^{(0)}) + C_{ij3\beta}\frac{\partial v_3^{(0)}}{\partial x_\beta'}, \tag{5.4.16}$$

$$\sigma_{ij}^{(2)} = C_{ijk\ell}\frac{\partial v_k^{(2)}}{\partial x_\ell} + C_{ij\alpha\beta}\varepsilon'_{\alpha\beta}(\boldsymbol{v}^{(1)}) + C_{ij3\beta}\frac{\partial v_3^{(1)}}{\partial x'_\beta}\,. \tag{5.4.17}$$

Using the expansion (Eq. (5.4.14)) in Cauchy's equations, we get at different orders of ϵ

$$\frac{\partial \sigma_{ij}^{(0)}}{\partial x_j} = 0\,, \quad \text{in } \Omega\,, \tag{5.4.18}$$

$$\frac{\partial \sigma_{ij}^{(1)}}{\partial x_j} + \frac{\partial \sigma_{i\alpha}^{(0)}}{\partial x'_\alpha} = 0\,, \quad \text{in } \Omega\,, \tag{5.4.19}$$

$$\frac{\partial \sigma_{ij}^{(2)}}{\partial x_j} + \frac{\partial \sigma_{i\alpha}^{(1)}}{\partial x'_\alpha} = 0\,, \quad \text{in } \Omega\,, \tag{5.4.20}$$

$$\frac{\partial \sigma_{ij}^{(3)}}{\partial x_j} + \frac{\partial \sigma_{i\alpha}^{(2)}}{\partial x'_\alpha} = 0\,, \quad \text{in } \Omega\,, \tag{5.4.21}$$

$$\frac{\partial \sigma_{ij}^{(4)}}{\partial x_j} + \frac{\partial \sigma_{i\alpha}^{(3)}}{\partial x'_\alpha} = 0\,, \quad \text{in } \Omega\,. \tag{5.4.22}$$

Based on earlier estimates we impose the stress boundary conditions on $S^\pm : x_3 = \pm h/2$ as follows:

$$\sigma_{ij}^{(0)} n_j = 0\,, \quad \sigma_{ij}^{(1)} n_j = 0\,, \quad \sigma_{ij}^{(2)} n_j = 0\,; \tag{5.4.23}$$

$$\sigma_{\alpha j}^{(3)} n_j = t_\alpha^\pm\,, \quad \sigma_{3j}^{(3)} n_j = 0\,; \tag{5.4.24}$$

$$\sigma_{ij}^{(4)} n_j = q^\pm \delta_{3i}\,, \quad (\text{i.e., } \sigma_{\alpha j}^{(4)} = 0, \sigma_{33}^{(4)} = q^\pm)\,. \tag{5.4.25}$$

From Eq. (5.4.18), the first term in the displacement expansion can be shown not to depend on the microscale variables

$$\boldsymbol{v}^{(0)} = \boldsymbol{v}^{(0)}(x'_1, x'_2)\,, \tag{5.4.26}$$

as in the case studied in Sec. 5.1.

At the order $\mathcal{O}(\epsilon^1)$, from Eq. (5.4.19), we have

$$\frac{\partial}{\partial x_j}\left(C_{ijk\ell}\frac{\partial v_\ell^{(1)}}{\partial x_k}\right) = -\frac{\partial}{\partial x_j}\left(C_{ij\alpha\beta}\varepsilon'_{\alpha\beta}(\boldsymbol{v}^{(0)}) + C_{ij3\beta}\frac{\partial v_3^{(0)}}{\partial x'_\beta}\right), \quad \text{in } \Omega\,, \tag{5.4.27}$$

$$C_{ijk\ell}\frac{\partial v_\ell^{(1)}}{\partial x_k} n_j = -\left(C_{ij\alpha\beta}\varepsilon'_{\alpha\beta}(\boldsymbol{v}^{(0)}) + C_{ij3\beta}\frac{\partial v_3^{(0)}}{\partial x'_\beta}\right) n_j\,, \quad \text{on } \Gamma\,, \tag{5.4.28}$$

$$\boldsymbol{v}^{(1)} \text{ is } (x_1, x_2)\text{-periodic}\,. \tag{5.4.29}$$

The above problem has the solution of the form

$$\boldsymbol{v}^{(1)} = \frac{\partial v_n^{(0)}}{\partial x'_\alpha} \boldsymbol{\chi}^{n\alpha} + \boldsymbol{U}(x'_\alpha), \tag{5.4.30}$$

where the vector $\boldsymbol{U}(\boldsymbol{x}')$ is unknown and represents the depth-averaged displacement. For fixed (n, α), $\boldsymbol{\chi}^{n\alpha}$ is a vector defined by the cell problem

$$\frac{\partial}{\partial x_j}\left(C_{ijk\ell}\frac{\partial \chi_\ell^{n\alpha}}{\partial x_k}\right) = -\frac{\partial C_{ijn\alpha}}{\partial x_j} \quad \text{in } \Omega, \tag{5.4.31}$$

$$C_{ijk\ell}\frac{\partial \chi_\ell^{n\alpha}}{\partial x_k} n_j = -C_{ijn\alpha} n_j \quad \text{on } \Gamma, \tag{5.4.32}$$

$$\boldsymbol{\chi}^{n\alpha} \text{ is } (x_1, x_2)\text{-periodic}. \tag{5.4.33}$$

It follows that

$$\boldsymbol{\chi}^{31} = (-x_3, 0, 0), \quad \boldsymbol{\chi}^{32} = (0, -x_3, 0). \tag{5.4.34}$$

Let

$$A_{ij}^{n\alpha} = C_{ijk\ell}\frac{\partial \chi_\ell^{n\alpha}}{\partial x_k} + C_{ijn\alpha}, \tag{5.4.35}$$

then

$$A_{ij}^{3\alpha} = C_{ijk\ell}\varepsilon_{k\ell}(\boldsymbol{\chi}^{3\alpha}) + C_{ij3\alpha} = -C_{ij3\alpha} + C_{ij3\alpha} = 0. \tag{5.4.36}$$

From Eqs. (5.4.16) and (5.4.30) we get

$$\sigma_{ij}^{(1)} = A_{ij}^{\alpha\beta}\varepsilon'_{\alpha\beta}(\boldsymbol{v}^{(0)}). \tag{5.4.37}$$

At order $\mathcal{O}(\epsilon^2)$, we get, after averaging Eq. (5.4.20) over Ω,

$$\frac{\partial}{\partial x'_\alpha}\langle \sigma_{i\alpha}^{(1)} \rangle = 0. \tag{5.4.38}$$

This yields

$$\varepsilon'_{\alpha\beta}(\boldsymbol{v}^{(0)}) = 0. \tag{5.4.39}$$

Thus the leading-order displacement is

$$v_1^{(0)} = v_2^{(0)} = 0, \quad \text{and} \quad v_3^{(0)} = W(x'_1, x'_2), \tag{5.4.40}$$

which is dominated by the transverse component W. It follows from Eq. (5.4.30) that the displacement components at the next order are

$$v_\alpha^{(1)} = -x_3\frac{\partial W}{\partial x'_\alpha} + U_\alpha(x'_1, x'_2), \quad v_3^{(1)} = U_3(x'_1, x'_2). \tag{5.4.41}$$

Note that the in-plane displacement is linear in x_3 as assumed in Kirchoff's approximation for simple plates, and that the depth-averaged in-plane displacements (U_1, U_2) are of higher order than the leading-order transverse displacement W, to which U_3 is the next order correction. From Eqs. (5.4.37) and (5.4.17) we have

$$\sigma_{ij}^{(1)} = 0\,, \tag{5.4.42}$$

$$\sigma_{ij}^{(2)} = C_{ijk\ell}\frac{\partial v_\ell^{(2)}}{\partial x_k} + C_{ijk\beta}\frac{\partial U_k}{\partial x'_\beta} - x_3 C_{ij\alpha\beta}\frac{\partial}{\partial x'_\beta}\left(\frac{\partial W}{\partial x'_\alpha}\right). \tag{5.4.43}$$

Because of Eq. (5.4.42), Eq. (5.4.20) reduces to

$$\frac{\partial \sigma_{ij}^{(2)}}{\partial x_j} = 0\,, \tag{5.4.44}$$

which implies that $\boldsymbol{v}^{(2)}$ satisfies

$$\frac{\partial}{\partial x_j}\left(C_{ijk\ell}\frac{\partial v_\ell^{(2)}}{\partial x_k}\right) = -\frac{\partial}{\partial x_j}\left(C_{ijk\beta}\frac{\partial U_k}{\partial x'_\beta}\right) + C_{i3\alpha\beta}\left(\frac{\partial^2 W}{\partial x'_\alpha \partial x'_\beta}\right) + x_3\frac{\partial}{\partial x_j}\left(C_{ij\alpha\beta}\frac{\partial^2 W}{\partial x'_\alpha \partial x'_\beta}\right). \tag{5.4.45}$$

Thus $\boldsymbol{v}^{(2)}$ can be written in the form

$$\boldsymbol{v}^{(2)} = \frac{\partial U_k}{\partial x'_\beta}\boldsymbol{\chi}^{k\beta} + s_{\alpha\beta}(W)\boldsymbol{\tau}^{\alpha\beta}\,, \tag{5.4.46}$$

where $\boldsymbol{\chi}^{k\ell}$ are solutions to the cell problems (Eqs. (5.4.31)–(5.4.33)), and $\boldsymbol{\tau}^{\alpha\beta}$ the solution to another cell problem governed by

$$\frac{\partial}{\partial x_j}\left(C_{ijk\ell}\frac{\partial \tau_\ell^{\alpha\beta}}{\partial x_k}\right) = -C_{i3\alpha\beta} - x_3\frac{\partial C_{ij\alpha\beta}}{\partial x_j} \quad \text{in } \Omega\,, \tag{5.4.47}$$

$$C_{ijk\ell}\frac{\partial \tau_\ell^{\alpha\beta}}{\partial x_k}n_j = x_3 C_{ij\alpha\beta}n_j \quad \text{on } \Gamma\,, \tag{5.4.48}$$

$$\boldsymbol{\tau}^{\alpha\beta} \text{ is } (x_1, x_2)\text{-periodic}\,. \tag{5.4.49}$$

In Eq. (5.4.46) $s_{\alpha\beta}(W)$ denotes the flexural curvature

$$\boxed{s_{\alpha\beta}(W) = -\frac{\partial^2 W}{\partial x'_\alpha \partial x'_\beta}}\,. \tag{5.4.50}$$

With the displacements $\chi^{k\ell}$ and $\tau^{\alpha\beta}$ determined, the stress (Eq. (5.4.43)) takes the form

$$\sigma_{ij}^{(2)} = A_{ij}^{\alpha\beta}\varepsilon'_{\alpha\beta}(U) + B_{ij}^{\alpha\beta}s_{\alpha\beta}(W)\,, \tag{5.4.51}$$

where $A_{ij}^{k\beta}$ are defined by Eq. (5.4.35) and

$$B_{ij}^{\alpha\beta} = C_{ijk\ell}\frac{\partial\tau_\ell^{\alpha\beta}}{\partial x_k} + x_3 C_{ij\alpha\beta}\,. \tag{5.4.52}$$

Using the two-cell problems (Eqs. (5.4.31)–(5.4.33)) and (Eqs. (5.4.47)–(5.4.49)) we obtain by averaging over Ω that

$$\langle A_{i3}^{\alpha\beta}\rangle = \langle x_3 A_{i3}^{\alpha\beta}\rangle = \langle B_{i3}^{\alpha\beta}\rangle = \langle x_3 B_{i3}^{\alpha\beta}\rangle = 0\,, \tag{5.4.53}$$

and

$$\langle\sigma_{i3}^{(2)}\rangle = \langle x_3\sigma_{i3}^{(2)}\rangle = 0\,. \tag{5.4.54}$$

To obtain the macroscopic equations, we first compute the average of the stress $\sigma_{\gamma\delta}^{(2)}$ and the moment $x_3\sigma_{\gamma\delta}^{(2)}$

$$\langle\sigma_{\gamma\delta}^{(2)}\rangle = \mathcal{C}_{\gamma\delta\alpha\beta}\varepsilon'_{\alpha\beta}(U) + \mathcal{C}^*_{\gamma\delta\alpha\beta}s_{\alpha\beta}(W)\,, \tag{5.4.55}$$

$$\langle x_3\sigma_{\alpha\beta}^{(2)}\rangle = \mathcal{M}_{\gamma\delta\alpha\beta}\varepsilon'_{\alpha\beta}(U) + \mathcal{M}^*_{\gamma\delta\alpha\beta}s_{\alpha\beta}(W)\,. \tag{5.4.56}$$

The effective coefficients are given by

$$\mathcal{C}_{\gamma\delta\alpha\beta} = \langle A_{\gamma\delta}^{\alpha\beta}\rangle = \left\langle C_{\gamma\delta\alpha\beta} + C_{\gamma\delta\ell n}\frac{\partial\chi_n^{\alpha\beta}}{\partial x_\ell}\right\rangle, \tag{5.4.57}$$

$$\mathcal{C}^*_{\gamma\delta\alpha\beta} = \langle B_{\gamma\delta}^{\alpha\beta}\rangle = \left\langle x_3 C_{\gamma\delta\alpha\beta} + C_{\gamma\delta k\ell}\frac{\partial\tau_\ell^{\alpha\beta}}{\partial x_k}\right\rangle, \tag{5.4.58}$$

$$\mathcal{M}_{\gamma\delta\alpha\beta} = \langle x_3 A_{\gamma\delta}^{\alpha\beta}\rangle\,, \tag{5.4.59}$$

$$\mathcal{M}^*_{\gamma\delta\alpha\beta} = \langle x_3 B_{\gamma\delta}^{\alpha\beta}\rangle\,, \tag{5.4.60}$$

and can be shown to satisfy the usual symmetry conditions. Thus, at the macroscopic level we get from the average of Eq. (5.4.21),

$$\frac{\partial}{\partial x'_\delta}\langle\sigma_{i\delta}^{(2)}\rangle + \frac{1}{\Omega}\iint_\Gamma \sigma_{ij}^{(3)}n_j d\Gamma = 0\,, \tag{5.4.61}$$

where $\Omega = S_o h$ denotes the cell volume and S_o the area of the horizontal cross-section. Now we apply Eq. (5.4.54), the boundary condition

(Eq. (5.4.24)) for $\sigma^{(3)}_{\gamma\beta}$ and the (x_1, x_2) periodicity to get,

$$\frac{\partial}{\partial x'_\delta}\left(C_{\gamma\delta\alpha\beta}\varepsilon'_{\alpha\beta}(\boldsymbol{U}) + C^*_{\gamma\delta\alpha\beta}s_{\alpha\beta}(W)\right) + \frac{1}{\Omega}\iint_S (t^+_\gamma + t^-_\gamma)dx_1 dx_2 = 0\,, \tag{5.4.62}$$

where $S = S^+ \cup S$, or

$$\boxed{\frac{\partial}{\partial x'_\delta}\left(C_{\gamma\delta\alpha\beta}\varepsilon'_{\alpha\beta}(\boldsymbol{U}) + C^*_{\gamma\delta\alpha\beta}s_{\alpha\beta}(W)\right) + \frac{T_\gamma}{h} = 0\,, \quad \gamma = 1, 2}\,, \tag{5.4.63}$$

where T_γ is the area-averaged tangential traction applied on the surfaces,

$$T_\gamma = \frac{1}{S_o}\iint_S (t^+_\gamma + t^-_\gamma)dx_1 dx_2\,. \tag{5.4.64}$$

Multiplying Eq. (5.4.21) by x_3 and averaging over Ω, we get

$$\frac{\partial}{\partial x'_\delta}\langle x_3\sigma^{(2)}_{\gamma\delta}\rangle + \frac{1}{\Omega}\iiint_\Omega x_3\frac{\partial\sigma^{(3)}_{\gamma j}}{\partial x_j}d\Omega = 0\,. \tag{5.4.65}$$

The last integral can be further expanded as

$$\begin{aligned}&\iiint_\Omega \frac{\partial}{\partial x_j}(x_3\sigma^{(3)}_{\gamma j})d\Omega - \iiint_\Omega \sigma^{(3)}_{\gamma 3}d\Omega\\ &\quad = \iint_S x_3(t^+_\gamma + t^-_\gamma)dx_1 dx_2 - \Omega\langle\sigma^{(3)}_{\gamma 3}\rangle\,,\end{aligned} \tag{5.4.66}$$

and thus

$$\frac{\partial}{\partial x'_\delta}\langle x_3\sigma^{(2)}_{\gamma\delta}\rangle + \frac{1}{\Omega}\iint_S x_3(t^+_\gamma + t^-_\gamma)dx_1 dx_2 - \langle\sigma^{(3)}_{\gamma 3}\rangle = 0\,. \tag{5.4.67}$$

From Eq. (5.4.22)

$$\frac{\partial}{\partial x'_\alpha}\langle\sigma^{(3)}_{i\alpha}\rangle + \frac{1}{\Omega}\iint_{\partial\Omega}\sigma^{(4)}_{ij}n_j dS = 0\,, \tag{5.4.68}$$

and $\langle\sigma^{(3)}_{\gamma 3}\rangle$ can be eliminated between Eqs. (5.4.67) and (5.4.68) to get

$$\begin{aligned}&\frac{\partial}{\partial x'_\gamma}\left(\frac{\partial}{\partial x'_\delta}\langle x_3\sigma^{(2)}_{\gamma\delta}\rangle + \frac{1}{\Omega}\iint_S x_3(t^+_\alpha + t^-_\alpha)dx_1 dx_2\right)\\ &\quad - \frac{1}{\Omega}\iint_S (q^+ + q^-)dx_1 dx_2 = 0\,.\end{aligned} \tag{5.4.69}$$

Use has been made of Eq. (5.4.23). In more explicit form, it reads,

$$\boxed{\frac{\partial}{\partial x'_\gamma}\left(\frac{\partial}{\partial x'_\delta}\left(\mathcal{M}_{\gamma\delta\alpha\beta}\varepsilon'_{\alpha\beta}(\boldsymbol{U})+\mathcal{M}^*_{\gamma\delta\alpha\beta}s_{\alpha\beta}(W)\right)+\frac{M_\gamma}{h}\right)-\frac{Q}{h}=0}\,. \tag{5.4.70}$$

The effective coefficients are defined in Eqs. (5.4.57)–(5.4.60) in terms of the solutions of the cell problems (Eqs. (5.4.31)–(5.4.33)) and (Eqs. (5.4.47)–(5.4.49)), where M is the area-averaged surface torque,

$$M_\gamma = \frac{1}{S_o}\iint_S x_3(t^+_\gamma + t^-_\gamma)dx_1dx_2\,, \tag{5.4.71}$$

and Q is the area-averaged transverse stress

$$Q = \frac{1}{S_o}\iint_S (q^+ + q^-)dx_1dx_2 \tag{5.4.72}$$

applied externally.

In summary, the panel is governed by the coupled system of three equations (5.4.63) and (5.4.70) for the three unknowns U_1, U_2, and W. They are first derived via homogenization theory by Caillerie (1984) (see also Lewinsky and Telega, 2000; Parton and Kudryavtsev, 1993).

These equations are extensions of Kirchoff's equations for a homogeneous plate. They must be supplemented by macroscale boundary conditions. For example if all panel edges are clamped, we must require

$$U_1 = U_2 = W = 0\,, \quad \frac{\partial W}{\partial x_\alpha}n_\alpha = 0\,, \tag{5.4.73}$$

along the panel edge of the plate, where $\boldsymbol{n} = \{n_\alpha\}$ is the two-dimensional unit normal to the edge. Effective moduli have been calculated by approximate analytical as well as finite element solutions of the cell problems (see, e.g., Xu and Qiao, 2002). For the case of non-planar panels, the derivation of the effective equations is similar but the results are complicated, and the solutions to the cell problems are more computationally involved (Parton and Kudryavtsev, 1993). A mathematical treatment of plates and shells can be found in Lewinsky and Telega (2000).

5.4.3. *Homogeneous Plate — A Limiting Case*

In the limiting case when the panel is just a plate formed by an isotropic elastic material with no microstructure, the macroscopic equations (5.4.63) and (5.4.70) reduce to the well known Kirchoff plate equations.

In the isotropic case, the elastic tensor has only the following nonzero entries:

$$C_{1111} = C_{2222} = C_{3333} = \lambda + 2G\,, \tag{5.4.74}$$

$$C_{1122} = C_{1133} = C_{2233} = \lambda\,, \tag{5.4.75}$$

$$C_{1212} = C_{1313} = C_{2323} = G\,, \tag{5.4.76}$$

as well as their symmetric counterparts.

Now the cell problems (Eqs. (5.4.31)–(5.4.33)) become

$$\frac{\partial}{\partial x_j}\left(C_{ijk\ell}\frac{\partial \chi_\ell^{\alpha\beta}}{\partial x_k}\right) = 0 \quad \text{in } \Omega\,, \tag{5.4.77}$$

$$C_{i3k\ell}\frac{\partial \chi_\ell^{\alpha\beta}}{\partial x_k} = -C_{i3\alpha\beta} \quad \text{on } S^{\pm}\,, \tag{5.4.78}$$

$$\boldsymbol{\chi}^{n\alpha} \text{ is } (x_1, x_2)\text{-periodic}\,. \tag{5.4.79}$$

Since the plate does not have a microstructure, the solutions do not depend on (x_1, x_2); thus Eqs. (5.4.77) and (5.4.78) become

$$\frac{\partial}{\partial x_3}\left(C_{i33\ell}\frac{\partial \chi_\ell^{\alpha\beta}}{\partial x_3}\right) = 0 \quad \text{in } \Omega\,, \tag{5.4.80}$$

$$C_{i33\ell}\frac{\partial \chi_\ell^{\alpha\beta}}{\partial x_3} = -C_{i3\alpha\beta} \quad \text{on } S^{\pm}\,. \tag{5.4.81}$$

The right-hand side of Eq. (5.4.81) is not zero only for $(\alpha, \beta) = (1, 1)$ or $(\alpha, \beta) = (2, 2)$, and $i = 3$ for which nonzero solutions are obtained

$$\chi_3^{11} = \chi_3^{22} = -\frac{C_{3311}}{C_{3333}}x_3 = -\frac{\lambda}{\lambda + 2G}x_3\,. \tag{5.4.82}$$

Similar conclusions can be drawn from Eqs. (5.4.47)–(5.4.49), and thus

$$\boldsymbol{\chi}^{11} = \boldsymbol{\chi}^{22} = \left(0, 0, -\frac{\lambda}{\lambda + 2G}x_3\right), \tag{5.4.83}$$

$$\boldsymbol{\tau}^{11} = \boldsymbol{\tau}^{22} = \left(0, 0, -\frac{\lambda}{2(\lambda + 2G)}x_3^2\right). \tag{5.4.84}$$

The effective coefficients can be readily deduced from Eqs. (5.4.57)–(5.4.60),

$$\mathcal{C}_{1111} = \mathcal{C}_{2222} = \frac{4G(\lambda + G)}{\lambda + 2G} = \frac{E}{1 - \nu^2}\,, \tag{5.4.85}$$

$$\mathcal{C}_{1122} = \mathcal{C}_{2211} = \frac{2G\lambda}{\lambda + 2G} = \frac{E\nu}{1 - \nu^2}\,, \tag{5.4.86}$$

$$\mathcal{C}_{1212} = \mathcal{C}_{2121} = G = \frac{E}{2(1+\nu)}\,, \tag{5.4.87}$$

$$\mathcal{C}^*_{\gamma\delta\alpha\beta} = \langle x_3 \rangle A^{\alpha\beta}_{\gamma\delta} = 0\,, \tag{5.4.88}$$

$$\mathcal{M}_{\gamma\delta\alpha\beta} = 0\,, \tag{5.4.89}$$

$$\mathcal{M}^*_{\gamma\delta\alpha\beta} = \langle x_3^2 \rangle \mathcal{C}_{\gamma\delta\alpha\beta} = \frac{h^2}{12}\mathcal{C}_{\gamma\delta\alpha\beta} \tag{5.4.90}$$

as, for instance,

$$\mathcal{C}_{1111} = \left\langle C_{1111} + C_{1133}\frac{\partial \chi_3^{11}}{\partial x_3} \right\rangle = \frac{4G(\lambda+G)}{\lambda+2G}\,. \tag{5.4.91}$$

E and ν are Young's modulus and Poisson's ratio, respectively.

Equations (5.4.63) and (5.4.70) are no longer coupled from Eqs. (5.4.88) and (5.4.89). From Eq. (5.4.70), we get

$$\frac{h^3}{12}\left\{\mathcal{C}_{1111}\frac{\partial^4 W}{\partial {x'_1}^4} + \mathcal{C}_{2222}\frac{\partial^4 W}{\partial {x'_2}^4} + 2(\mathcal{C}_{1122}+\mathcal{C}_{1212}+\mathcal{C}_{2121}+\mathcal{C}_{2211})\frac{\partial^4 W}{\partial {x'_1}^2 \partial {x'_2}^2}\right\}$$
$$+\frac{\partial M_\gamma}{\partial x_\gamma} - Q = 0\,, \tag{5.4.92}$$

which becomes, after using Eqs. (5.4.85)–(5.4.87)

$$\boxed{\frac{Eh^3}{12(1-\nu^2)}\left\{\frac{\partial^4 W}{\partial {x'_1}^4} + 2\frac{\partial^4 W}{\partial {x'_1}^2 \partial {x'_2}^2} + \frac{\partial^4 W}{\partial {x'_2}^4}\right\} + \frac{\partial M_\gamma}{\partial x_\gamma} - Q = 0}\,. \tag{5.4.93}$$

From Eq. (5.4.63), we get

$$\mathcal{C}_{1111}\frac{\partial^2 U_1}{\partial {x'_1}^2} + \mathcal{C}_{1122}\frac{\partial^2 U_2}{\partial x'_1 \partial x'_2} + \mathcal{C}_{1212}\frac{\partial}{\partial x'_2}\left(\frac{\partial U_1}{\partial x'_2} + \frac{\partial U_2}{\partial x'_1}\right) + T_1 = 0\,, \tag{5.4.94}$$

and

$$\mathcal{C}_{2121}\frac{\partial}{\partial x'_1}\left(\frac{\partial U_1}{\partial x'_2} + \frac{\partial U_2}{\partial x'_1}\right) + \mathcal{C}_{2211}\frac{\partial}{\partial x'_2}\frac{\partial U_1}{\partial x'_1} + \mathcal{C}_{2222}\frac{\partial^2 U_2}{\partial {x'_2}^2} + T_2 = 0\,, \tag{5.4.95}$$

which becomes

$$\boxed{Eh\left\{\frac{\partial^2 U_1}{\partial {x'_1}^2} + \frac{1-\nu}{2}\frac{\partial U_1}{\partial {x'_2}^2} + \frac{1+\nu}{2}\frac{\partial^2 U_2}{\partial x'_1 \partial x'_2}\right\} + (1-\nu^2)T_1 = 0}\,, \tag{5.4.96}$$

and

$$\boxed{Eh\left\{\frac{\partial^2 U_2}{\partial {x'_2}^2}+\frac{1-\nu}{2}\frac{\partial U_2}{\partial {x'_1}^2}+\frac{1+\nu}{2}\frac{\partial^2 U_1}{\partial x'_1\partial x'_2}\right\}+(1-\nu^2)T_2=0}\,. \tag{5.4.97}$$

Equations (5.4.93), (5.4.96), and (5.4.97) are the classical results of Kirchoff (see Fung, 1965).

5.5. Variational Principles and Bounds for the Elastic Moduli

In addition to numerical solution of cell problems, a large theoretical literature has been dedicated to estimating the effective properties of elastic composites. Voigt (1889) was the first to estimate the effective properties by assuming that the average strain of each phase is equal to the applied strain, which yields the arithmetic average estimate for the effective moduli. Later, Reuss (1929) assumed that the average stress of each phase is equal to the applied stress, which yields the harmonic mean estimate for the effective moduli. It is due to Hill (1952) that these estimates were proven to be the upper and lower bounds, respectively, for the elastic moduli.

In this section we will deduce two variational principles: the potential energy minimization principle and the complementary energy minimization principle. These will imply the elementary bounds given by the harmonic and arithmetic means for the effective elasticity tensor $\boldsymbol{C} \equiv \{C_{ijk\ell}\}$:

$$\langle \boldsymbol{C}^{-1}\rangle^{-1} \le \boldsymbol{C} \le \langle \boldsymbol{C}\rangle\,, \tag{5.5.1}$$

which means, in component form,

$$\left\langle {C_{ijk\ell}}^{-1}\right\rangle^{-1}\xi_{ij}\xi_{k\ell} \le C_{ijk\ell}\xi_{ij}\xi_{k\ell} \le \langle C_{ijk\ell}\rangle\xi_{ij}\xi_{k\ell} \tag{5.5.2}$$

for any second-order tensor $\boldsymbol{\xi} = \{\xi_{ij}\}$. For convenience the inverse and other properties of the fourth-rank tensor are discussed in Appendix 5A.

5.5.1. *First Variational Principle and the Upper Bound*

Let us define the following functional:

$$\mathcal{F}(\boldsymbol{v}) = \frac{1}{\Omega}\iiint_\Omega C_{pq\ell m}(\xi_{pq}+\varepsilon_{pq}(\boldsymbol{v}))(\xi_{\ell m}+\varepsilon_{\ell m}(\boldsymbol{v}))d\Omega\,, \tag{5.5.3}$$

where ξ_{ij} is any constant tensor of the second rank. We will show that

$$\boxed{\mathcal{C}_{ijk\ell}\xi_{ij}\xi_{k\ell} = \min_{\boldsymbol{v}} \mathcal{F}(\boldsymbol{v})}\,. \tag{5.5.4}$$

One inequality is immediate; indeed from Eq. (5.2.7) we have

$$\mathcal{C}_{ijk\ell}\xi_{ij}\xi_{k\ell} = \mathcal{F}(\boldsymbol{\chi}^{ij}\xi_{ij}) \geq \min_{\boldsymbol{v}} \mathcal{F}(\boldsymbol{v})\,, \tag{5.5.5}$$

and therefore we only need to prove the opposite inequality:

$$\mathcal{C}_{ijk\ell}\xi_{ij}\xi_{k\ell} \leq \min_{\boldsymbol{v}} \mathcal{F}(\boldsymbol{v})\,. \tag{5.5.6}$$

Let us denote by $\boldsymbol{\delta} = \boldsymbol{\chi}^{ij}\xi_{ij} - \boldsymbol{v}$, then

$$\begin{aligned}\mathcal{F}(\boldsymbol{v}) = \mathcal{F}(\boldsymbol{\chi}^{ij}\xi_{ij} - \boldsymbol{\delta}) &= \langle C_{pq\ell m}(\xi_{pq} + \varepsilon_{pq}(\boldsymbol{\chi}^{ij}\xi_{ij}))(\xi_{\ell m} + \varepsilon_{\ell m}(\boldsymbol{\chi}^{ij}\xi_{ij}))\rangle \\ &\quad + \langle C_{pq\ell m}\varepsilon_{pq}(\boldsymbol{\delta})\varepsilon_{\ell m}(\boldsymbol{\delta})\rangle \\ &\quad - 2\langle C_{pq\ell m}(\xi_{pq} + \varepsilon_{pq}(\boldsymbol{\chi}^{ij}\xi_{ij}))\varepsilon_{\ell m}(\boldsymbol{\delta})\rangle\,.\end{aligned} \tag{5.5.7}$$

The last term in Eq. (5.5.7) can be written as

$$\begin{aligned}&2\langle C_{pq\ell m}(\xi_{pq} + \varepsilon_{pq}(\boldsymbol{\chi}^{ij}\xi_{ij}))\varepsilon_{\ell m}(\boldsymbol{\delta})\rangle \\ &\quad = 2\xi_{ij}\langle C_{pq\ell m}(\delta_{pi}\delta_{qj} + \varepsilon_{pq}(\boldsymbol{\chi}^{ij}))\varepsilon_{\ell m}(\boldsymbol{\delta})\rangle \\ &\quad = \xi_{ij}\iint_{\partial\Omega} C_{pq\ell m}(\delta_{pi}\delta_{qj} + \varepsilon_{pq}(\boldsymbol{\chi}^{ij}))\delta_\ell n_m d\Gamma \\ &\qquad - \left\langle \delta_\ell \frac{\partial}{\partial x_m}(C_{pq\ell m}(\delta_{pi}\delta_{qj} + \varepsilon_{pq}(\boldsymbol{\chi}^{ij})))\right\rangle,\end{aligned} \tag{5.5.8}$$

where the integral on Γ is canceled due to Eq. (5.1.27). The surface integral above is zero because of periodicity on opposite faces of Ω and the second term is zero from the cell problem (Eq. (5.1.25)). Using the positivity property (Eq. (5.1.5)) in Eq. (5.5.7), we get

$$\begin{aligned}\mathcal{C}_{pq\ell m}\xi_{pq}\xi_{\ell m} &= \langle C_{pq\ell m}(\xi_{pq} + \varepsilon_{pq}(\boldsymbol{\chi}^{ij}\xi_{ij}))(\xi_{\ell m} + \varepsilon_{\ell m}(\boldsymbol{\chi}^{ij}\xi_{ij}))\rangle \\ &= \mathcal{F}(\boldsymbol{v}) - \langle C_{pq\ell m}\varepsilon_{pq}(\boldsymbol{\delta})\varepsilon_{\ell m}(\boldsymbol{\delta})\rangle \leq \mathcal{F}(\boldsymbol{v})\,.\end{aligned} \tag{5.5.9}$$

Since the above inequality holds for any $\boldsymbol{v}$, the inequality (Eq. (5.5.6)) holds.

An upper bound on the effective elastic tensor can be obtained now, from the variational principle, using suitable choices for the field $\boldsymbol{v}$ in the variational principle (Eq. (5.5.4)). Indeed the simplest choice is a constant displacement field $\boldsymbol{v}$ that yields the upper bound:

$$\mathcal{C}_{pq\ell m}\xi_{pq}\xi_{\ell m} \leq \langle C_{pq\ell m}\rangle\xi_{pq}\xi_{\ell m}\,, \tag{5.5.10}$$

where $\langle C_{pq\ell m}\rangle$ is the arithmetic average of the elasticity tensor, and thus the arithmetic mean upper bound is obtained. For simplicity we write inequality (Eq. (5.5.10)) as

$$\mathcal{C} \leq \langle C\rangle\,. \tag{5.5.11}$$

5.5.2. *Second Variational Principle and the Lower Bound*

Using the characterization (Eq. (5.2.6)) for the effective elasticity tensor, we will develop a variational principle for its inverse, denoted by $\mathcal{C}^{-1}$.

Consider the functional

$$\mathcal{J}(\boldsymbol{\tau}) = \frac{1}{\Omega}\iiint_{\Omega} C^{-1}_{ijkl}\tau_{ij}\tau_{k\ell}d\Omega \tag{5.5.12}$$

defined for all solenoidal, symmetric tensors $\boldsymbol{\tau}$, with given average $\boldsymbol{\sigma}$

$$\nabla\cdot\boldsymbol{\tau} = 0\,,\quad \tau_{ij} = \tau_{ji}\,,\quad \langle\tau_{ij}\rangle = \sigma_{ij}\,. \tag{5.5.13}$$

The second (or complementary) variational principle is:

$$\boxed{\mathcal{C}^{-1}_{ijk\ell}\sigma_{ij}\sigma_{k\ell} = \min_{\boldsymbol{\tau}}\mathcal{J}(\boldsymbol{\tau})}\,. \tag{5.5.14}$$

Let us start by showing, in a similar way as in Sec. 2.6.2, that in general for any positive, fourth-order tensor $(A_{ijk\ell})$, the following is true

$$\frac{1}{2}A_{ijk\ell}\xi_{ij}\xi_{k\ell} = \max_{\boldsymbol{\tau}}\left(\xi_{ij}\tau_{ij} - \frac{1}{2}A^{-1}_{ijk\ell}\tau_{ij}\tau_{k\ell}\right)\,. \tag{5.5.15}$$

First we show that the inequality

$$\frac{1}{2}A_{ijk\ell}\xi_{ij}\xi_{k\ell} \geq \left(\xi_{ij}\tau_{ij} - \frac{1}{2}A^{-1}_{ijk\ell}\tau_{ij}\tau_{k\ell}\right) \tag{5.5.16}$$

holds for any $\boldsymbol{\xi}$ and $\boldsymbol{\tau}$. For that we choose $\boldsymbol{\eta}$ in such a way that $\xi_{ij} = \eta_{ij} + A^{-1}_{ijpq}\tau_{pq}$ and compute

$$\begin{aligned}
\frac{1}{2}A_{ijk\ell}\xi_{ij}\xi_{kl} &= \frac{1}{2}A_{ijk\ell}\eta_{ij}\eta_{k\ell} + A_{ijk\ell}A^{-1}_{k\ell st}\eta_{ij}\tau_{st} + \frac{1}{2}A_{ijk\ell}A^{-1}_{ijpq}A^{-1}_{k\ell st}\tau_{pq}\tau_{st}\\
&= \frac{1}{2}A_{ijk\ell}\eta_{ij}\eta_{k\ell} + \eta_{ij}\tau_{ij} + \frac{1}{2}A^{-1}_{ijpq}\tau_{ij}\tau_{pq}\\
&= \frac{1}{2}A_{ijk\ell}\eta_{ij}\eta_{k\ell} + \xi_{ij}\tau_{ij} - \frac{1}{2}A^{-1}_{ijpq}\tau_{ij}\tau_{pq}\,.
\end{aligned} \tag{5.5.17}$$

Because of this result, Eq. (5.5.16) follows immediately, as the first term in the equality above is positive.

Second, we see from Eq. (5.5.17), that for $\boldsymbol{\tau}$ chosen such that $\xi_{ij} = A^{-1}_{ijpq}\tau_{pq}$ (i.e., when $\eta_{ij} = 0$) the equality in Eq. (5.5.16) holds. This completes the proof of Eq. (5.5.15).

We now use the characterization of the effective elasticity tensor (Eq. (5.2.6)) and the inequality (Eq. (5.5.16)) with ξ_{ij} replaced by $\xi_{pq} + \varepsilon_{pq}(\chi^{ij}\xi_{ij})$, and obtain

$$\begin{aligned}\frac{1}{2}\mathcal{C}_{ijkl}\xi_{ij}\xi_{kl} &= \frac{1}{2}\langle C_{pqmn}(\xi_{pq} + \varepsilon_{pq}(\chi^{ij}\xi_{ij}))(\xi_{mn} + \varepsilon_{mn}(\chi^{kl}\xi_{kl}))\rangle \\ &\geq \left\langle (\xi_{mn} + \varepsilon_{mn}(\chi^{ij}\xi_{ij}))\tau_{mn} - \frac{1}{2}C^{-1}_{pqmn}\tau_{pq}\tau_{mn}\right\rangle \\ &= \langle \xi_{mn}\tau_{mn}\rangle + \langle \varepsilon_{mn}(\chi^{ij}\xi_{ij})\tau_{mn}\rangle - \frac{1}{2}\left\langle C^{-1}_{pqmn}\tau_{pq}\tau_{mn}\right\rangle .\end{aligned} \tag{5.5.18}$$

The second term above cancels due to the choice (Eq. (5.5.13)) of admissible tensors $\boldsymbol{\tau}$. Equality holds only for $\xi_{pq} + \varepsilon_{pq}(\chi^{ij}\xi_{ij}) = C^{-1}_{pqk\ell}\tau_{k\ell}$, or, in other words for $C_{mnpq}(\xi_{pq} + \varepsilon_{pq}(\chi^{ij}\xi_{ij})) = \tau_{mn}$ which implies that

$$\langle \tau_{mn}\rangle = \langle C_{mnpq}(\delta_{pi}\delta_{qj} + \varepsilon_{pq}(\chi^{ij}))\rangle \xi_{ij} = \mathcal{C}_{mnij}\xi_{ij} \,. \tag{5.5.19}$$

The inequality (Eq. (5.5.18)) can then be written as

$$\begin{aligned}\frac{1}{2}\mathcal{C}_{ijk\ell}\xi_{ij}\xi_{k\ell} &= \max_{\boldsymbol{\tau}}\left(\mathcal{C}_{mnij}\xi_{mn}\xi_{ij} - \frac{1}{2}\left\langle C^{-1}_{pqmn}\tau_{pq}\tau_{mn}\right\rangle\right) \\ &= \mathcal{C}_{mnij}\xi_{mn}\xi_{ij} - \min_{\boldsymbol{\tau}}\frac{1}{2}\left\langle C^{-1}_{pqmn}\tau_{pq}\tau_{mn}\right\rangle ,\end{aligned} \tag{5.5.20}$$

and thus

$$\mathcal{C}_{ijk\ell}\xi_{ij}\xi_{k\ell} = \min_{\boldsymbol{\tau}}\left\langle C^{-1}_{pqmn}\tau_{pq}\tau_{mn}\right\rangle . \tag{5.5.21}$$

From Eq. (5.5.19) we find $\xi_{ij} = \mathcal{C}^{-1}_{ijk\ell}\langle \tau_{k\ell}\rangle = \mathcal{C}^{-1}_{ijk\ell}\sigma_{k\ell}$ which can be substituted in Eq. (5.5.20) to conclude the proof of the variational principle (Eq. (5.5.14)).

Choosing constant tensors $\boldsymbol{\tau}$ in the variational principle (Eq. (5.5.14)) we obtain the inequality

$$(\boldsymbol{\mathcal{C}})^{-1}_{ijk\ell}\sigma_{ij}\sigma_{k\ell} \leq \langle (\boldsymbol{C}^{-1})_{ijk\ell}\rangle \sigma_{ij}\sigma_{k\ell} \,, \tag{5.5.22}$$

which implies the harmonic mean lower bound

$$\langle \boldsymbol{C}^{-1}\rangle^{-1} \leq \boldsymbol{\mathcal{C}} \,. \tag{5.5.23}$$

5.6. Hashin–Shtrikman Bounds

As in the diffusion case of Chapter 2, the upper and lower bounds for the effective elasticity tensor derived in the previous section can be improved. In this section, we will derive the celebrated Hashin–Shtrikman variational principles and bounds for the effective bulk modulus and the effective shear modulus of a composite formed by two isotropic elastic materials. One important question in structural optimization, outside the scope of this chapter, is the optimality of the bounds: for any tensor within the bounds, can one find a composite with it as the effective tensor? Starting with the results of Hashin and Shtrikman (1963), the literature related to these bounds and their optimality is vast: Lurie and Cherkaev (1984), Murat and Tartar (1985), Francfort and Murat (1986), Avellaneda (1987), Kohn and Lipton (1988), Lipton (1988), Milton and Kohn (1988), Milton (1990), to mention just a few (see also Cherkaev (2000), Allaire (2001) and Milton (2002) and the references therein).

We assume that the two isotropic elastic materials in the composite are totally ordered, which means $\boldsymbol{C}_1 < \boldsymbol{C}_2$, or equivalently $(C_1)_{ijk\ell}\xi_{ij}\xi_{k\ell} < (C_2)_{ijk\ell}\xi_{ij}\xi_{k\ell}$, for any second-order tensor $\boldsymbol{\xi}$. In the special case of isotropic materials, this inequality implies that the bulk and shear moduli are well-ordered ($K_1 < K_2$ and $G_1 < G_2$). We choose $\boldsymbol{C}_0$ to be a fourth-order tensor such that $\boldsymbol{C}_0 < \boldsymbol{C}_1 < \boldsymbol{C}_2$.

From the first variational principle (Eq. (5.5.4)) we obtain

$$\begin{aligned}\mathcal{C}_{ijk\ell}\xi_{ij}\xi_{k\ell} = \min_{\boldsymbol{v}} \frac{1}{\Omega}\bigg(&\iiint_\Omega (C_{ijk\ell} - (C_0)_{ijk\ell})(\xi_{ij} + \varepsilon_{ij}(\boldsymbol{v}))(\xi_{k\ell} + \varepsilon_{k\ell}(\boldsymbol{v}))d\Omega \\ &+ \iiint_\Omega (C_0)_{ijk\ell}(\xi_{ij} + \varepsilon_{ij}(\boldsymbol{v}))(\xi_{k\ell} + \varepsilon_{k\ell}(\boldsymbol{v}))d\Omega\bigg). \qquad (5.6.1)\end{aligned}$$

Using Eq. (5.5.15) we can write

$$\begin{aligned}&(\boldsymbol{C} - \boldsymbol{C_0})_{ijk\ell}(\xi_{ij} + \varepsilon_{ij}(\boldsymbol{v}))(\xi_{k\ell} + \varepsilon_{k\ell}(\boldsymbol{v})) \\ &\quad = \max_{\boldsymbol{\tau}} \left\langle 2(\xi_{ij} + \varepsilon_{ij}(\boldsymbol{v}))\tau_{ij} - (\boldsymbol{C} - \boldsymbol{C_0})^{-1}_{ijk\ell}\tau_{ij}\tau_{k\ell}\right\rangle, \qquad (5.6.2)\end{aligned}$$

and thus Eq. (5.6.1) becomes

$$\begin{aligned}\mathcal{C}_{ijk\ell}\xi_{ij}\xi_{k\ell} = \min_{\boldsymbol{v}} \max_{\boldsymbol{\tau}} \big\langle &2(\xi_{ij} + \varepsilon_{ij}(\boldsymbol{v}))\tau_{ij} - (\boldsymbol{C} - \boldsymbol{C_0})^{-1}_{ijk\ell}\tau_{ij}\tau_{k\ell} \\ &+ (C_0)_{ijk\ell}(\xi_{ij} + \varepsilon_{ij}(\boldsymbol{v}))(\xi_{k\ell} + \varepsilon_{k\ell}(\boldsymbol{v}))\big\rangle,\end{aligned}$$

which yields (see footnote in Sec. 2.7.2)

$$C_{ijk\ell}\xi_{ij}\xi_{k\ell} \geq \max_{\boldsymbol{\tau}} \Big(\min_{\boldsymbol{v}} \langle (C_0)_{ijk\ell}\varepsilon_{ij}(\boldsymbol{v})\varepsilon_{k\ell}(\boldsymbol{v}) + 2\tau_{ij}\varepsilon_{ij}(\boldsymbol{v})\rangle + \big\langle 2\xi_{ij}\tau_{ij} - (\boldsymbol{C}-\boldsymbol{C}_0)^{-1}_{ijk\ell}\tau_{ij}\tau_{k\ell} + (C_0)_{ijk\ell}\xi_{ij}\xi_{k\ell}\big\rangle \Big) . \quad (5.6.3)$$

We will use Eq. (5.6.3) to derive the Hashin–Shtrikman lower bounds. For the reader's convenience, the derivation will be made in four separate steps.

Step 1. Let us first find the minimum above. Consider the functional

$$\mathcal{H}(\boldsymbol{v}) = \langle (C_0)_{ijk\ell}\varepsilon_{ij}(\boldsymbol{v})\varepsilon_{k\ell}(\boldsymbol{v}) + 2\tau_{ij}\varepsilon_{ij}(\boldsymbol{v})\rangle , \quad (5.6.4)$$

whose first variation is

$$\delta\mathcal{H}(\boldsymbol{v}) = 2\langle (C_0)_{ijk\ell}\varepsilon_{ij}(\boldsymbol{v})\varepsilon_{k\ell}(\delta\boldsymbol{v}) + \tau_{ij}\varepsilon_{ij}(\delta\boldsymbol{v})\rangle . \quad (5.6.5)$$

By applying Gauss' theorem and canceling the surface integral by periodicity, we have for any vector (δv_i)

$$\left\langle \frac{\partial}{\partial x_j}((C_0)_{ijk\ell}\varepsilon_{k\ell}(\boldsymbol{v}) + \tau_{ij})\delta v_i \right\rangle = 0 . \quad (5.6.6)$$

Thus for a given tensor $\boldsymbol{\tau}$, the vector $\boldsymbol{v}_{\boldsymbol{\tau}}$ that minimizes $\mathcal{H}$ is governed by the following cell problem,

$$-\frac{\partial}{\partial x_j}((C_0)_{ijk\ell}\varepsilon_{k\ell}(\boldsymbol{v}_{\boldsymbol{\tau}})) = \frac{\partial}{\partial x_j}(\tau_{ij}) , \quad \text{in } \Omega , \quad (5.6.7)$$

$$\boldsymbol{v}_{\boldsymbol{\tau}1} = \boldsymbol{v}_{\boldsymbol{\tau}2} , \quad \text{on } \Gamma , \quad (5.6.8)$$

$$(C_0)_{ijk\ell}\varepsilon_{k\ell}(\boldsymbol{v}_{\boldsymbol{\tau}1})n_j = (C_0)_{ijk\ell}\varepsilon_{k\ell}(\boldsymbol{v}_{\boldsymbol{\tau}2})n_j , \quad \text{on } \Gamma , \quad (5.6.9)$$

$$\boldsymbol{v}_{\boldsymbol{\tau}} \text{ is } \Omega\text{-periodic} , \quad (5.6.10)$$

which implies

$$\langle (C_0)_{ijk\ell}\varepsilon_{ij}(\boldsymbol{v}_{\boldsymbol{\tau}})\varepsilon_{k\ell}(\boldsymbol{v}_{\boldsymbol{\tau}}) + \tau_{ij}\varepsilon_{ij}(\boldsymbol{v}_{\boldsymbol{\tau}})\rangle = 0 . \quad (5.6.11)$$

Therefore, the minimum value for $\mathcal{H}$ is

$$\mathcal{H}(\boldsymbol{v}_{\boldsymbol{\tau}}) = \langle \tau_{ij}\varepsilon_{ij}(\boldsymbol{v}_{\boldsymbol{\tau}})\rangle . \quad (5.6.12)$$

The inequality (Eq. (5.6.3)) becomes

$$(\boldsymbol{C}-\boldsymbol{C}_0)_{ijk\ell}\xi_{ij}\xi_{k\ell} \geq \max_{\boldsymbol{\tau}} \big\langle \tau_{ij}\varepsilon_{ij}(\boldsymbol{v}_{\boldsymbol{\tau}}) + 2\xi_{ij}\tau_{ij} - (\boldsymbol{C}-\boldsymbol{C}_0)^{-1}_{ijk\ell}\tau_{ij}\tau_{k\ell}\big\rangle . \quad (5.6.13)$$

Step 2. Next for a particular choice of $\boldsymbol{\tau}$ we will explicitly compute the solution $\boldsymbol{v}_{\boldsymbol{\tau}}$ of the cell problem Eqs. (5.6.7)–(5.6.10). We choose $\boldsymbol{\tau} = \chi_2\boldsymbol{\eta}$

where χ_2 is the characteristic function of Phase 2 (i.e., $\chi_2 = 1$ in material 2 and $\chi_2 = 0$ in material 1) and $\boldsymbol{\eta}$ is a constant symmetric tensor.

The solution of Eqs. (5.6.7)–(5.6.10) is periodic in $\boldsymbol{x} = (x_1, x_2, x_3)$, and hence it can be expressed as a Fourier series (for simplicity the subscript $\boldsymbol{\tau}$ is omitted)

$$\boldsymbol{v} = \sum_{\boldsymbol{m}\in\mathbb{Z}^3} \hat{\boldsymbol{v}}(\boldsymbol{m}) e^{2\pi i \boldsymbol{m}\cdot\boldsymbol{x}}, \tag{5.6.14}$$

where $\hat{\boldsymbol{v}}(\boldsymbol{m})$ denote the Fourier coefficients and the summation is over all wave vectors $\boldsymbol{m} = (m_1, m_2, m_3) \in \mathbb{Z}^3$ (i.e., triplets of integers). Since the solution is defined up to an additive constant, we can choose $\hat{\boldsymbol{v}}(\boldsymbol{0}) = 0$ so that the sum (Eq. (5.6.14)) does not include the term corresponding to $\boldsymbol{m} = (0, 0, 0)$, i.e.,

$$\boldsymbol{v} = \sum_{\boldsymbol{m}\neq\boldsymbol{0}} \hat{\boldsymbol{v}}(\boldsymbol{m}) e^{2\pi i \boldsymbol{m}\cdot\boldsymbol{x}}. \tag{5.6.15}$$

To compute the left-hand side of Eq. (5.6.7) we note that

$$\varepsilon_{ij}(\boldsymbol{v}) = \pi i \sum_{\boldsymbol{m}\neq\boldsymbol{0}} (\hat{v}_i m_j + \hat{v}_j m_i) e^{2\pi i \boldsymbol{m}\cdot\boldsymbol{x}}, \tag{5.6.16}$$

and thus

$$\frac{\partial}{\partial x_j}((C_0)_{ijk\ell}\varepsilon_{k\ell}(\boldsymbol{v})) = -2\pi^2 \sum_{\boldsymbol{m}\neq\boldsymbol{0}} (C_0)_{ijk\ell}(\hat{v}_k(\boldsymbol{m}) m_\ell + \hat{v}_\ell(\boldsymbol{m}) m_k) m_j e^{2\pi i \boldsymbol{m}\cdot\boldsymbol{x}}. \tag{5.6.17}$$

Next, we consider the Fourier expansion for the characteristic function χ_2

$$\chi_2 = \sum_{\boldsymbol{m}\in\mathbb{Z}^3} \hat{\chi}_2(\boldsymbol{m}) e^{2\pi i \boldsymbol{m}\cdot\boldsymbol{x}} = \theta_2 + \sum_{\boldsymbol{m}\neq\boldsymbol{0}} \hat{\chi}_2(\boldsymbol{m}) e^{2\pi i \boldsymbol{m}\cdot\boldsymbol{x}}. \tag{5.6.18}$$

The right-hand side of Eq. (5.6.7) is

$$\frac{\partial}{\partial x_j}(\eta_{ij}\chi_2) = 2\pi i \sum_{\boldsymbol{m}\neq\boldsymbol{0}} \eta_{ij} m_j \hat{\chi}_2(\boldsymbol{m}) e^{2\pi i \boldsymbol{m}\cdot\boldsymbol{x}}, \tag{5.6.19}$$

where $\hat{\chi}_2(\boldsymbol{m})$ are the Fourier coefficients for the characteristic function χ_2. Thus Eq. (5.6.7) reduces to solving the system

$$\pi (C_0)_{ijk\ell}(\hat{v}_k(\boldsymbol{m}) m_\ell + \hat{v}_\ell(\boldsymbol{m}) m_k) m_j = i \eta_{ij} m_j \hat{\chi}_2(\boldsymbol{m}). \tag{5.6.20}$$

Taking now an isotropic elasticity tensor

$$(C_0)_{ijk\ell} = \lambda_0 \delta_{ij}\delta_{kl} + G_0(\delta_{ik}\delta_{jl} + \delta_{il}\delta_{jk}), \tag{5.6.21}$$

where $\lambda_0 = K_0 - (2/3)G_0$ is the Lamè coefficient, Eq. (5.6.20) becomes

$$(G_0 k^2 \delta_{ik} + (\lambda_0 + G_0) m_i m_k)\hat{v}_k(\boldsymbol{m}) = \frac{i}{2\pi}\eta_{ij} m_j \hat{\chi}_2(\boldsymbol{m}) . \tag{5.6.22}$$

The inverse of the matrix on the left-hand side is

$$\frac{1}{|\boldsymbol{m}|^4}\left(\frac{1}{G_0}|\boldsymbol{m}|^2\delta_{mi} - \frac{\lambda_0 + G_0}{G_0(\lambda_0 + 2G_0)} m_m m_i\right), \tag{5.6.23}$$

where $|\boldsymbol{m}|^2 = \boldsymbol{m}\cdot\boldsymbol{m}$. Equation (5.6.22) can be solved to give

$$\hat{v}_m(\boldsymbol{m}) = \frac{i}{2\pi|\boldsymbol{m}|^4}\left(\frac{1}{G_0}|\boldsymbol{m}|^2\eta_{mj} - \frac{\lambda_0 + G_0}{G_0(\lambda_0 + 2G_0)} m_m m_i \eta_{ij}\right) m_j \hat{\chi}_2(\boldsymbol{m}) . \tag{5.6.24}$$

Step 3. We can now compute $\mathcal{H}$ in Eq. (5.6.12) and the right-hand side in Eq. (5.6.13) for the chosen $\boldsymbol{\tau} = \chi_2\boldsymbol{\eta}$.

From Eq. (5.6.24) $\varepsilon_{pq}(\boldsymbol{v})$ has the Fourier coefficients

$$\begin{aligned}\pi i(m_q\hat{v}_p + m_p\hat{v}_q) = -\frac{1}{2G_0}\Bigg(&\frac{m_j}{|\boldsymbol{m}|}\frac{m_q}{|\boldsymbol{m}|}\delta_{pi} + \frac{m_j}{|\boldsymbol{m}|}\frac{m_p}{|\boldsymbol{m}|}\delta_{qi} \\ &- \frac{2(\lambda_0 + G_0)}{\lambda_0 + 2G_0}\frac{m_i}{|\boldsymbol{m}|}\frac{m_j}{|\boldsymbol{m}|}\frac{m_p}{|\boldsymbol{m}|}\frac{m_q}{|\boldsymbol{m}|}\Bigg)\eta_{ij}\hat{\chi}_2(\boldsymbol{m}) .\end{aligned} \tag{5.6.25}$$

Let (L_{ijpq}) denote the fourth-order tensor on the right-hand side

$$L_{ijpq}(\boldsymbol{m}) = \frac{1}{2G_0}\left(\frac{m_j}{|\boldsymbol{m}|}\frac{m_q}{|\boldsymbol{m}|}\delta_{pi} + \frac{m_j}{|\boldsymbol{m}|}\frac{m_p}{|\boldsymbol{m}|}\delta_{qi} - \frac{2(\lambda_0 + G_0)}{\lambda_0 + 2G_0}\frac{m_i}{|\boldsymbol{m}|}\frac{m_j}{|\boldsymbol{m}|}\frac{m_p}{|\boldsymbol{m}|}\frac{m_q}{|\boldsymbol{m}|}\right). \tag{5.6.26}$$

Using Parseval's theorem, Eq. (5.6.12) becomes

$$\mathcal{H}(\chi_2\boldsymbol{\eta}) = \iiint_\Omega \varepsilon_{pq}((\chi_2\boldsymbol{\eta}))\hat{\chi}_2\eta_{pq}d\Omega = -\sum_{\boldsymbol{m}\neq\boldsymbol{0}} L_{ijpq}(\boldsymbol{m})\hat{\chi}_2^2(\boldsymbol{m})\eta_{ij}\eta_{pq} . \tag{5.6.27}$$

If we introduce the tensor $\boldsymbol{\Gamma}$ with components

$$\Gamma_{ijpq} = \sum_{\boldsymbol{m}\neq\boldsymbol{0}} L_{ijpq}(\boldsymbol{m})\hat{\chi}_2^2(\boldsymbol{m}) , \tag{5.6.28}$$

then Eq. (5.6.13) becomes

$$\begin{aligned}&(\boldsymbol{C} - \boldsymbol{C}_0)_{ijk\ell}\xi_{ij}\xi_{k\ell} \\ &\quad\geq \max_{\boldsymbol{\eta}}\big(-(\Gamma_{ijk\ell} + \theta_2(\boldsymbol{C}_2 - \boldsymbol{C}_0)^{-1}_{ijk\ell})\eta_{ij}\eta_{k\ell} + 2\theta_2\xi_{ij}\eta_{ij}\big) .\end{aligned} \tag{5.6.29}$$

The quadratic form on the right-hand side of Eq. (5.6.29) has the maximum

$$\theta_2^2\big(\theta_2(\boldsymbol{C}_2 - \boldsymbol{C}_0)^{-1}_{ijk\ell} + \Gamma_{ijk\ell}\big)^{-1}\xi_{ij}\xi_{k\ell} , \tag{5.6.30}$$

and thus from here we obtain the tensor inequality

$$(\mathcal{C} - \boldsymbol{C}_0)^{-1} \leq \theta_2^{-2}(\theta_2(\boldsymbol{C}_2 - \boldsymbol{C}_0)^{-1} + \boldsymbol{\Gamma}) . \tag{5.6.31}$$

By making the particular choice $\boldsymbol{C}_0 = \boldsymbol{C}_1$ we get

$$(\mathcal{C} - \boldsymbol{C}_1)^{-1} \leq \theta_2^{-2}(\theta_2(\boldsymbol{C}_2 - \boldsymbol{C}_1)^{-1} + \boldsymbol{\Gamma}) . \tag{5.6.32}$$

Step 4. In order to obtain the bounds for the effective moduli we next apply to Eq. (5.6.32) the projections P_b and P_s defined in Eqs. (A.3) and (A.4) in Appendix 5A. Thus

$$\frac{1}{3(\mathcal{K} - K_1)} P_b \leq \theta_2^{-2} \left(\theta_2 \frac{1}{3(K_2 - K_1)} P_b + P_b \Gamma \right) , \tag{5.6.33}$$

and

$$\frac{1}{2(\mathcal{G} - G_1)} P_s \leq \theta_2^{-2} \left(\theta_2 \frac{1}{2(G_2 - G_1)} P_s + P_s \Gamma \right) , \tag{5.6.34}$$

where the effective elasticity tensor $\mathcal{C}$ is characterized by the effective bulk modulus $\mathcal{K}$ and effective shear modulus $\mathcal{G}$.

In order to compute the projections of Γ we note that

$$\begin{aligned} (P_b)_{ijk\ell}\Gamma_{k\ell mn} &= \sum_{\boldsymbol{m}\neq \boldsymbol{0}} (P_b)_{ijk\ell} L_{k\ell mn} \hat{\chi}_2^2(\boldsymbol{m}) \quad \text{and} \\ (P_s)_{ijk\ell}\Gamma_{k\ell mn} &= \sum_{\boldsymbol{m}\neq \boldsymbol{0}} (P_s)_{ijk\ell} L_{k\ell mn} \hat{\chi}_2^2(\boldsymbol{m}) . \end{aligned} \tag{5.6.35}$$

Let us first compute the bulk projection of L

$$(P_b)_{ijk\ell} L_{k\ell pq} = \frac{1}{3} L_{kkpq} \delta_{ij} = \frac{1}{3(\lambda_1 + 2G_1)} \frac{m_p}{|\boldsymbol{m}|} \frac{m_q}{|\boldsymbol{m}|} \delta_{ij} , \tag{5.6.36}$$

and its trace

$$Tr(P_b L) = \frac{1}{3(\lambda_1 + 2G_1)} = \frac{1}{3K_1 + 4G_1} . \tag{5.6.37}$$

Similarly, the trace of the projection $P_s L$ is

$$Tr(P_s L) = \frac{2\lambda_1 + 8G_1}{3G_1(\lambda_1 + 2G_1)} = \frac{3(K_1 + 2G_1)}{G_1(3K_1 + 4G_1)} . \tag{5.6.38}$$

Therefore, by taking the trace of the projections of Γ in Eq. (5.6.35) we obtain

$$\begin{aligned} Tr(P_b \Gamma) &= \frac{1}{3K_1 + 4G_1} \sum_{\boldsymbol{m}\neq \boldsymbol{0}} \hat{\chi}_2^2(\boldsymbol{m}) \quad \text{and} \\ Tr(P_s \Gamma) &= \frac{3(K_1 + 2G_1)}{G_1(3K_1 + 4G_1)} \sum_{\boldsymbol{m}\neq \boldsymbol{0}} \hat{\chi}_2^2(\boldsymbol{m}) . \end{aligned} \tag{5.6.39}$$

Now applying again Parseval's theorem we have

$$\sum_{m \neq 0} \hat{\chi}_2^2(m) = \iiint_\Omega (\chi_2 - \theta_2)^2 d\Omega = \theta_1 \theta_2 \,. \tag{5.6.40}$$

Thus the lower bounds (Eqs. (5.6.33) and (5.6.34)) become

$$\frac{\theta_2}{3(\mathcal{K} - K_1)} \leq \frac{1}{3(K_2 - K_1)} + \frac{\theta_1}{3K_1 + 4G_1} \tag{5.6.41}$$

and

$$\frac{5\theta_2}{2(\mathcal{G} - G_1)} \leq \frac{5}{2(G_2 - G_1)} + \frac{3(K_1 + 2G_1)\theta_1}{G_1(3K_1 + 4G_1)} \,. \tag{5.6.42}$$

These are the Hashin–Shtrikman lower bounds for the effective bulk modulus $\mathcal{K}$ and effective shear modulus $\mathcal{G}$, which can be equivalently rewritten as

$$\boxed{\mathcal{K} \geq \theta_1 K_1 + \theta_2 K_2 - \frac{\theta_1 \theta_2 (K_1 - K_2)^2}{\theta_2 K_1 + \theta_1 K_2 + (4/3)G_1}} \,, \tag{5.6.43}$$

$$\boxed{\mathcal{G} \geq \theta_1 G_1 + \theta_2 G_2 - \frac{\theta_1 \theta_2 (G_1 - G_2)^2}{\theta_2 G_1 + \theta_1 G_2 + ((G_1(9K_1 + 8G_1))/(6(K_1 + 2G_1)))}} \,. \tag{5.6.44}$$

The upper bounds, which can be obtained in a similar fashion from the dual variational principles, are

$$\boxed{\mathcal{K} \leq \theta_1 K_1 + \theta_2 K_2 - \frac{\theta_1 \theta_2 (K_1 - K_2)^2}{\theta_2 K_1 + \theta_1 K_2 + (4/3)G_2}} \,, \tag{5.6.45}$$

$$\boxed{\mathcal{G} \geq \theta_1 G_1 + \theta_2 G_2 - \frac{\theta_1 \theta_2 (G_1 - G_2)^2}{\theta_2 G_1 + \theta_1 G_2 + ((G_2(9K_2 + 8G_2))/(6(K_2 + 2G_2)))}} \,. \tag{5.6.46}$$

5.7. Partially Cohesive Composites

In many industrial materials (e.g., fiber-reinforced composites) perfect bonding between phases is not a good assumption. In the modeling of imperfect bonding two approaches have been considered.

One approach is to assume the presence of a thin layer of different materials sandwiched between two phases. This has been considered by Theocaris *et al.* (1985), Sideridis (1988), Bigoni and Movchan (2002), among other authors.

A second approach is to allow for discontinuities of the displacements along the interface between the constituent materials. In these models the tractions are continuous and proportional to the jump of the displacement across the interface. The constant of proportionality represents the stiffness of the interface. This approach is taken in Benveniste (1985), Aboudi (1987), Achenbach and Zhu (1989), Bigoni and Movchan (2002), and others.

We will first derive the effective equations for the partially cohesive composites, using periodic homogenization. A similar derivation for the case allowing only for tangential jump of the displacement can be found in Lenè and Leguillon (1982).

Let $\boldsymbol{n}$ be the unit normal to the interface Γ, and the normal and tangential components of a vector $\boldsymbol{v}$ be denoted, respectively, by $v_N = \boldsymbol{v} \cdot \boldsymbol{n} = v_i n_i$ and $\boldsymbol{v}_T = \boldsymbol{v} - v_N \boldsymbol{n}$. The dot product of two vectors $\boldsymbol{u}$ and $\boldsymbol{v}$ can be written as

$$\boldsymbol{u} \cdot \boldsymbol{v} = u_N v_N + \boldsymbol{u}_T \cdot \boldsymbol{v}_T \,. \tag{5.7.1}$$

Similarly, considering the stress tensor $\boldsymbol{\sigma}$ and the orthonormal basis $\{\boldsymbol{e}_i\}$, the normal stress vector

$$\boldsymbol{\sigma n} = \sigma_{ij} n_j \boldsymbol{e}_i \tag{5.7.2}$$

can be projected along the normal

$$(\boldsymbol{\sigma n})_N = \sigma_{ij} n_i n_j \,, \tag{5.7.3}$$

and in the tangent plane

$$(\boldsymbol{\sigma n})_T = \boldsymbol{\sigma n} - \sigma_{ij} n_i n_j \boldsymbol{n} \,. \tag{5.7.4}$$

The jump of a field ϕ on the surface will be denoted by

$$[\phi] = \phi_2|_\Gamma - \phi_1|_\Gamma \tag{5.7.5}$$

with $\phi_1|_\Gamma$ and $\phi_2|_\Gamma$ the values of ϕ of the two materials on Γ.

Using the same notations as in Sec. 5.1, the problem for the partially cohesive composites is described by

$$\frac{\partial (\sigma_\alpha)_{ij}}{\partial x_j} = 0 \quad \text{in } \Omega_\alpha, \quad \alpha = 1, 2 \tag{5.7.6}$$

with the stress continuous across the interface

$$(\sigma_1)_{ij} n_j = (\sigma_2)_{ij} n_j \quad \text{on } \Gamma \,, \tag{5.7.7}$$

and a Hooke-type law satisfied on the interface

$$(\boldsymbol{\sigma n})_N = -\alpha [v_N] \,, \quad (\boldsymbol{\sigma n})_T = -\beta [\boldsymbol{v}_T] \,, \quad \text{on } \Gamma \,. \tag{5.7.8}$$

Here the spring constants α and β model the stiffness of the interface.

5.7.1. *Effective Equations on the Macroscale*

Introducing the multiple scale expansion for the displacement field (5.1.10) we obtain the same expressions for the stress (Eqs. (5.1.11)–(5.1.14)). At order $\mathcal{O}(\epsilon^0)$ the problem will be similar to Eqs. (5.1.15)–(5.1.18) with the displacement continuity condition replaced by

$$(\boldsymbol{\sigma}^{(0)}\boldsymbol{n})_N = -\alpha\big[v_N^{(0)}\big]\,, \quad (\boldsymbol{\sigma}^{(0)}\boldsymbol{n})_T = -\beta\big[\boldsymbol{v}_T^{(0)}\big]\,. \tag{5.7.9}$$

The problem has again only constant solutions, so $\boldsymbol{v}^0$ depends only on the macroscopic coordinate:

$$\boldsymbol{v}^{(0)} = \boldsymbol{v}^{(0)}(\boldsymbol{x}')\,. \tag{5.7.10}$$

At the order $\mathcal{O}(\epsilon')$ the problem is

$$\frac{\partial}{\partial x_j}\big(\sigma_\alpha^{(1)}\big)_{ij} = 0\,, \quad \text{in } \Omega_\alpha\,, \tag{5.7.11}$$

$$\boldsymbol{\sigma}_1^{(1)}\boldsymbol{n} = \boldsymbol{\sigma}_2^{(1)}\boldsymbol{n}\,, \quad \text{on } \Gamma\,, \tag{5.7.12}$$

$$\big(\boldsymbol{\sigma}_1^{(1)}\boldsymbol{n}\big)_N = -\alpha\big[v_N^1\big]\,, \quad \text{on } \Gamma\,, \tag{5.7.13}$$

$$\big(\boldsymbol{\sigma}_1^{(1)}\boldsymbol{n}\big)_T = -\beta\big[\boldsymbol{v}_T^1\big]\,, \quad \text{on } \Gamma\,. \tag{5.7.14}$$

Let the solution be of the form

$$\boldsymbol{v}^{(1)} = \boldsymbol{\chi}^{k\ell}\varepsilon'_{k\ell}(\boldsymbol{v}^0) + \text{const}\,. \tag{5.7.15}$$

Then the third-rank tensor $\boldsymbol{\chi}^{k\ell}$ is defined by the cell problem

$$\frac{\partial}{\partial x_j}\big(C_{ijmn}\varepsilon_{mn}(\boldsymbol{\chi}_\alpha^{k\ell})\big) = -\frac{\partial C_{ijk\ell}}{\partial x_j}\,, \quad \text{in } \Omega_\alpha\,, \tag{5.7.16}$$

$$C_{ijmn}\varepsilon_{mn}(\boldsymbol{\chi}_1^{k\ell})n_j = C_{ijmn}\varepsilon_{mn}(\boldsymbol{\chi}_2^{k\ell})n_j\,, \quad \text{on } \Gamma\,, \tag{5.7.17}$$

$$C_{ijmn}\varepsilon_{mn}(\boldsymbol{\chi}^{k\ell})n_i n_j = -\alpha\big[\chi_N^{k\ell}\big]\,, \quad \text{on } \Gamma\,, \tag{5.7.18}$$

$$C_{ijmn}\varepsilon_{mn}(\boldsymbol{\chi}^{k\ell})n_j - C_{hjmn}\varepsilon_{mn}(\boldsymbol{\chi}^{k\ell})n_h n_j n_i = -\beta\big[\boldsymbol{\chi}_T^{k\ell}\big]_i\,, \quad \text{on } \Gamma\,, \tag{5.7.19}$$

$$\boldsymbol{\chi}^{k\ell} \text{ is } \Omega\text{-periodic}\,, \tag{5.7.20}$$

and $\boldsymbol{\chi}^{k\ell}$ has zero average on Ω

$$\langle \boldsymbol{\chi}_\alpha^{k\ell} \rangle = 0\,. \tag{5.7.21}$$

At the order $\mathcal{O}(\epsilon^2)$ we have

$$\frac{\partial \sigma_{ij}^{(2)}}{\partial x_j} = -\frac{\partial \sigma_{ij}^{(1)}}{\partial x'_j}, \quad \text{in } \Omega_\alpha, \tag{5.7.22}$$

$$\boldsymbol{\sigma}_1^{(2)} \boldsymbol{n} = \boldsymbol{\sigma}_2^{(2)} \boldsymbol{n}, \quad \text{on } \Gamma, \tag{5.7.23}$$

$$\left(\boldsymbol{\sigma}_1^{(2)} \boldsymbol{n}\right)_N = -\alpha \left[v_N^{(2)}\right], \quad \text{on } \Gamma, \tag{5.7.24}$$

$$\left(\boldsymbol{\sigma}_1^{(2)} \boldsymbol{n}\right)_N = -\beta \left[\boldsymbol{v}_T^{(2)}\right], \quad \text{on } \Gamma. \tag{5.7.25}$$

By averaging Eq. (5.7.22) over Ω_1 and Ω_2, and using Gauss' theorem with the boundary condition (Eq. (5.7.23)) we get the macroscale governing equation,

$$\boxed{\frac{\partial}{\partial x'_j} \langle \sigma_{ij}^{(1)} \rangle = 0}. \tag{5.7.26}$$

By using Eqs. (5.1.12) and (5.7.15) we get from Eq. (5.7.26) the macroscopic Hooke's law

$$\boxed{\left\langle \sigma_{ij}^{(1)} \right\rangle = \mathcal{C}_{ijk\ell} \varepsilon'_{k\ell}(\boldsymbol{v}^{(0)})}, \tag{5.7.27}$$

where the homogenized coefficients are given by

$$\boxed{\mathcal{C}_{ijk\ell} = \langle C_{ijmn}(\delta_{mk}\delta_{n\ell} + \varepsilon_{mn}(\boldsymbol{\chi}^{k\ell})) \rangle}. \tag{5.7.28}$$

In order to study the symmetry and positivity properties of the effective coefficients, it is more convenient to rewrite Eq. (5.7.28) making use of the cell problems (Eqs. (5.7.16)–(5.7.21)). Indeed by multiplying the cell problem (Eq. (5.7.16)) by χ_i^{pq} in each domain Ω_1 and Ω_2 and by applying Gauss' formula we obtain

$$- \iiint_\Omega C_{ijmn} \varepsilon_{mn}(\boldsymbol{\chi}^{k\ell}) \varepsilon_{ij}(\boldsymbol{\chi}^{pq}) d\Omega + \iint_\Gamma \left[C_{ijmn} \varepsilon_{mn}(\boldsymbol{\chi}^{k\ell}) n_j \chi_i^{pq}\right] d\Gamma$$
$$= \iiint_\Omega C_{ijk\ell} \varepsilon_{ij}(\boldsymbol{\chi}^{pq}) d\Omega, \tag{5.7.29}$$

where we have used the periodicity to cancel the integrals on opposite faces of $\partial\Omega$. Due to the continuity condition (Eq. (5.7.17)), the surface term becomes

$$\iint_\Gamma C_{ijmn} \varepsilon_{mn}(\boldsymbol{\chi}^{k\ell}) n_j \left[\chi_i^{pq}\right] d\Gamma, \tag{5.7.30}$$

and using the jump conditions (Eqs. (5.7.18) and (5.7.19)) it can be rewritten as

$$-\alpha \iint_{\Gamma} [\chi_N^{k\ell}] [\chi_N^{pq}] d\Gamma - \beta \iint_{\Gamma} [\chi_T^{k\ell}] [\chi_T^{pq}] d\Gamma . \tag{5.7.31}$$

Thus Eq. (5.7.29) becomes

$$\begin{aligned} &\langle C_{ijmn}(\delta_{mk}\delta_{nl} + \varepsilon_{mn}(\boldsymbol{\chi}^{k\ell}))\varepsilon_{ij}(\boldsymbol{\chi}^{pq})\rangle \\ &\quad + \alpha \iint_{\Gamma} [\chi_N^{k\ell}] [\chi_N^{pq}] d\Gamma + \beta \iint_{\Gamma} [\chi_T^{k\ell}] [\chi_T^{pq}] d\Gamma = 0 . \end{aligned} \tag{5.7.32}$$

By adding Eq. (5.7.32) to the definition of the effective coefficients (Eq. (5.7.28)) we get a new expression for the effective coefficients

$$\begin{aligned} \mathcal{C}_{ijk\ell} &= \langle C_{pqmn}(\delta_{pi}\delta_{qj} + \varepsilon_{pq}(\boldsymbol{\chi}^{ij}))(\delta_{mk}\delta_{n\ell} + \varepsilon_{mn}(\boldsymbol{\chi}^{k\ell}))\rangle \\ &\quad + \alpha \iint_{\Gamma} [\chi_N^{ij}] [\chi_N^{k\ell}] d\Gamma + \beta \iint_{\Gamma} [\chi_T^{ij}] [\chi_T^{k\ell}] d\Gamma . \end{aligned} \tag{5.7.33}$$

From here the symmetry $\mathcal{C}_{ijk\ell} = \mathcal{C}_{k\ell ij}$ is immediate using the symmetry of $C_{ijk\ell}$.

The Eq. (5.7.33) also implies the positivity of $\mathcal{C}_{ijk\ell}$. Indeed if (ξ_{ij}) is a symmetric matrix, then

$$\begin{aligned} \mathcal{C}_{ijk\ell}\xi_{ij}\xi_{k\ell} &= \langle C_{pqmn}(\xi_{pq} + \varepsilon_{pq}(\boldsymbol{\chi}^{ij}\xi_{ij}))(\xi_{mn} + \varepsilon_{mn}(\boldsymbol{\chi}^{k\ell}\xi_{k\ell}))\rangle \\ &\quad + \alpha \iint_{\Gamma} [\chi_N^{ij}\xi_{ij}] [\chi_N^{k\ell}\xi_{k\ell}] d\Gamma + \beta \iint_{\Gamma} [\chi_T^{ij}\xi_{ij}] [\chi_T^{k\ell}\xi_{k\ell}] d\Gamma , \end{aligned} \tag{5.7.34}$$

and the positivity of C_{pqmn} implies the positivity of $\mathcal{C}_{ijk\ell}$. In order to show Eq. (5.2.2) we need to prove that if $\mathcal{C}_{ijk\ell}\xi_{ij}\xi_{k\ell} = 0$ then $\xi_{ij} = 0$. Indeed if Eq. (5.7.34) is zero, then from the positive definiteness of C_{pqmn} we get

$$\xi_{pq} + \varepsilon_{pq}(\boldsymbol{\chi}^{ij}\xi_{ij}) = 0 \quad \text{and} \quad [\boldsymbol{\chi}^{ij}] = 0 . \tag{5.7.35}$$

Integrating Eq. (5.7.35) on a cell Ω we obtain $\xi_{pq} = 0$.

In summary, the macroscopic behavior of the elastic composite is similar to an elastic material having the effective elasticity coefficients given by Eq. (5.7.28) in terms of the solutions $\boldsymbol{\chi}^{k\ell}$ of the cell problems (Eqs. (5.7.16)–(5.7.21)). In general these cell problems must be solved numerically.

5.7.2. *Variational Principles*

Variational principles for the effective elasticity of composites with imperfect interfaces have been obtained by Hashin (1992) and for the case of continuous normal displacements on the interface by Lenè and Leguillon (1982). According to these principles the effective elasticity is given by

$$\boxed{\begin{aligned} C_{ijk\ell}\xi_{ij}\xi_{k\ell} = \min_{\boldsymbol{v}} \Big\{ \iiint_\Omega C_{ijk\ell}(x)(\varepsilon_{ij}(\boldsymbol{v}) + \xi_{ij})(\varepsilon_{k\ell}(\boldsymbol{v}) + \xi_{k\ell})d\Omega \\ + \alpha \iint_\Gamma ([v_n])^2 d\Gamma + \beta \iint_\Gamma ([\boldsymbol{v}_T]^2) d\Gamma \Big\} . \end{aligned}} \tag{5.7.36}$$

For the complementary variational principle we consider as the admissible set of tensors, the second-order symmetric tensors $\boldsymbol{\tau}$ that satisfy

$$\frac{\partial \tau_{ij}}{\partial x_j} = 0 \text{ in } \Omega_1 \text{ and } \Omega_2 , \quad [\tau_{ij} n_j] = 0 \quad \text{on } \Gamma , \quad \Omega\text{-periodic} , \tag{5.7.37}$$

and denote the cell volume average on Ω by $\langle \boldsymbol{\tau} \rangle$.

With these notations, the effective compliance is given by

$$\boxed{\begin{aligned} (\boldsymbol{C}^{-1})_{ijk\ell}\langle \tau_{ij}\rangle\langle \tau_{k\ell}\rangle = \min_{\boldsymbol{\tau}} \Big(\iiint_\Omega (\boldsymbol{C}^{-1})_{ijk\ell}\tau_{ij}\tau_{k\ell} d\Omega + \alpha^{-1} \iint_\Gamma |(\boldsymbol{\tau n})_N|^2 d\Gamma \\ + \beta^{-1} \iint_\Gamma |(\boldsymbol{\tau n})_T|^2 d\Gamma \Big) , \end{aligned}} \tag{5.7.38}$$

where $\boldsymbol{\tau n}$ is the vector with components $(\boldsymbol{\tau n})_i = \tau_{ij} n_j$.

We note that the above variational principles (Eqs. (5.7.36) and (5.7.38)) are generalizations of the variational principles presented in Sec. 5.3. Indeed, if we make α and β tend to ∞ in Eq. (5.7.36) and respectively in Eq. (5.7.38), we recover Eq. (5.5.4) and respectively Eq. (5.5.14).

Next we derive two other variational principles describing the effective elastic tensor, introduced by Lipton and Vernescu (1995). These will be used to provide tighter bounds for the elastic moduli and to describe size effects in particulate composites.

Let us start with Eq. (5.7.36) and subtract from both sides a reference energy, corresponding to a fourth-order tensor $\boldsymbol{C}_0$

$$\begin{aligned} &(\boldsymbol{C} - \boldsymbol{C}_0)_{ijk\ell}\xi_{ij}\xi_{k\ell} \\ &\quad = \min_{\boldsymbol{v}} \Big(\iiint_\Omega (\boldsymbol{C}(x) - \boldsymbol{C}_0)_{ijk\ell}(\varepsilon_{ij}(\boldsymbol{v}) + \xi_{ij})(\varepsilon_{k\ell}(\boldsymbol{v}) + \xi_{k\ell})d\Omega \end{aligned}$$

$$+\iiint_\Omega (C_0)_{ijk\ell}\varepsilon_{ij}(\boldsymbol{v})\varepsilon_{k\ell}(\boldsymbol{v})d\Omega + 2\iiint_\Omega (C_0)_{ijk\ell}\varepsilon_{ij}(\boldsymbol{v})\xi_{k\ell}d\Omega$$

$$+\alpha\iint_\Gamma ([v_N])^2 d\Gamma + \beta\iint_\Gamma ([\boldsymbol{v}_T])^2 d\Gamma\Bigg). \tag{5.7.39}$$

The third term above can be rewritten by partial integration

$$2\iiint_\Omega (C_0)_{ijk\ell}\varepsilon_{ij}(\boldsymbol{v})\xi_{k\ell}d\Omega = 2\iint_\Gamma [v_i](C_0)_{ijk\ell}\xi_{k\ell}n_j d\Gamma\,. \tag{5.7.40}$$

Denoting by $\boldsymbol{c}$ the vector with components

$$c_i = (C_0)_{ijk\ell}\xi_{k\ell}n_j\,, \tag{5.7.41}$$

and using Eq. (5.7.1), Eq. (5.7.40) becomes

$$2\iiint_\Omega (C_0)_{ijk\ell}\varepsilon_{ij}(\boldsymbol{v})\xi_{k\ell}d\Omega$$

$$= 2\iint_\Gamma [v_i]c_i d\Gamma = 2\iint_\Gamma ([v_N]c_N + [\boldsymbol{v}_T]\cdot \boldsymbol{c}_T)d\Gamma\,. \tag{5.7.42}$$

Applying Eq. (5.7.42) and completing the squares in the surface terms of Eq. (5.7.39) gives:

$$(\boldsymbol{C}-\boldsymbol{C}_0)_{ijk\ell}\xi_{ij}\xi_{k\ell} + \alpha^{-1}\iint_\Gamma |c_N|^2 d\Gamma + \beta^{-1}\iint_\Gamma |\boldsymbol{c}_T|^2 d\Gamma$$

$$= \min_{\boldsymbol{v}}\Bigg(\iiint_\Omega (\boldsymbol{C}-\boldsymbol{C}_0)_{ijk\ell}(\varepsilon_{ij}(\boldsymbol{v})+\xi_{ij})(\varepsilon_{k\ell}(\boldsymbol{v})+\xi_{k\ell})d\Omega$$

$$+\iiint_\Omega (C_0)_{ijk\ell}\varepsilon_{ij}(\boldsymbol{v})\varepsilon_{k\ell}(\boldsymbol{v})d\Omega + \alpha\iint_\Gamma |[v_N]+\alpha^{-1}c_N|^2 d\Gamma$$

$$+\beta\iint_\Gamma |[\boldsymbol{v}_T]+\beta^{-1}\boldsymbol{c}_T|^2 d\Gamma\Bigg). \tag{5.7.43}$$

Introducing a second-order tensor $\boldsymbol{\tau}$ and a vector $\boldsymbol{u}$ defined on Γ one can use Eq. (5.5.15) to rewrite the last two terms in the right-hand side of Eq. (5.7.43) as

$$\alpha\iint_\Gamma |[v_N]+\alpha^{-1}c_N|^2 d\Gamma + \beta\iint_\Gamma |[\boldsymbol{v}_T]+\beta^{-1}\boldsymbol{c}_T|^2 d\Gamma$$

$$= \max_{\boldsymbol{u}}\Bigg(\iint_\Gamma (2u_N([v_N]+\alpha^{-1}c_N) - \alpha^{-1}|u_N|^2)d\Gamma$$

$$+\iint_\Gamma (2\boldsymbol{u}_T\cdot([\boldsymbol{v}_T]+\beta^{-1}\boldsymbol{c}_T) - \beta^{-1}|\boldsymbol{u}_T|^2)d\Gamma\Bigg), \tag{5.7.44}$$

and the first term as

$$\iiint_\Omega (\boldsymbol{C}-\boldsymbol{C}_0)_{ijk\ell}(\varepsilon_{ij}(\boldsymbol{v})+\xi_{ij})(\varepsilon_{k\ell}(\boldsymbol{v})+\xi_{k\ell})d\Omega$$
$$= \max_{\boldsymbol{\tau}}\left(2\iiint_\Omega \tau_{ij}(\varepsilon_{ij}(\boldsymbol{v})+\xi_{ij})d\Omega - \iiint_\Omega (\boldsymbol{C}-\boldsymbol{C}_0)^{-1}_{ijk\ell}\tau_{ij}\tau_{k\ell}d\Omega\right), \tag{5.7.45}$$

for any $\boldsymbol{\tau}$ and $\boldsymbol{v}$. Introducing the Lagrangian $\underline{\mathcal{L}}(\boldsymbol{\tau},\boldsymbol{u},\boldsymbol{v})$ defined by:

$$\begin{aligned}\underline{\mathcal{L}}(\boldsymbol{\tau},\boldsymbol{u},\boldsymbol{v}) &= 2\iiint_\Omega \tau_{ij}\xi_{ij}d\Omega + 2\alpha^{-1}\iint_\Gamma u_N c_N d\Gamma + 2\beta^{-1}\iint_\Gamma \boldsymbol{u}_T\cdot\boldsymbol{c}_T d\Gamma\\ &\quad - \iiint_\Omega (\boldsymbol{C}(x)-\boldsymbol{C}_0)^{-1}_{ijk\ell}\tau_{ij}\tau_{k\ell}d\Omega - \alpha^{-1}\iint_\Gamma (u_N)^2 d\Gamma\\ &\quad - \beta^{-1}\iint_\Gamma |\boldsymbol{u}_T|^2 d\Gamma + 2\iiint_\Omega \tau_{ij}\varepsilon_{ij}(\boldsymbol{v})d\Omega + 2\iint_\Gamma u_N[v_N]d\Gamma\\ &\quad + 2\iint_\Gamma (\boldsymbol{u}_T)_i[(\boldsymbol{v}_T)_i]d\Gamma + \iiint_\Omega (C_0)_{ijk\ell}\varepsilon_{ij}(\boldsymbol{v})\varepsilon_{k\ell}(\boldsymbol{v})d\Omega\,,\end{aligned} \tag{5.7.46}$$

and applying Eqs. (5.7.44) and (5.7.45) to Eq. (5.7.43), we get

$$\begin{aligned}(\boldsymbol{C}-\boldsymbol{C}_0)_{ijk\ell}\xi_{ij}\xi_{k\ell} + \alpha^{-1}\iint_\Gamma |c_N|^2 d\Gamma + \beta^{-1}\iint_\Gamma |\boldsymbol{c}_T|^2 d\Gamma\\ = \min_{\boldsymbol{v}}\max_{\boldsymbol{\tau},\boldsymbol{u}}\mathcal{L}(\boldsymbol{\tau},\boldsymbol{u},\boldsymbol{v})\,.\end{aligned} \tag{5.7.47}$$

Observe next that

$$\begin{aligned}(\boldsymbol{C}-\boldsymbol{C}_0)_{ijk\ell}\xi_{ij}\xi_{k\ell} + \alpha^{-1}\iint_\Gamma |c_N|^2 d\Gamma + \beta^{-1}\iint_\Gamma |\boldsymbol{c}_T|^2 d\Gamma\\ = \min_{\boldsymbol{v}}\max_{\boldsymbol{\tau},\boldsymbol{u}}\mathcal{L}(\boldsymbol{\tau},\boldsymbol{u},\boldsymbol{v}) \geq \max_{\boldsymbol{\tau},\boldsymbol{u}}\min_{\boldsymbol{v}}\mathcal{L}(\boldsymbol{\tau},\boldsymbol{u},\boldsymbol{v})\,.\end{aligned} \tag{5.7.48}$$

Denoted by $\overset{*}{\boldsymbol{v}}$ the vector function that attains the minimum of

$$\begin{aligned}\min_{\boldsymbol{v}}\Bigg\{2\iiint_\Omega \tau_{ij}\varepsilon_{ij}(\boldsymbol{v})d\Omega + 2\iint_\Gamma u_N[v_N]d\Gamma\\ + 2\iint_\Gamma (\boldsymbol{u}_T)_i[(\boldsymbol{v}_T)_i]d\Gamma + \iiint_\Omega (C_0)_{ijk\ell}\varepsilon_{ij}(\boldsymbol{v})\varepsilon_{k\ell}(\boldsymbol{v})d\Omega\Bigg\}.\end{aligned} \tag{5.7.49}$$

By requiring the first variation of the above functional to be zero, the equation

$$\iiint_\Omega (C_0)_{ijk\ell}\varepsilon_{ij}(\boldsymbol{v})\varepsilon_{k\ell}(\delta\boldsymbol{v})d\Omega + \iint_\Gamma (\boldsymbol{u})_i[(\delta\boldsymbol{v})_i]d\Gamma + \iiint_\Omega \tau_{ij}\varepsilon_{ij}(\delta\boldsymbol{u})d\Omega = 0 \qquad (5.7.50)$$

implies that the minimum is attained for $\overset{*}{\boldsymbol{v}}$, which satisfies

$$\frac{\partial}{\partial x_j}((C_0)_{ijk\ell}\varepsilon_{k\ell}(\overset{*}{\boldsymbol{v}})) = -\frac{\partial}{\partial x_j}\tau_{ij}\,, \quad \text{in } \Omega_1 \text{ and } \Omega_2\,, \qquad (5.7.51)$$

$$((C_0)_{ijk\ell}\varepsilon_{k\ell}(\overset{*}{\boldsymbol{v}}) + \tau_{k\ell})_2 n_j = ((C_0)_{ijk\ell}\varepsilon_{k\ell}(\overset{*}{\boldsymbol{v}}) + \tau_{k\ell})_1 n_j\,, \quad \text{on } \Gamma\,, \qquad (5.7.52)$$

$$\left[\overset{*}{v}_i\right] = -u_i\,, \quad \text{on } \Gamma\,, \qquad (5.7.53)$$

and the minimum value for Eq. (5.7.49) is obtained taking $\boldsymbol{v} = \overset{*}{\boldsymbol{v}}$

$$-\iiint_\Omega (C_0)_{ijk\ell}\varepsilon_{ij}(\overset{*}{\boldsymbol{v}})\varepsilon_{k\ell}(\overset{*}{\boldsymbol{v}})d\Omega\,. \qquad (5.7.54)$$

We next observe that $\overset{*}{\boldsymbol{v}}$ can be written as $\overset{*}{\boldsymbol{v}} = \boldsymbol{v}^\tau + \boldsymbol{v}^u$ where $\boldsymbol{v}^\tau$ and $\boldsymbol{v}^u$ solve, respectively, the problems

$$\frac{\partial}{\partial x_j}((C_0)_{ijk\ell}\varepsilon_{k\ell}(\boldsymbol{v}^\tau)) = -\frac{\partial \tau_{ij}}{\partial x_j}\,, \quad \text{in } \Omega_1 \text{ and } \Omega_2\,, \qquad (5.7.55)$$

$$((C_0)_{ijk\ell}\varepsilon_{k\ell}(\boldsymbol{v}^\tau) + \tau_{k\ell})_2 n_j = ((C_0)_{ijk\ell}\varepsilon_{k\ell}(\boldsymbol{v}^\tau) + \tau_{k\ell})_1 n_j\,, \quad \text{on } \Gamma\,, \qquad (5.7.56)$$

$$\left[v_i^\tau\right] = 0\,, \quad \text{on } \Gamma\,, \qquad (5.7.57)$$

and

$$\frac{\partial}{\partial x_j}((C_0)_{ijk\ell}\varepsilon_{k\ell}(\boldsymbol{v}^u)) = 0, \quad \text{in } \Omega_1 \text{ and } \Omega_2\,, \qquad (5.7.58)$$

$$((C_0)_{ijk\ell}\varepsilon_{k\ell}(\boldsymbol{v}^u) + \tau_{k\ell})_2 n_j = ((C_0)_{ijk\ell}\varepsilon_{k\ell}(\boldsymbol{v}^u) + \tau_{k\ell})_1 n_j\,, \quad \text{on } \Gamma\,, \qquad (5.7.59)$$

$$\left[v_i^u\right] = -u_i\,, \quad \text{on } \Gamma\,. \qquad (5.7.60)$$

Now Eq. (5.7.48) becomes

$$(\boldsymbol{C} - \boldsymbol{C}_0)_{ijk\ell}\xi_{ij}\xi_{k\ell} + \alpha^{-1}\iint_\Gamma |c_N|^2 d\Gamma + \beta^{-1}\iint_\Gamma |\boldsymbol{c}_T|^2 d\Gamma \geq \max_{\boldsymbol{\tau},\boldsymbol{u}} \mathcal{L}(\boldsymbol{\tau},\boldsymbol{u},\overset{*}{\boldsymbol{v}}) = \max_{\boldsymbol{\tau},\boldsymbol{u}} \mathcal{L}(\boldsymbol{\tau},\boldsymbol{u},\boldsymbol{v}^\tau + \boldsymbol{v}^u)\,, \qquad (5.7.61)$$

where

$$\mathcal{L}(\boldsymbol{\tau}, \boldsymbol{u}, \overset{*}{\boldsymbol{v}}) = 2\left(\iiint_\Omega \tau_{ij}\xi_{ij}d\Omega + \alpha^{-1}\iint_\Gamma u_N c_N + \beta^{-1}\iint_\Gamma (\boldsymbol{u}_T)_i(\boldsymbol{c}_T)_i \, d\Gamma\right)$$
$$-\left(\iiint_\Omega (\boldsymbol{C}(x) - \boldsymbol{C}_0)^{-1}_{ijk\ell}\tau_{ij}\tau_{k\ell}d\Omega + \alpha^{-1}\iint_\Gamma |u_N|^2 d\Gamma\right.$$
$$+\beta^{-1}\iint_\Gamma |\boldsymbol{u}_T|^2 d\Gamma + \iiint_\Omega (C_0)_{ijk\ell}(\varepsilon_{ij}(\boldsymbol{v}^\tau) + \varepsilon_{ij}(\boldsymbol{v}^u))(\varepsilon_{k\ell}(\boldsymbol{v}^\tau)$$
$$\left. + \varepsilon_{k\ell}(\boldsymbol{v}^u))d\Omega\right). \tag{5.7.62}$$

We will now split the right-hand side of Eq. (5.7.62) into a linear part

$$\underline{L}(\boldsymbol{\tau}, \boldsymbol{u}) = \iiint_\Omega \tau_{ij}\xi_{ij}d\Omega + \alpha^{-1}\iint_\Gamma u_N c_N d\Gamma + \beta^{-1}\iint_\Gamma (\boldsymbol{u}_T)_i(\boldsymbol{c}_T)_i d\Gamma, \tag{5.7.63}$$

and a quadratic part $\underline{Q}(\boldsymbol{\tau}, \boldsymbol{v})$ is given by

$$\underline{Q}(\boldsymbol{\tau}, \boldsymbol{u}) = \iiint_\Omega (\boldsymbol{C} - \boldsymbol{C}_0)^{-1}_{ijk\ell}\tau_{ij}\tau_{k\ell}d\Omega + \alpha^{-1}\iint_\Gamma |u_N|^2 d\Gamma + \beta^{-1}\iint_\Gamma |\boldsymbol{u}_T|^2 d\Gamma$$
$$+ \iiint_\Omega (C_0)_{ijk\ell}(\varepsilon_{ij}(\boldsymbol{v}^\tau) + \varepsilon_{ij}(\boldsymbol{v}^u))(\varepsilon_{k\ell}(\boldsymbol{v}^\tau) + \varepsilon_{k\ell}(\boldsymbol{v}^u))d\Omega. \tag{5.7.64}$$

Equation (5.7.61) then becomes

$$(\boldsymbol{C} - \boldsymbol{C}_0)_{ijk\ell}\xi_{ij}\xi_{k\ell} + \alpha^{-1}\iint_\Gamma |c_N|^2 d\Gamma + \beta^{-1}\iint_\Gamma |\boldsymbol{c}_T|^2 d\Gamma$$
$$\geq \max_{\boldsymbol{\tau}, \boldsymbol{u}}(2\underline{L}(\boldsymbol{\tau}, \boldsymbol{u}) - \underline{Q}(\boldsymbol{\tau}, \boldsymbol{u})). \tag{5.7.65}$$

One observes that the choice of bulk and surface polarizations, $\boldsymbol{\tau}$ and $\boldsymbol{v}$, is consistent with the actual displacements inside the composite, i.e.,

$$\tau_{ij} = (\boldsymbol{C} - \boldsymbol{C}_0)_{ijk\ell}(\varepsilon_{k\ell}(\boldsymbol{v}) + \xi_{k\ell}), \tag{5.7.66}$$
$$u_N = [v_N] + \alpha^{-1}c_N, \tag{5.7.67}$$
$$\boldsymbol{u}_T = [\boldsymbol{v}_T] + \beta^{-1}\boldsymbol{c}_T \tag{5.7.68}$$

makes Eq. (5.7.65) hold with equality.

To conclude, we rewrite the left-hand side of Eq. (5.7.65), use the definition of the vector $\boldsymbol{c}$ in Eq. (5.7.41) and expand c_N and $\boldsymbol{c}_T$ to find

$$c_N^2 = (\Gamma_h(\boldsymbol{n}))_{ijk\ell}(C_0)_{k\ell mn}\xi_{mn}(C_0)_{ijps}\xi_{ps}\,, \tag{5.7.69}$$

$$|\boldsymbol{c}_T|^2 = \frac{1}{2}(\Gamma_s(\boldsymbol{n}))_{ijk\ell}(C_0)_{k\ell mn}\xi_{mn}(C_0)_{ijps}\xi_{ps}\,, \tag{5.7.70}$$

where $\Gamma_h(\boldsymbol{n})$ and $\Gamma_s(\boldsymbol{n})$ are tensor-valued functions of the unit normal given by

$$(\Gamma_h(\boldsymbol{n}))_{ijk\ell} = n_i n_j n_k n_\ell\,. \tag{5.7.71}$$

$$(\Gamma_s(\boldsymbol{n}))_{ijk\ell} = \frac{1}{2}(n_i n_\ell \delta_{jk} + n_i n_k \delta_{j\ell} + n_j n_\ell \delta_{ik} + n_j n_k \delta_{i\ell}) - 2n_i n_j n_k n_\ell\,. \tag{5.7.72}$$

From these results we obtain the variational principle which states

$$\boxed{\begin{aligned} &(\boldsymbol{C} - \boldsymbol{C}_0)_{ijk\ell}\xi_{ij}\xi_{k\ell} + \\ &\quad + \left(\alpha^{-1}\iint_\Gamma (\Gamma_h(\boldsymbol{n}))_{ijk\ell}d\Gamma + (2\beta)^{-1}\iint_\Gamma (\Gamma_s(\boldsymbol{n}))_{ijk\ell}d\Gamma\right) \\ &\quad \times (C_0)_{k\ell mn}\xi_{mn}(C_0)_{ijps}\xi_{ps} \\ &\quad = \max_{\boldsymbol{\tau},\boldsymbol{u}}\left(2\underline{L}(\boldsymbol{\tau},\boldsymbol{u}) - \underline{Q}(\boldsymbol{\tau},\boldsymbol{u})\right). \end{aligned}} \tag{5.7.73}$$

In a similar way, a complementary variational principle can be obtained starting from Eq. (5.7.38). Only the results are stated here. For details the reader is referred to Lipton and Vernescu (1995).

Let us introduce an isotropic comparison material such that $\boldsymbol{C}_0 > \boldsymbol{C}_2$ and for any second-order tensor $\boldsymbol{\tau}$ defined on Ω and any vector $\boldsymbol{u}$ defined on Γ, we formulate two auxiliary elasticity-type problems: (i) find the potential $\boldsymbol{\psi}^\tau$, periodic in Ω, which solves:

$$\frac{\partial}{\partial x_j}((C_0)_{ijk\ell}\varepsilon_{k\ell}(\boldsymbol{\psi}^\tau)) = \frac{\partial}{\partial x_j}((C_0)_{ijk\ell}\tau_{k\ell}) \quad \text{in } \Omega_1 \text{ and } \Omega_2\,, \tag{5.7.74}$$

$$[(C_0)_{ijk\ell}\varepsilon_{k\ell}(\boldsymbol{\psi}^\tau) - \tau_{k\ell}]n_j = 0\,, \quad [\psi_i^\tau] = 0 \text{ on } \Gamma\,, \tag{5.7.75}$$

and (ii) find the potential $\boldsymbol{\psi}^u$, periodic in Ω, solution to

$$\frac{\partial}{\partial x_j}((C_0)_{ijk\ell}\varepsilon_{k\ell}(\boldsymbol{\psi}^u)) = 0 \quad \text{in } \Omega_1 \text{ and } \Omega_2\,, \tag{5.7.76}$$

$$[(C_0)_{ijk\ell}\varepsilon_{k\ell}(\boldsymbol{\psi}^u)]n_j = 0\,, \quad [\psi_i^u] = -u_i \quad \text{on } \Gamma\,. \tag{5.7.77}$$

The complementary variational principle states that

$$\boxed{(\boldsymbol{C}^{-1} - \boldsymbol{C}_0^{-1})_{ijk\ell}\sigma_{ij}\sigma_{k\ell} = \max_{\boldsymbol{\tau},\boldsymbol{u}}(2\bar{L}(\boldsymbol{\tau},\boldsymbol{u}) - \bar{Q}(\boldsymbol{\tau},\boldsymbol{u}))}\,, \tag{5.7.78}$$

where the max is taken over all $\boldsymbol{\tau}$, Ω-periodic second-order tensors with

$$\frac{\partial \tau_{ij}}{\partial x_j} = 0\,, \quad \text{in } \Omega\,, \quad \langle \tau_{ij} \rangle = \sigma_{ij}\,, \quad [\tau_{ij} n_j] = 0\,, \quad \text{on } \Gamma\,, \tag{5.7.79}$$

the linear form $\bar{L}$ is given by:

$$\bar{L}(\boldsymbol{\tau},\boldsymbol{u}) = 2\iiint_\Omega \tau_{ij}\sigma_{ij}d\Omega + 2\iint_\Gamma \sigma_{ij}u_i n_j d\Gamma\,, \tag{5.7.80}$$

and the quadratic form $\bar{Q}$ is given by:

$$\begin{aligned}\bar{Q}(\boldsymbol{\tau},\boldsymbol{u}) = &\iiint_\Omega (\boldsymbol{C}^{-1} - \boldsymbol{C}_0^{-1})^{-1}_{ijk\ell}\tau_{ij}\tau_{k\ell}d\Omega + \alpha\iint_\Gamma |u_N|^2 d\Gamma + \beta\iint_\Gamma |\boldsymbol{u}_T|^2 d\Gamma \\ &+ \iiint_\Omega (C_0)_{ijk\ell}(\zeta_{ij} - \langle\zeta_{ij}\rangle)(\zeta_{k\ell} - \langle\zeta_{k\ell}\rangle)d\Omega\,,\end{aligned} \tag{5.7.81}$$

with $\zeta_{ij} = \varepsilon_{ij}(\boldsymbol{\psi}^u + \boldsymbol{\psi}^\tau) - \tau_{ij}$.

5.7.3. Bounds for Particulate Composites

Bounds on the effective elasticity can be obtained by taking particular choices of admissible functions in the variational principles. In this section we will consider particulate composites formed by stiffer particles, of elasticity $\boldsymbol{C}_2$ in a more compliant matrix of elasticity $\boldsymbol{C}_1$ ($\boldsymbol{C}_1 < \boldsymbol{C}_2$).

In the variational principle, Eq. (5.7.73), we choose

$$\boldsymbol{C}_0 = \boldsymbol{C}_1\,, \quad \tau_{ij} = \chi_2\eta_{ij}\,, \quad u_i = r_{ij}n_j\,, \tag{5.7.82}$$

with $\boldsymbol{\eta} = (\eta_{ij})$ and $\boldsymbol{r} = (r_{ij})$ constant, symmetric second-order tensors, χ_2 the characteristic function of material 2 and $\boldsymbol{n} = (n_i)$ the normal vector on the interface Γ pointing toward material 2. With these choices the linear form in Eq. (5.7.63) becomes

$$\begin{aligned}\underline{L}(\boldsymbol{r},\boldsymbol{n}) = &\,\theta_2\eta_{ij}\xi_{ij} \\ &+ (C_1)_{ijk\ell}\left(\alpha^{-1}\iint_\Gamma (\Gamma_h(\boldsymbol{n}))_{ijpq}d\Gamma\right. \\ &\left. + (2\beta)^{-1}\iint_\Gamma (\Gamma_s(\boldsymbol{n}))_{ijpq}d\Gamma\right) r_{pq}\xi_{k\ell}\,.\end{aligned} \tag{5.7.83}$$

The first three terms in the definition of the quadratic form $\underline{Q}$ in Eq. (5.7.64) can be easily computed to be

$$\theta_2(\boldsymbol{C}_2-\boldsymbol{C}_1)^{-1}_{ijk\ell}\eta_{ij}\eta_{k\ell} \\ +\left(\alpha^{-1}\iint_\Gamma(\Gamma_h(\boldsymbol{n}))_{ijk\ell}d\Gamma+(2\beta)^{-1}\iint_\Gamma(\Gamma_s(\boldsymbol{n}))_{ijk\ell}d\Gamma\right)r_{ij}r_{k\ell}\,. \tag{5.7.84}$$

In order to compute the fourth term of the form $\underline{Q}$ we need to compute the solutions of the problems (Eqs. (5.7.55)–(5.7.57)) and (Eqs. (5.7.58)–(5.7.60)) corresponding to the choices of fields in Eq. (5.7.82). It is easy to check from Eqs. (5.7.55)–(5.7.57) that $C_1\boldsymbol{\varepsilon}(\boldsymbol{v}^\tau)=-\chi_2\boldsymbol{\eta}$. Also, from Eqs. (5.7.58)–(5.7.60) we see that $C_1\boldsymbol{\varepsilon}(\boldsymbol{v}^u)=-\boldsymbol{r}$ in Ω_2. Therefore

$$\iiint_{\Omega_2}(\boldsymbol{C}_1)_{ijk\ell}\varepsilon_{ij}(\boldsymbol{v}^u)\varepsilon_{k\ell}(\boldsymbol{v}^u)d\Omega=\theta_2(\boldsymbol{C}_1)^{-1}_{ijk\ell}r_{ij}r_{k\ell}\,.$$

In Ω_1 the problems (Eqs. (5.7.58)–(5.7.60)) become

$$\frac{\partial}{\partial x_j}((C_1)_{ijk\ell}\varepsilon_{k\ell}(\boldsymbol{v}^u))=0 \quad \text{in } \Omega_1\,, \tag{5.7.85}$$

$$(C_1)_{ijk\ell}\varepsilon_{k\ell}(\boldsymbol{v}^u)n_j=-r_{ij}n_j \quad \text{on } \Gamma\,. \tag{5.7.86}$$

After multiplying Eq. (5.7.85) by v_i^u and using the periodicity and the boundary condition (Eq. (5.7.86)) we get

$$\iiint_{\Omega_1}(C_1)_{ijk\ell}\varepsilon_{ij}(\boldsymbol{v}^u)\varepsilon_{k\ell}(\boldsymbol{v}^u)d\Omega=\iiint_{\Omega_1}r_{ij}\varepsilon_{ij}(\boldsymbol{v}^u)d\Omega\,. \tag{5.7.87}$$

Let us now use Eq. (5.7.87) to compute

$$\iiint_{\Omega_1}(C_1)_{ijk\ell}(\varepsilon_{ij}(\boldsymbol{v}^u)+(C_1)^{-1}_{ijpq}r_{pq})(\varepsilon_{k\ell}(\boldsymbol{v}^u)+(C_1)^{-1}_{k\ell pq}r_{pq})d\Omega \\ =\iiint_{\Omega_1}((C_1)_{ijk\ell}\varepsilon_{ij}(\boldsymbol{v}^u)\varepsilon_{k\ell}(\boldsymbol{v}^u)+2r_{ij}\varepsilon_{ij}(\boldsymbol{v}^u))d\Omega+\theta_1(C_1)^{-1}_{ijk\ell}r_{ij}r_{k\ell} \\ =-\iiint_{\Omega_1}(C_1)_{ijk\ell}\varepsilon_{ij}(\boldsymbol{v}^u)\varepsilon_{k\ell}(\boldsymbol{v}^u)d\Omega+\theta_1(C_1)^{-1}_{ijk\ell}r_{ij}r_{k\ell}\,. \tag{5.7.88}$$

By comparison to Eq. (5.2.7), the left-hand side of the above equation is seen to represent the effective elasticity tensor C^* of a composite with the same geometry but with soft material (i.e., holes) instead of material 2.

Thus the left-hand side can be written as $C^*_{ijk\ell}(C_1)^{-1}_{ijpq}r_{pq}(C_1)^{-1}_{k\ell mn}r_{mn}$ and

$$\iiint_{\Omega_1}(C_1)_{ijk\ell}\varepsilon_{ij}(\boldsymbol{v}^u)\varepsilon_{k\ell}(\boldsymbol{v}^u)d\Omega$$
$$= -C^*_{ijk\ell}(C_1)^{-1}_{ijpq}r_{pq}(C_1)^{-1}_{k\ell mn}r_{mn} + \theta_1(C_1)^{-1}_{ijk\ell}r_{ij}r_{k\ell}\,. \quad (5.7.89)$$

Before computing the quadratic form $\underline{Q}$ let us introduce the fourth-order tensor F defined by

$$F_{ijk\ell} = \alpha^{-1}\iint_{\Gamma}(\Gamma_h(\boldsymbol{n}))_{ijk\ell}d\Gamma + (2\beta)^{-1}\iint_{\Gamma}(\Gamma_s(\boldsymbol{n}))_{ijk\ell}d\Gamma\,. \quad (5.7.90)$$

Now from Eqs. (5.7.84), (5.7.85), and (5.7.89) we have

$$\underline{Q}(\boldsymbol{\eta},\boldsymbol{r}) = \theta_2(\boldsymbol{C}_2-\boldsymbol{C}_1)^{-1}_{ijk\ell}\eta_{ij}\eta_{k\ell} + F_{ijk\ell}r_{ij}r_{k\ell} + \theta_2(C_1)^{-1}_{ijk\ell}\eta_{ij}\eta_{k\ell}$$
$$+ 2\theta_2(C_1)^{-1}_{ijk\ell}\eta_{ij}r_{k\ell} - C^*_{ijk\ell}(C_1)^{-1}_{ijpq}r_{pq}(C_1)^{-1}_{k\ell mn}r_{mn}$$
$$+ (C_1)^{-1}_{ijk\ell}r_{ij}r_{k\ell}\,. \quad (5.7.91)$$

The bound equation (5.7.73) becomes

$$(\boldsymbol{C}-\boldsymbol{C}_1)_{ijk\ell}\xi_{ij}\xi_{k\ell} + F_{ijk\ell}(C_1)_{ijmn}\xi_{mn}(C_1)_{k\ell pq}\xi_{pq}$$
$$\geq \max_{\boldsymbol{\eta},\boldsymbol{r}}(2\underline{L}(\boldsymbol{\eta},\boldsymbol{r}) - \underline{Q}(\boldsymbol{\eta},\boldsymbol{r}))\,, \quad (5.7.92)$$

which can be written as:

$$(\boldsymbol{C}-\boldsymbol{C}_1)_{ijk\ell}\xi_{ij}\xi_{k\ell} + F_{ijk\ell}(C_1)_{ijmn}\xi_{mn}(C_1)_{k\ell pq}\xi_{pq}$$
$$\geq \max_{\boldsymbol{\eta},\boldsymbol{r}}\left\{2\begin{pmatrix}\theta_2\boldsymbol{\xi}\\ \boldsymbol{\zeta}\end{pmatrix} - \begin{pmatrix}\theta_2((C_2-C_1)^{-1}+C_1^{-1}) & \theta_2 C_1^{-1}\\ \theta_2 C_1^{-1} & F+C_1^{-1}-C_1^{-1}C^*C_1^{-1}\end{pmatrix}\right.$$
$$\left.\times\begin{pmatrix}\boldsymbol{\eta}\\ \boldsymbol{r}\end{pmatrix}\right\}\cdot\begin{pmatrix}\boldsymbol{\eta}\\ \boldsymbol{r}\end{pmatrix} \quad (5.7.93)$$

after using Eqs. (5.7.83) and (5.7.91), where $\boldsymbol{\zeta}$ is the second-order tensor with components $\zeta_{ij} = (C_1)_{ijk\ell}F_{k\ell mn}\xi_{mn}$. The fourth-order tensor $C_1^{-1}C^*C_1^{-1}$ has the $(ijk\ell)$ component given by $(C_1)^{-1}_{ijmn}C^*_{mnpq}(C_1)^{-1}_{pqk\ell}$. The lower bound is obtained after computing the maximum over all pairs of symmetric tensors $(\boldsymbol{\eta}, \boldsymbol{r})$, yielding

$$
\begin{aligned}
&(\mathcal{C}_{ijk\ell} - (C_1)_{ijk\ell} + F_{mnpq}(C_1)_{mnij}(C_1)_{pqk\ell})\xi_{ij}\xi_{k\ell} \\
&\quad \geq \begin{pmatrix} \theta_2\boldsymbol{\xi} \\ \boldsymbol{\zeta} \end{pmatrix} \cdot \begin{pmatrix} \theta_2((\boldsymbol{C}_2 - \boldsymbol{C}_1)^{-1} + \boldsymbol{C}_1^{-1}) & \theta_2\boldsymbol{C}_1^{-1} \\ \theta_2\boldsymbol{C}_1^{-1} & F + \boldsymbol{C}_1^{-1} - \boldsymbol{C}_1^{-1}\boldsymbol{C}^*\boldsymbol{C}_1^{-1} \end{pmatrix}^{-1} \\
&\qquad \times \begin{pmatrix} \theta_2\boldsymbol{\xi} \\ \boldsymbol{\zeta} \end{pmatrix}.
\end{aligned} \tag{5.7.94}
$$

From this one gets the tensor inequality:

$$
\begin{aligned}
&\boldsymbol{\mathcal{C}} - \boldsymbol{C}_1 + \boldsymbol{C}_1 F \boldsymbol{C}_1 \\
&\quad \geq ((F + \boldsymbol{C}_1^{-1} - \boldsymbol{C}_1^{-1}\boldsymbol{C}^* \boldsymbol{C}_1^{-1})((\boldsymbol{C}_2 - \boldsymbol{C}_1)^{-1} + \boldsymbol{C}_1^{-1}) - \theta_2 \boldsymbol{C}_1^{-1}\boldsymbol{C}_1^{-1})^{-1} \\
&\qquad \times (\theta_2(\boldsymbol{C}_1^{-1} - \boldsymbol{C}_1^{-1}\boldsymbol{C}^*\boldsymbol{C}_1^{-1} - \boldsymbol{C}_1 F \boldsymbol{C}_1^{-1}) \\
&\qquad + \boldsymbol{C}_1 F((\boldsymbol{C}_2 - \boldsymbol{C}_1)^{-1} + \boldsymbol{C}_1^{-1})\boldsymbol{C}_1 F).
\end{aligned} \tag{5.7.95}
$$

This tensor inequality yields the lower bounds for the bulk modulus $\mathcal{K}$ and shear modulus $\mathcal{G}$, of the effective elasticity tensor $\boldsymbol{\mathcal{C}}$ in terms of the bulk moduli K_1, K_2, K^* and shear moduli G_1, G_2, G^* of the elasticity tensors $\boldsymbol{C}_1$, $\boldsymbol{C}_2$ and $\boldsymbol{C}^*$ respectively. We first need to compute the following traces

$$
\begin{aligned}
&Tr P_b \boldsymbol{C}_1 = 3K_1, \quad Tr P_b \boldsymbol{C}_2 = 3K_2, \\
&Tr P_b \boldsymbol{C}^* = 3K^*, \quad Tr P_b F = \frac{s}{3\alpha},
\end{aligned} \tag{5.7.96}
$$

$$
\begin{aligned}
&Tr P_s \boldsymbol{C}_1 = 2G_1, \quad Tr P_s \boldsymbol{C}_2 = 2G_2, \\
&Tr P_s \boldsymbol{C}^* = 2G^*, \quad Tr P_s F = \frac{2s}{3\alpha} + \frac{s}{\beta},
\end{aligned} \tag{5.7.97}
$$

where s is the interfacial surface area of the particles. This can be immediately obtained by using the definitions (Eqs. (A.3), (A.4) of Appendix 5A). Indeed the first identity in Eq. (5.7.96) can be immediately obtained

$$
(P_b C_1)_{ijmn} = \frac{1}{d}\delta_{ij}\delta_{k\ell}(C_1)_{k\ell mn} = \frac{1}{d}\delta_{ij}(C_1)_{kkmn} = \frac{1}{d}\delta_{ij}(K_1 d\delta_{mn}), \tag{5.7.98}
$$

$$
Tr P_h C_1 = (P_b C_1)_{ijij} = K_1 \delta_{ij}\delta_{ij} = 3K_1, \tag{5.7.99}
$$

and similarly for the others.

If we apply on both sides of Eq. (5.7.95) the trace of the bulk projection TrP_b the lower bound for the bulk modulus is obtained to be

$$3\mathcal{K} \geq 3K_1 - \cfrac{3K_1}{\cfrac{1}{1-\cfrac{K^*}{K_1}} + \cfrac{1}{\theta_2 K_1\left(\cfrac{1}{K_2} - \cfrac{1}{K_1} + \cfrac{s}{\alpha\theta_2}\right)}}. \tag{5.7.100}$$

Similarly, applying the trace of the shear projection TrP_s the lower bound for the shear trace is obtained

$$2\mathcal{G} \geq 2G_1 - \cfrac{2G_1}{\cfrac{1}{1-\cfrac{G^*}{G_1}} + \cfrac{1}{G_1\theta_2\left(\cfrac{1}{G_2} - \cfrac{1}{G_1} + \cfrac{2}{5}\left(\cfrac{2s}{3\alpha\theta_2} + \cfrac{s}{\beta\theta_2}\right)\right)}}. \tag{5.7.101}$$

Both lower bounds depend on K^* and G^* that represent the bulk and shear moduli of the effective elasticity tensor C^* of a composite with the same geometry but material 2 replaced by voids (holes). These are parameters that incorporate the geometric information on the composite's microstructure. The bounds are monotone increasing in these two parameters; therefore the case when K^* and G^* are taken to be zero, also provide a lower bound

$$\begin{aligned} 3\mathcal{K} &\geq 3K_1 - \cfrac{3K_1}{\cfrac{1}{1-\cfrac{K^*}{K_1}} + \cfrac{1}{\theta_2 K_1\left(\cfrac{1}{K_2} - \cfrac{1}{K_1} + \cfrac{s}{\alpha\theta_2}\right)}} \\ &\geq \left(\frac{\theta_1}{3K_1} + \frac{\theta_2}{3K_2} + \frac{s}{3\alpha}\right)^{-1}, \end{aligned} \tag{5.7.102}$$

$$\begin{aligned} 2G &\geq 2G_1 - \cfrac{2G_1}{\cfrac{1}{1-\cfrac{G^*}{G_1}} + \cfrac{1}{G_1\theta_2\left(\cfrac{1}{G_2} - \cfrac{1}{G_1} + \cfrac{2}{5}\left(\cfrac{2s}{3\alpha\theta_2} + \cfrac{s}{\beta\theta_2}\right)\right)}} \\ &\geq \left(\frac{\theta_1}{2G_2} + \frac{\theta_2}{2G_2} + \frac{s}{5}\left(\frac{2}{3\alpha} + \frac{1}{\beta}\right)\right)^{-1}. \end{aligned} \tag{5.7.103}$$

These latter bounds represent the harmonic mean lower bounds for the case of elastic partially cohesive composites bonding and were first obtained by Hashin (1992).

In a similar way, starting from the complementary variational principle (Eq. (5.7.78)) one can get a tensor inequality that provides the upper bound for the effective tensor. We denote the region occupied by the mth particle by Y_m and its boundary by ∂Y_m, the position vector of its center of mass by $\boldsymbol{r}^{(m)}$. A point on the surface ∂Y_m with position vector $\boldsymbol{x}$ has the position vector from the center of the particle denoted by $\boldsymbol{y}^{(m)} = \boldsymbol{x} - \boldsymbol{r}^{(m)}$. The fourth-order tensors $\boldsymbol{A}$, $\boldsymbol{B}$, $\boldsymbol{M}$, and $\boldsymbol{R}$ are defined as:

$$A_{ijk\ell} = \theta_1\theta_2(C_2)_{ijk\ell} - 2\sum_{\boldsymbol{m}\neq\boldsymbol{0}} |\hat{\chi}_1(\boldsymbol{m})|^2 (C_2)_{ijmn}\omega_{mnpq}(C_2)_{pqk\ell}$$

$$- \sum_{\boldsymbol{m}\neq\boldsymbol{0}} |\hat{\chi}_1(\boldsymbol{m})|^2 \omega_{ijmn}\omega_{mnpq}(C_2)_{pqk\ell}\,, \tag{5.7.104}$$

$$B_{ijk\ell} = \sum_m \int_{\Gamma_m} \frac{1}{4}(n_i y_k^{(m)}\delta_{j\ell} + n_j y_k^{(m)}\delta_{i\ell} + n_i y_\ell^{(m)}\delta_{jk} + n_j y_\ell^{(m)}\delta_{ik})\,d\Gamma\,, \tag{5.7.105}$$

$$M_{ijk\ell} = \sum_m \int_{\Gamma_m} \frac{1}{4}(y_i^{(m)}y_\ell^{(m)}\delta_{kj} + y_i^{(m)}y_k^{(m)}\delta_{\ell j}$$

$$+ y_j^{(m)}y_\ell^{(m)}\delta_{ki} + y_j^{(m)}y_k^{(m)}\delta_{\ell i})\,d\Gamma\,, \tag{5.7.106}$$

$$R_{ijk\ell} = \sum_m \int_{\Gamma_m} \frac{1}{4}(y_i^{(m)}n_j y_k^{(m)}n_\ell + y_i^{(m)}n_j y_\ell^{(m)}n_k$$

$$+ y_j^{(m)}n_i y_k^{(m)}n_\ell + y_j^{(m)}n_i y_\ell^{(m)}n_k)\,d\Gamma\,, \tag{5.7.107}$$

where the fourth-order tensor $\boldsymbol{\omega}$ has components $\omega_{mnpq} = (2G_2)^{-1} (\Gamma_s(\boldsymbol{m}))_{mnpq} + r(\Gamma_h(\boldsymbol{m}))_{mnpq}$.

The tensor inequality which provides the upper bound is

$$(C^{-1}_{ijk\ell} - (C_2^{-1})_{ijk\ell})\sigma_{ij}\sigma_{k\ell}$$

$$\geq \begin{pmatrix}\theta_1\boldsymbol{\sigma}\\ \boldsymbol{B\sigma}\end{pmatrix}^T \cdot \begin{pmatrix}\theta_1(\boldsymbol{C}_1^{-1} - \boldsymbol{C}_2^{-1})^{-1} + A & -A\\ -A & \beta M + (\alpha-\beta)R + A\end{pmatrix}^{-1}$$

$$\times \begin{pmatrix}\theta_1\boldsymbol{\sigma}\\ \boldsymbol{B\sigma}\end{pmatrix}, \tag{5.7.108}$$

where $\boldsymbol{\sigma}$ is any second-order symmetric tensor. By taking the trace of the bulk projection in Eq. (5.7.108) we obtain the upper bound on the bulk trace of the effective compliance

$$3\mathcal{K} \leq \left((3K_1)^{-1} + \frac{(\theta_2^2 - g_b\theta_2\Delta_b)(1+\Delta_b(1-t_b)3K_2)}{g_b(1+\theta_2\Delta_b(1-t_b)3K_2)+\theta_1\theta_2(1-t_b)3K_2}\right)^{-1}, \tag{5.7.109}$$

where $\Delta_b = ((3K_1)^{-1} - (3K_2)^{-1})$, $t_b = 3K_2/(3K_2+4G_2)$, and

$$g_b = \frac{\beta}{3}\sum_m \int_{\Gamma_m} |y^{(m)}|^2 d\Gamma + \frac{(\alpha-\beta)}{3}\sum_m \int_{\Gamma_m} (y^{(m)}\cdot n)^2 d\Gamma\,. \tag{5.7.110}$$

Similarly, by taking the shear projection in Eq. (5.7.108) we obtain the upper bound on the shear trace of the effective compliance

$$2\mathcal{G} \leq \left((2G_1)^{-1} + \frac{(\theta_2^2 - g_s\theta_2\Delta_s)(1+\Delta_s(1-t_s)2G_2)}{g_s(1+\theta_2\,\Delta_s(1-t_s)2G_2)+\theta_1\theta_2(1-t_s)2G_2}\right)^{-1}, \tag{5.7.111}$$

where $\Delta_s = ((2G_1)^{-1} - (2G_2)^{-1})$, $t_s = (6/5)(K_2+2G_2)/(3K_2+4G_2)$, and

$$\begin{aligned} g_s = &\left(\frac{\beta}{5}\right)\left(\frac{7}{6}\sum_m \int_{\Gamma_m} |\boldsymbol{y}^{(m)}|^2 d\Gamma - \frac{1}{6}\sum_m \int_{\Gamma_m} (y_i^{(m)} n_i)^2 d\Gamma\right) \\ &+ \left(\frac{\alpha}{5}\right)\left(\frac{1}{2}\sum_m \int_{\Gamma_m} |\boldsymbol{y}^{(m)}|^2 d\Gamma + \frac{1}{6}\sum_m \int_{\Gamma_m} (y_i^{(m)} n_i)^2 d\Gamma\right). \end{aligned} \tag{5.7.112}$$

The upper bounds (Eqs. (5.7.111) and (5.7.112)) are increasing in the parameters g_b and g_s, respectively. As a limit case, when both α and β tend to ∞, that is when the composite has perfect bonding between phases, g_b and g_s also tend to ∞ and the corresponding upper bounds become

$$3\mathcal{K} \leq 3K_2 + \frac{\theta_1}{(1/(3K_1-3K_2)) + (\theta_2/(3K_2+4G_2))}, \tag{5.7.113}$$

and

$$2\mathcal{G} \leq 2G_2 + \frac{\theta_1}{(1/(2G_1-2G_2)) + ((3\theta_2(K_2+2G_2)/(5G_2(3K_2+4G_2)))}\,. \tag{5.7.114}$$

The limit cases above are precisely the upper bulk and shear modulus bounds for isotropic composites, discussed in Sec. 5.5 and due to Hashin and Shtrikman (1963). However, these bounds are not appropriate for the partly bonded composites, as the properties of the interface are lost by taking g_b and g_s to tend to ∞.

Without losing the interfacial effect, simpler upper bounds, however not as tight, can be obtained by straightforward computation from Eqs. (5.7.109) and (5.7.111). Indeed one can show that

$$3\mathcal{K} \leq \theta_1 3K_1 + \theta_2 3K_2 \left(\frac{1}{1+\alpha_b}\right), \tag{5.7.115}$$

$$2\mathcal{G} \leq \theta_1 2G_1 + \theta_2 2G_2 \left(\frac{1}{1+\alpha_s}\right), \tag{5.7.116}$$

where

$$\alpha_b = \frac{\theta_2 3K_2}{g_b}, \quad \alpha_s = \frac{\theta_2 2G_2}{g_s}. \tag{5.7.117}$$

The bounds in Eqs. (5.7.103) and (5.7.104) and Eqs. (5.7.115) and (5.7.116) can also be directly obtained from the simpler variational principles given by Eqs. (5.7.39) and (5.7.43). These are bounds that generalize the harmonic and arithmetic mean bounds for the partial bonding case

$$\boxed{\left(\frac{\theta_1}{3K_1} + \frac{\theta_2}{2K_2} + \frac{s}{3\alpha}\right)^{-1} \leq 3\mathcal{K} \leq \theta_1 3K_1 + \theta_2 3K_2 \left(\frac{1}{1+\alpha_b}\right)}, \tag{5.7.118}$$

$$\boxed{\left(\frac{\theta_1}{2G_1} + \frac{\theta_2}{2G_2} + \frac{s}{5}\left(\frac{1}{\beta} + \frac{2}{3\alpha}\right)\right)^{-1} \leq 2\mathcal{G} \leq \theta_1 2G_1 + \theta_2 2G_2 \left(\frac{1}{1+\alpha_s}\right)}. \tag{5.7.119}$$

5.7.4. *Size Effects for Particulate Composites*

In the case of partially cohesive elastic composites, size effects are present, in a similar way as in the conductivity problem. We consider here, as in Lipton and Vernescu (1995), a monodisperse suspension of relatively stiff isotropic elastic spheres with elasticity $\boldsymbol{C}_2$ embedded in a softer matrix of elasticity $\boldsymbol{C}_1$. We will see that the composite can get softer than the matrix, although stiffer particles are present; depending on the size of the particles, increasing the volume fraction might not make the material stiffer, as there is an interlay between the particles' elasticity and the interface properties.

For a monodisperse suspension of graphite spheres of radius a in an epoxy matrix, we plot the shear modulus upper bound (US) (Eq. (5.7.111)) and the lower and upper mean bounds (LMS and UMS) (Eq. (5.7.118))

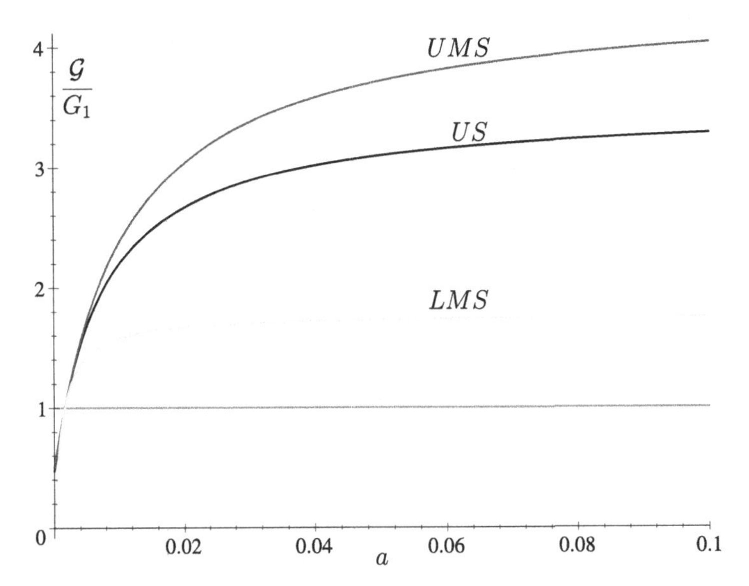

Fig. 5.4. Comparison between the effective shear trace bounds US, the upper and lower mean bounds UMS and LMS for a monodisperse suspension of graphite spheres in an epoxy matrix. The volume fraction of spheres is fixed at 50% and the sphere radius is less than 0.1 m. From Lipton and Vernescu (1995), *Math. Models Methods Appl. Sci.*

(Fig. 5.4). The parameters are $g_s = (1/5)(3\beta + 2\alpha)\theta_2 a$ and $s/\theta_2 = 3/a$. The bounds are plotted for radii between 0 and 0.1×10^{-2} m, with sphere volume fraction fixed at 50%. The interfacial stiffnesses are chosen to be $\alpha = \beta = 1 \times 10^6$ MPa/m. Here, the shear moduli and Poisson's ratios of the spheres are $\mu_1 = 0.7 \times 10^3$ MPa and $\nu_1 = 0.33$; the moduli for the matrix are $\mu_2 = 5.5 \times 10^3$ MPa, $\nu_2 = 0.25$. Note that for sufficiently small sphere radii the bounds indicate that the shear trace drops below that of the matrix.

In Fig. 5.5 the shear trace bounds are plotted for the monodisperse suspension of graphite spheres in epoxy for small sphere radii, i.e., $a < 0.12 \times 10^{-4}$ m. Here $\alpha = \beta = 1 \times 10^6$ MPa/m. Note that the shear trace upper bounds lie far below that of the matrix shear trace.

5.7.5. *Critical Radii for Particulate Composites*

In the case of partially cohesive elastic composites, critical radii for which the composite has the same elasticity as the elasticity of the matrix, independent of the volume fraction of stiffer particles. In some sense the particles are "invisible" or "cloaked". Indeed if we consider again the example

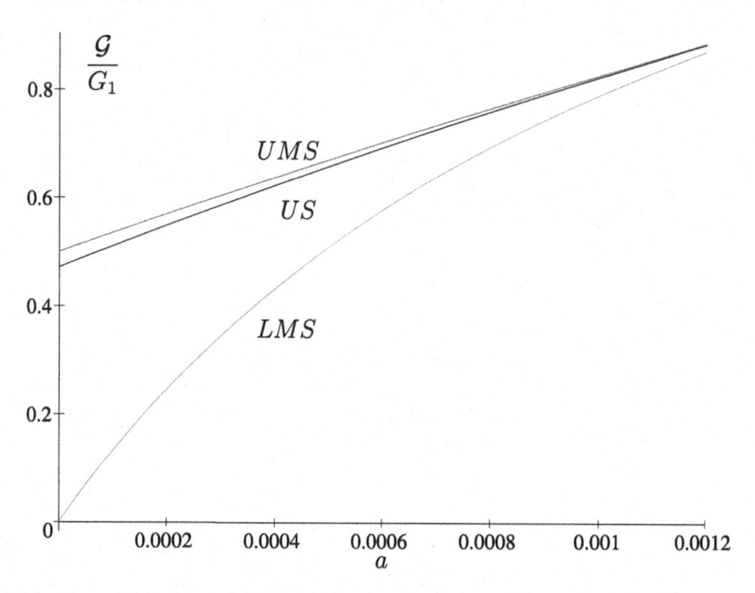

Fig. 5.5. Shear trace bounds US and the mean upper and lower bounds UMS and LMS for monodisperse suspensions of graphite in an epoxy matrix for particle radius less than critical: 0.0012 m. From Lipton and Vernescu (1995), *Math. Models Methods Appl. Sci.*

in the previous section and plot (Fig. 5.6) the previous bounds for the shear modulus, this time in a neighborhood of $a = 0.0016\,\mathrm{m}$, a value for a appears where the upper and lower bounds coincide. This occurs exactly when $\mathcal{G}/G_1 = 1$.

In what follows we shall consider, as in Lipton and Vernescu (1995), a monodisperse suspension of relatively stiff isotropic elastic spheres with elasticity $\boldsymbol{C}_2$ embedded in a softer matrix with isotropic elasticity $\boldsymbol{C}_1$. Let us find the critical radii for which the effect of the interface is balanced by the larger elastic stiffness of the particles, in such a way that the composite has the same moduli as the matrix.

Indeed, we will show that for prescribed interfacial spring constant α, at the critical particle radius R_b^{cr}

$$R_b^{cr} = (\alpha\Delta_b)^{-1}, \tag{5.7.120}$$

the effective bulk trace TrP_bC of the composite is identical to that of the matrix.

Similarly, when the interfacial spring constants α and β are equal, a critical particle radius

$$R_s^{cr} = (\beta\Delta_s)^{-1}, \tag{5.7.121}$$

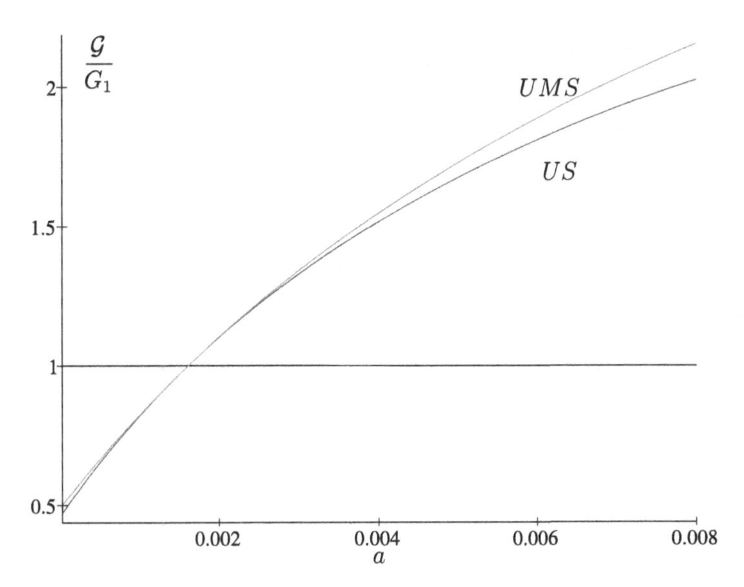

Fig. 5.6. Comparison between the bounds US and UMS for radius a close to the critical value 0.0016 m. From Lipton and Vernescu (1995), *Math. Models Methods Appl. Sci.*

exists so that the effective shear trace TrP_sC equals that of the matrix material.

Let us consider a constant tensor $\boldsymbol{\xi}$ and multiply by $\xi_{k\ell}$ the cell problems (Eqs. (5.7.16)–(5.7.21)). If the sum $\boldsymbol{\chi}^{k\ell}\xi_{k\ell}$ is denoted by $\boldsymbol{\phi}$, the cell problem becomes

$$\frac{\partial}{\partial x_j}(C_{ijmn}(\boldsymbol{\varepsilon}_{mn}(\boldsymbol{\phi}) + \xi_{mn})) = 0\,, \quad \text{in } \Omega_\alpha\,, \tag{5.7.122}$$

$$[C_{ijmn}(\boldsymbol{\varepsilon}_{mn}(\boldsymbol{\phi}) + \xi_{mn})n_j] = 0\,, \quad \text{on } \Gamma\,, \tag{5.7.123}$$

$$C_{ijmn}(\boldsymbol{\varepsilon}_{mn}(\boldsymbol{\phi}) + \xi_{mn})n_i n_j = -\alpha[\phi_N]\,, \quad \text{on } \Gamma\,, \tag{5.7.124}$$

$$C_{ijmn}(\varepsilon_{mn}(\boldsymbol{\phi}) + \xi_{mn})n_j - C_{hjmn}(\varepsilon_{mn}(\boldsymbol{\phi}) + \xi_{mn})n_h n_j n_i$$

$$= -\beta[\boldsymbol{\phi}_T]_i\,, \quad \text{on } \Gamma\,, \tag{5.7.125}$$

$$\boldsymbol{\phi} \text{ is } \Omega\text{-periodic } \langle\boldsymbol{\phi}\rangle_\Omega = 0\,, \tag{5.7.126}$$

and $\boldsymbol{\phi}$ has zero average on Ω, $\langle\boldsymbol{\phi}\rangle_\Omega = 0$.

From Eq. (5.7.28), the effective elasticity tensor is then given by

$$\mathcal{C}_{ijk\ell}\xi_{k\ell} = \langle C_{ijk\ell}(\varepsilon_{kl}(\boldsymbol{\phi}) + \xi_{k\ell})\rangle\,. \tag{5.7.127}$$

Now consider a distribution of N nontouching spheres of common radius a inside the unit cube, with centers located at $\boldsymbol{r}^{(m)}$.

Next let us look for a solution of Eqs. (5.7.122)–(5.7.126) of the form

$$\phi_i + \xi_{ij}x_j = \begin{cases} (\xi_1)_{ij}x_j \text{ in the matrix,} \\ (\xi_2)_{ij}x_j + V_i^m \text{ in the } m\text{th particle,} \end{cases} \tag{5.7.128}$$

where $\boldsymbol{V}^m$ is a constant vector. Note that Eq. (5.7.122) is identically satisfied and from Eqs. (5.7.123)–(5.7.125) we get a system for the unknowns $\boldsymbol{\xi}_1$, $\boldsymbol{\xi}_2$, and $\boldsymbol{V}^m$

$$(C_1)_{ijmn}(\xi_1)_{mn}n_j = (C_2)_{ijmn}(\xi_2)_{mn}n_j\,, \tag{5.7.129}$$

$$(C_1)_{ijmn}(\xi_1)_{mn}n_jn_i = -\alpha(((\xi_2)_{ij} - (\xi_1)_{ij})x_j + V_i^M)n_i\,, \tag{5.7.130}$$

$$\begin{aligned}(C_1)_{\ell jmn}(\xi_1)_{mn}n_j - (C_1)_{ijmn}(\xi_1)_{mn}n_jn_in_\ell \\ = -\beta(((\xi_2)_{\ell j} - (\xi_1)_{\ell j})x_j + V_\ell^M) - (((\xi_2)_{ij} - (\xi_1)_{ij})x_j + V_i^M)n_in_\ell\,.\end{aligned} \tag{5.7.131}$$

From Eq. (5.7.129) it follows that $(C_1^{-1})_{ijk\ell}(C_2)_{k\ell mn}(\xi_2)_{mn} = (\xi_1)_{ij}$. On the surface of the ith sphere the unit normal is written $\boldsymbol{n} = (\boldsymbol{x} - \boldsymbol{r}^i)/a$, thus $\boldsymbol{x} = a\boldsymbol{n} + \boldsymbol{r}^i$ on the surface and

$$\begin{aligned}((\xi_2)_{ij} - (\xi_1)_{ij})x_j = a\big(1 - (C_1^{-1})_{ijk\ell}(C_2)_{k\ell mn}\big)(\xi_2)_{mn}n_j \\ + \big(1 - (C_1^{-1})_{ijk\ell}(C_2)_{k\ell mn}\big)(\xi_2)_{mn}r_j\,.\end{aligned} \tag{5.7.132}$$

Thus Eqs. (5.7.130) and (5.7.131) can be written as

$$\begin{aligned}\left(\left(3K_2 + \alpha a\left(1 - \frac{K_2}{K_1}\right)\right)P_b\boldsymbol{\xi}_2\boldsymbol{n} + \left(2G_2 + \alpha a\left(1 - \frac{G_2}{G_1}\right)\right)P_s\boldsymbol{\xi}_2\boldsymbol{n}\right)_N \\ = -\alpha\left(\left(1 - \frac{K_2}{K_1}\right)P_b\boldsymbol{\xi}_2\boldsymbol{r}^i + \left(1 - \frac{G_2}{G_1}\right)P_s\boldsymbol{\xi}_2\boldsymbol{r}^i + \boldsymbol{V}^i\right)_N\,,\end{aligned} \tag{5.7.133}$$

and

$$\begin{aligned}\left(\left(3K_2 + \beta a\left(1 - \frac{K_2}{K_1}\right)\right)P_b\boldsymbol{\xi}^2\boldsymbol{n} + \left(2G_2 + \beta a\left(1 - \frac{G_2}{G_1}\right)\right)P_s\boldsymbol{\xi}_2\boldsymbol{n}\right)_T \\ = -\beta\left(\left(1 - \frac{K_2}{K_1}\right)P_b\boldsymbol{\xi}_2\boldsymbol{r}^i + \left(1 - \frac{G_2}{G_1}\right)P_s\boldsymbol{\xi}_2\boldsymbol{r}^i + \boldsymbol{V}^i\right)_T\,.\end{aligned} \tag{5.7.134}$$

It may be observed that Eq. (5.7.133) is of the form:

$$h + \boldsymbol{q}\cdot\boldsymbol{n} + L\boldsymbol{n}\cdot\boldsymbol{n} = 0 \tag{5.7.135}$$

for all unit vectors $\boldsymbol{n}$, where h is a scalar, $\boldsymbol{q}$ is a constant vector and $TrL = 0$. It can be shown (see Lipton and Vernescu, 1995) that this is possible only if

$$L = 0\,, \quad \boldsymbol{q} = 0\,, \text{ and } h = 0\,. \tag{5.7.136}$$

Applying Eq. (5.7.136) to Eq. (5.7.133) we obtain

$$\left(3K_2 + \alpha a\left(1 - \frac{K_2}{K_1}\right)\right) P_b\boldsymbol{\xi}_2 = 0\,, \tag{5.7.137}$$

$$\left(2G_2 + \alpha a\left(1 - \frac{G_2}{G_1}\right)\right) P_s\boldsymbol{\xi}_2 = 0\,, \tag{5.7.138}$$

$$\left(1 - \frac{K_2}{K_1}\right) P_b\boldsymbol{\xi}_2\boldsymbol{r}^i + \left(1 - \frac{G_2}{G_1}\right) P_s\boldsymbol{\xi}_2\boldsymbol{r}^i + \boldsymbol{V}^i = 0\,. \tag{5.7.139}$$

Next we observe that Eq. (5.7.134) is of the form

$$(L\boldsymbol{n})_T = \boldsymbol{q}_T\,, \tag{5.7.140}$$

where L is a constant symmetric matrix and $\boldsymbol{q}$ is a constant vector, which needs to hold for all unit vectors $\boldsymbol{n}$. One can show (see Lipton and Vernescu, 1995) that such an equation holds for all unit vectors $\boldsymbol{n}$ only if

$$\boldsymbol{q} = 0 \quad \text{and} \quad L = cI \tag{5.7.141}$$

with c an arbitrary constant and I the identity tensor. Thus from Eq. (5.7.134) we have that there exists a constant c such that:

$$\left(3K_2 + \beta a\left(1 - \frac{K_2}{K_1}\right)\right) P_b\boldsymbol{\xi}_2 = cI\,, \tag{5.7.142}$$

$$\left(2G_2 + \beta a\left(1 - \frac{G_2}{G_1}\right)\right) P_s\boldsymbol{\xi}_2 = 0\,, \tag{5.7.143}$$

$$\left(1 - \frac{K_2}{K_1}\right) P_b\boldsymbol{\xi}_2\boldsymbol{r}^i + \left(1 - \frac{G_2}{G_1}\right) P_s\boldsymbol{\xi}_2\boldsymbol{r}^i + \boldsymbol{V}^i = 0\,. \tag{5.7.144}$$

Since $P_b\boldsymbol{\xi}_2 = (1/3Tr\boldsymbol{\xi}_2)I$, it is evident that Eq. (5.7.142) is satisfied for all values of β, a, K_1, and K_2. Also Eqs. (5.7.139) and (5.7.144) are identical and provide a value for $\boldsymbol{V}^i$.

Next we choose a spherical tensor $\boldsymbol{\xi}_2$ (i.e., a tensor with a zero shear projection $(P_s)_{ijk\ell}(\xi_2)_{k\ell} = 0$) and both Eqs. (5.7.138) and (5.7.143) are satisfied. The remaining condition (Eq. (5.7.137)) implies

$$3K_2 + \alpha a \left(1 - \frac{K_2}{K_1}\right) = 0\,, \tag{5.7.145}$$

which defines the critical radius

$$a = R_b^{cr} = \frac{1}{\alpha\left((1/3K_1) - (1/3K_2)\right)} = (\alpha\Delta_b)^{-1}\,. \tag{5.7.146}$$

We find from Eq. (5.7.127) that the effective bulk modulus is

$$\mathcal{K} = K_1\,. \tag{5.7.147}$$

Therefore, the composite has the same bulk modulus as the matrix, independent of the volume fraction of the stiffer particles.

By choosing a deviatoric tensor $\boldsymbol{\xi}_2$ (i.e., a tensor with a zero bulk projection $(P_b)_{ijk\ell}(\xi_2)_{k\ell} = 0$), then Eq. (5.7.137) is satisfied. If $\alpha = \beta$ then Eqs. (5.7.138) and (5.7.143) are identical. Under the condition

$$2G_2 + \alpha a \left(1 - \frac{G_2}{G_1}\right) = 0\,, \tag{5.7.148}$$

the critical radius exists and is

$$a = R_s^{cr} = \frac{1}{\alpha\left((1/2G_1) - (1/2G_2)\right)} = (\alpha\Delta_s)^{-1}\,. \tag{5.7.149}$$

The effective shear modulus then follows from Eq. (5.7.127)

$$\mathcal{G} = G_1\,. \tag{5.7.150}$$

Therefore the composite has the same shear modulus as the matrix, independent of the volume fraction of the stiffer particles.

The above results also imply that the elastic field in the matrix remains undisturbed when the spheres are at critical radius.

In Fig. 5.7, the upper-bound US for the shear modulus is compared to the mean upper bound UMS for a monodisperse suspension of graphite spheres in an epoxy matrix in terms of particle volume fraction θ_2 and radius a. For the same composite material, the bulk modulus upper bound UB is compared in Fig. 5.8 to mean upper bound UMB and the mean lower bound LMB for composites. The volume fraction of spheres is fixed at 50% and the sphere radii range from 0.04 m to infinitesimal. One sees that the bounds are tangent and touch precisely at the critical radius $R_b^{cr} = 0.00683\,\mathrm{m}$.

Bounds for the torsional rigidity of a cylindrical shaft, containing a number of cylindrically orthotropic fibers, have been derived in Chen and

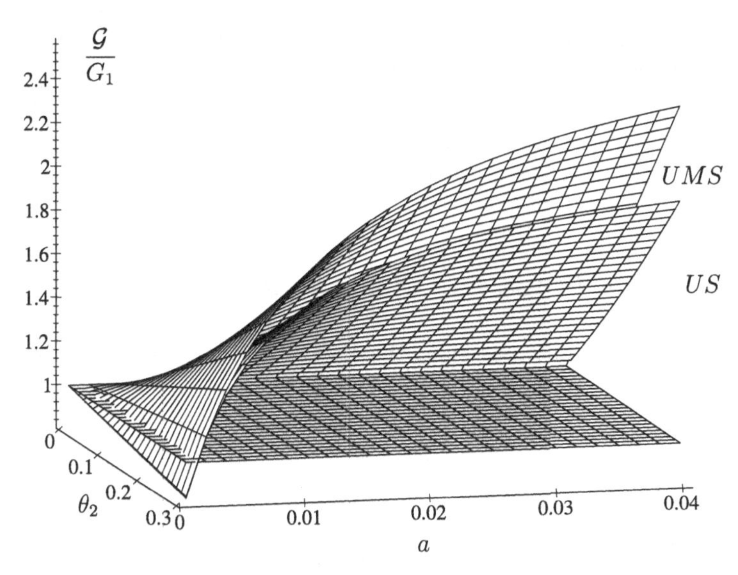

Fig. 5.7. Upper bound US for the shear modulus, compared to the mean upper bound UMS for a monodisperse suspension of graphite spheres in an epoxy matrix in terms of particle volume fraction θ_2 and radius a. From Lipton and Vernescu (1995), *Math. Models and Methods Appl. Sci.*

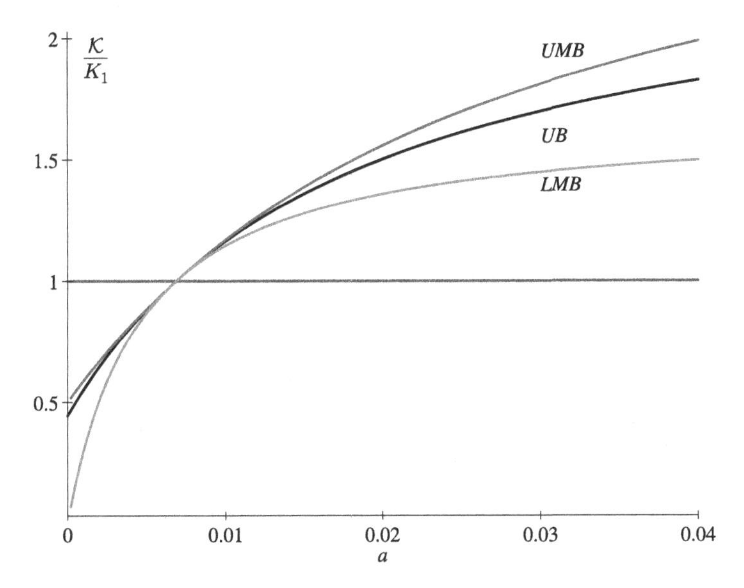

Fig. 5.8. Upper bound UB for the bulk modulus compared with the mean upper and lower bulk modulus bounds, UMB and LMB respectively.

Lipton (2007) and Chen and Chan (2008) and conditions for which the fibers become neutral under torsion have been identified. For nonlinear hyperelastic materials bounds have been obtained by Bisegna and Luciano (1998). Partially bonded composites have been used in the modeling of crack propagation in composites by Krasuki and Lenci (2000), and Lenci (2001).

Appendix 5A. Properties of a Tensor of Fourth Rank

In the case of isotropic materials, the elasticity tensor takes the form (Jeffreys (1931), Jog (2006))

$$C_{ijk\ell} = K\delta_{ij}\delta_{kl} + G\left(\delta_{ik}\delta_{j\ell} + \delta_{i\ell}\delta_{jk} - \frac{2}{3}\delta_{ij}\delta_{kl}\right) \tag{A.1}$$

or

$$C_{ijk\ell} = \lambda\delta_{ij}\delta_{kl} + G(\delta_{ik}\delta_{j\ell} + \delta_{i\ell}\delta_{jk}), \tag{A.2}$$

where K and G are the bulk and the shear modulus, respectively, and λ is the Lamè constant, $\lambda = K - (2/3)G$.

For isotropic fourth-order tensors one can introduce (Milton (2002)) two orthogonal projections P_b, the projection on hydrostatic fields (i.e., matrix-valued fields proportional to the identity) and P_s the projection on shear fields (i.e., fields characterized by trace-free matrices) (see also Rychlewski (1984)). These projections have the components:

$$(P_b)_{ijk\ell} = \frac{1}{d}\delta_{ij}\delta_{kl}, \tag{A.3}$$

$$(P_s)_{ijk\ell} = \frac{1}{2}(\delta_{ik}\delta_{j\ell} + \delta_{i\ell}\delta_{jk}) - \frac{1}{d}\delta_{ij}\delta_{k\ell}, \tag{A.4}$$

where d represents the dimension of the space. Thus the elasticity tensor, Eq. (A.1), for isotropic materials can be written as

$$C_{ijk\ell} = dK(P_b)_{ijk\ell} + 2G(P_s)_{ijk\ell}. \tag{A.5}$$

One can easily check that P_b and P_s satisfy the properties of orthogonal projections:

$$(P_b)_{ijk\ell}(P_b)_{k\ell mn} = (P_b)_{ijmn}, \quad (P_s)_{ihk\ell}(P_s)_{k\ell mn} = (P_s)_{ijmn}, \tag{A.6}$$

$$(P_b)_{ijk\ell}(P_s)_{k\ell mn} = (P_s)_{ijk\ell}(P_b)_{k\ell mn} = 0. \tag{A.7}$$

Denoting the trace of a fourth-order tensor A

$$TrA = A_{ijij}, \tag{A.8}$$

we note that

$$TrP_b = 1\,, \quad \text{and } TrP_s = \frac{1}{2}(d^2 + d) - 1\,. \tag{A.9}$$

The inverse $\boldsymbol{C}^{-1}$ of a fourth-order tensor $\boldsymbol{C}$ is defined by,

$$C_{ijk\ell}C^{-1}_{k\ell pq} = C^{-1}_{ijk\ell}C_{k\ell pq} = \delta_{ip}\delta_{jq}\,. \tag{A.10}$$

Using the properties of the projections, it can be easily seen that for isotropic materials the inverse of the elasticity tensor, Eq. (A.1), is

$$C^{-1}_{ijk\ell} = \frac{1}{dK}(P_b)_{ijk\ell} + \frac{1}{2G}(P_s)_{ijk\ell}\,. \tag{A.11}$$

References

Aboudi, J. (1987). Damage in composites modeling of imperfect bonding. *Composites Science and Technology* **28**: 103–128.

Achenbach, J. D. and H. Zhu (1989). Effect of interfacial zone on mechanical behavior and failure of fiber-reinforced composites. *J. Mech. Phys. Solids* **37**: 381–393.

Allaire, G. (2001). *Shape Optimization by the Homogenization Method*, Springer Verlag, Berlin.

Avellaneda, M. (1987). Optimal bounds and microgeometries for elastic two-phase composites. *SIAM J. Appl. Math.* **47**: 1216–1228.

Bakhvalov, N. and G. Panasenko (1989). *Homogenization: Averaging Processes in Periodic Media*, Kluwer, Dordrecht.

Benveniste, Y. (1985). The effective mechanical behaviour of composite materials with imperfect contact between the constituents. *Mech. Mat.* **4**: 197–208.

Bigoni, D. and A. B. Movchan (2002). Statics and dynamics of structural interfaces in elasticity. *Int. J. Solids Structures* **39**: 4843–4865.

Bisegna, L. and R. Luciano (1997). Bounds on the overall properties of composites with debonded frictionless interfaces. *Mech. Mat.* **28**: 23–32.

Caillerie, D. (1984). Thin elastic and periodic plates. *Math. Meth. Appl. Sci.* **6**: 159–191.

Chen, T. and I. T. Chan (2008). Rigorous bounds on the torsional rigidity of composite shafts with imperfect interfaces. *J. Elasticity* **98**: 91–108.

Chen, T. and R. Lipton (2007). Bounds for the torsional rigidity of shafts with arbitrary cross-sections containing cylindrically orthotropic fibres or coated fibres. *Proc. R. Soc. A* **463**: 3291–3309.

Cherkaev, A. (2000). *Variational Methods for Structural Optimization*, Springer-Verlag, New York.

Ciarlet, P. G. and P. Destuynder (1979). A justification of the two-dimensional linear plate model. *J. Mechanique* **18**: 315–344.

Cioranescu, D. and J. Saint Jean Paulin (1999). *Homogenization of Reticulated Structures*, Springer-Verlag, New York.

Cristescu, N., E. M. Craciun and E. Soós (2004). *Mechanics of Elastic Composites*, Chapman and Hall/CRC, Boca Raton.

Francfort, G. A. and F. Murat (1986). Homogenization and optimal bounds in linear elasticity. *Arch. Rat. Mech. Anal.* **94**: 197–212.

Fung, Y. C. (1965). *Foundations of Solid Mechanics*, Prentice-Hall.

Gibson, L. and M. E. Ashley (1997). *Cellular Solids: Structural Properties*, Cambridge University Press, Cambridge.

Gilbert, R. P. and K. Hackl (1995). *Asymptotic Theories for Plates and Shells*, CRC Press.

Hashin, Z. (1992). Extremum principles for elastic heterogeneous media with imperfect interfaces and their application to bounding of effective moduli. *J. Mech. Phys. Solids* **40**: 767–781.

Hashin, Z. and S. Shtrikman (1963). A variational approach to the theory of the elastic behaviour of multiphase materials. *J. Mech. Phys. Solids* **11**: 127–140.

Hill, R. (1952). The elastic behavior of a crystalline aggregate. *The Proceedings of the Physical Society* **A65**: 349–354.

Hill, R. (1972). An invariant treatment of interfacial discontinuities in elastic composites. *Continuum Mechanics and Related Problems of Analysis*, Moscow, pp. 597–604.

Höhe, J. and W. Becker (2002). Effective stress–strain relation for two-dimensional cellular sandwich cores: Homogenization, material models, and properties. *Appl. Mech. Rev.* **55**: 61–87.

Jeffreys, H. (1931). *Cartesian Tensors*, Cambridge University Press, Cambridge.

Jog, C. S. (2006). A concise proof of the representation theorem for fourth-order isotropic tensors. *J. Elasticity* **85**: 119–124.

Kohn, R. V. and R. Lipton (1988). Optimal bounds for the effective energy of a mixture of isotropic, incompressible, elastic materials. *Arch. Rat. Mech. Anal.* **102**: 331–350.

Kohn, R. V. and G. Strang (1986). Optimal design and relaxation of a variational problem in optimal design. *Comm. Pure Appl. Math.* **39**.

Kohn, R. V. and M. Vogelius (1984). A new model for thin plates with rapidly varying thickness. *I.J.E.S* **20**: 333–350.

Krasuki, F. and S. Lenci (2000). Analysis of interfaces of variable stiffness. *Int. J. Sol. Struc.* **37**: 3619–3632.

Lenci, S. (2001). Analysis of a crack at a weak interface. *Int. J. Fracture* **108**: 275–290.

Lenè, F. and D. Leguillon (1982). Homogenized constitutive law for a partially cohesive composite material. *Int. J. Solids Struc.* **18**: 413–458.

Lewinsky, T. and J. J. Telega (2000). *Plates, Laminates and Shells. Asymptotic Analysis and Homogenization*, World Scientific, River Edge, New York.

Lipton, R. (1988) On the effective elasticity of a two-dimensional homogenized incompressible elastic composite. *Proc. Roy. Soc. Edinburgh* **100A**: 45–61.

Lipton, R. and B. Vernescu (1995). Variational methods, size effects and extremal microgeometries for elastic composites with imperfect interface. *Math. Models Methods Appl. Sci.* **5**: 1139–1173.

Lipton, R. and B. Vernescu (1996). Two-phase elastic composites with interfacial slip. *ZAMM* **452**: 329–358.

Lurie, K. A. and A. V. Cherkaev (1984). G-closure of some particular sets of admissible material characteristics for the problem of bending of thin elastic plates. *J. Optimiz. Th. Appl.* **42**: 305–316.

Lurie, K. A. and A. V. Cherkaev (1986). Effective characteristics of Composite Materials and optimal design of structural elements, Advances in Mechanics, **9**, 2, Translated in English in: *Topics on the Mathematical Modeling of Composite Materials*, eds. A. Cherkaev and R. Kohn, Birkhauser, Boston, pp. 175–258.

Lurie, K. A., A. V. Fedorov and A. V. Cherkaev (1980). On the existence of solutions of certain optimal design problems for bars and plates. *Ioffe Physico-Technical Institute Report* **668**.

Milton, G. W. (1990). On characterizing the set of possible effective tensors of composites: The variational method and the translation method. *Comm. Pure Appl. Math.* **43**: 63–125.

Milton, G. W. (2002). *The Theory of Composites*, Cambridge University Press, Cambridge.

Milton, G. and R. V. Kohn (1988). Variational bounds on the effective moduli of anisotropic composites. *J. Mech. Phys. Solids* **36**: 597–629.

Murat, F. and L. Tartar (1985). Optimality conditions and homogenization. *Nonlinear Variational Problems*, Pitman Publishing Ltd., London, pp. 1–8.

Nemat-Nasser, S. and M. Hori (1999). *Micromechanics: Overall Properties of Heterogeneous Materials*, Elsevier, Amsterdam.

Oleinik, O. A., A. S. Shamaev and G. A. Yosifian (1992). *Mathematical Problems in Elasticity and Homogenization*, North Holland, Amsterdam.

Parton, V. Z. and B. A. Kudryavtsev (1993). *Engineering Mechanics of Composite Structures*, CRC Press, Boca Raton.

Reuss, A. (1929). Berechnung der Fliessgrenze von Mischkristallen auf Grund der Plastizitäts bedingung für Einkristalle. *ZAMM* **9**: 49–58.

Rychlewski, J. (1984). On Hooke's law. *Prikl. Matem. Mekhan.* **48**: 303–314.

Sanchez-Palencia, E. (1980). *Non-homogeneous Media and Vibration Theory, Lecture Notes in Physics*, Springer, Berlin.

Shi, G. and P. Tong (1995). The derivation of equivalent constitutive equations of honeycomb structures by a two scale method. *Computational Mechanics* **15**: 395–407.

Shi, G. and P. Tong (1995). Equivalent transverse shear stiffness of honeycomb cores. *Int. J. Solids Struct.* **32**: 1383–1393.

Sideridis, E. (1988). The inplane shear modulus of fiber reinforced composites as defined by the concept of interphase. *Composite Science and Technology* **31**: 35–53.

Theocaris, P. S., E. P. Sideridis and G. C. Papanicolaou (1985). The elastic longitudinal modulus and Poisson's ratio of fiber composites. *Journal of Reinforced Plastics and Composites* **4**: 396–418.

Tong, P. and C. C. Mei (1992). Mechanics of composites of multiple scales. *Computational Mechanics* **9**: 195–210.

Torquato, S. (2002). *Random Heterogeneous Materials*, Springer-Verlag, New York.

Tsai, S. W. (1987). *Composite Design*, 3rd edn., Think Composites, Dayton, Ohio.

Vinson, J. R. and R. I. Sierakowski (1990). *The Behavior of Structures Composed of Composite Materials*, Kluwer Academic.

Voigt, W. (1889). Über die Beziehung zwischen den beiden Elastizitäts Konstanten Isotroper Körper. *Wiedemanns Annalen der Physik und Chemie (Leipzig)* **38**: 573–587.

Xu, F. X. and P. Qiao (2002). Homogenized elastic properties of honeycomb sandwich with skin effect. *Int. J. Solids Struct.* **39**: 2135–2188.

Deformable Porous Media

6

Quantitative understanding of deformable porous media, fully or partially saturated by fluids, is of great importance in several branches of engineering and science. For foundations on land, soil response to the dead weight of buildings or to vibrations due to earthquakes is of crucial considerations in a safe design. For off-shore oil-drilling platforms of gravity type, persistent attacks by ocean waves cause severe strains to the seabed supporting the structure. Rational prediction of the ground response under quasi-static or dynamic loading is essential. In earth sciences, understanding of wave propagation through a poroelastic medium is fundamental in acoustic tomography of the seafloor or seismic exploration of the earth interior. Medical ultrasonics applied to bones and muscles is yet another area where the dynamics of poroelastic media is germaine.

The mechanistic basis of poroelasticity was pioneered by the soil engineer Terzaghi (1943) who developed the first theory of one-dimensional soil consolidation under surface weight. Extensions to three dimensions and to dynamical problems were subsequently made in a series of landmark papers by Biot (1941, 1956, 1962a, b). Treating the pore fluid and solid matrix as two interacting continua with the latter being linearly elastic, he derived the equations coupling the fluid pressure and the solid displacement on the macroscale without microscale considerations. In recent decades Biot's theory has been used widely in soil and rock mechanics (Jaeger and Cook, 1969; Wang, 2000), seismology (Deresiewicz, 1962; Deresiewicz and Rice, 1962; Rice and Cleary, 1976), response of the seabed to ocean waves (Mei and Foda, 1981; Mei and McTigue, 1984; Mei, 1990), as well as the mechanics of bones and muscles (Cowan and Doty, 2007), etc.

Systematic derivation of Biot's celebrated theory of three-dimensional poroelasticity from microscale considerations was started by Auriault and

Sanchez-Palencia (1977), Auriault (1991), Burridge and Keller (1981), and Levy (1979). Further extensions can be found in Santos *et al.* (2006) and the references cited therein. The starting points for both the solid matrix and the pore fluid are the full sets of differential equations coupling the two media. By homogenization one can then derive Biot's macroscale equations of poroelasticity for quasi-static as well as dynamic problems. The scheme for calculating the constitutive coefficients from the known properties of the component materials will be explained and some numerical results will be given. Since Biot's equations for two coupled continua are still quite complicated in general, an effective method by boundary-layer approximation is also described for efficient calculation as well as physical understanding of certain classes of macroscale problems.

6.1. Basic Equations for Fluid and Solid Phases

Ignoring gravity and assuming low Reynolds numbers as in Sec. 3.2, we recall that convective inertia has negligible effect on the macroscale. For brevity we therefore start from the linearized conservation laws. It is known that the bulk modulus of the pore fluid can be strongly affected by the presence of gas bubbles, often present due to chemical or biological processes. If the degree of saturation is S, with S slightly less than unity, the bulk modulus E_f is reduced from that of saturated water $\overline{E}_f$ according to

$$\frac{1}{E_f} = \frac{1}{\overline{E}_f} + \frac{1-S}{\overline{p}}, \qquad (6.1.1)$$

where $\overline{p}$ is the mean pressure (Veruijt, 1969). To have some quantitative ideas, let us take $\overline{p} = \mathcal{O}(1)$ atm. and $1 - S = 0.99$. If $\overline{E}_f = 2 \times 10^9\,\text{N/m}^2$, then E_f is reduced drastically to $10^6\,\text{N/m}^2$.

Allowing for compressibility, mass conservation of the pore fluid requires

$$\frac{1}{\rho_f}\frac{\partial \rho_f}{\partial t} + \frac{\partial u_i}{\partial x_i} = 0\,, \quad \boldsymbol{x} \in \Omega_f\,, \qquad (6.1.2)$$

where ρ_f is the density of the pore fluid. For small density changes the fluid pressure and density are linearly related by

$$\frac{dp}{d\rho_f} = \overline{\rho}_f E_f\,, \qquad (6.1.3)$$

where $\overline{\rho}_f$ is the density of the unperturbed fluid, hence

$$\frac{1}{\overline{\rho}_f E_f}\frac{\partial p}{\partial t} + \frac{\partial u_i}{\partial x_i} = 0\,, \quad \boldsymbol{x} \in \Omega_f\,. \qquad (6.1.4)$$

The linearized conservation law of fluid momentum reads

$$\bar{\rho}_f \frac{\partial u_i}{\partial t} = \frac{\partial \Sigma_{ij}}{\partial x_j}, \quad \boldsymbol{x} \in \Omega_f, \tag{6.1.5}$$

where Σ_{ij} denotes the fluid stress tensor, which is the sum of pore pressure and the viscous stress[1]

$$\Sigma_{ij} = -p\delta_{ij} + 2\mu\varepsilon_{ij}(\boldsymbol{u}), \tag{6.1.6}$$

with $\varepsilon_{ij}(\boldsymbol{u})$ being the rate of strain tensor:

$$\varepsilon_{ij}(\boldsymbol{u}) = \frac{1}{2}\left(\frac{\partial u_i}{\partial x_j} + \frac{\partial u_j}{\partial x_i}\right). \tag{6.1.7}$$

Recall that ε_{ij} is a tensorial operator with the dimension (length)$^{-1}$. When it operates on a tensor of rank n, the result is a tensor of rank $n+1$.

Let the solid displacement vector be $\boldsymbol{v} = \{v_i\}$ and the solid stress tensor be σ_{ij}. Momentum conservation requires that

$$\rho_s \frac{\partial^2 v_i}{\partial t^2} = \frac{\partial \sigma_{ij}}{\partial x_j}, \quad \boldsymbol{x} \in \Omega_s. \tag{6.1.8}$$

From the linear theory of elasticity (Fung, 1965), Hookes' law holds for the solid phase when strains are infinitesimally small:

$$\sigma_{ij} = C_{ijk\ell}\varepsilon_{k\ell}(\boldsymbol{v}), \tag{6.1.9}$$

where $\boldsymbol{v} = \{v_i\}$ is the solid displacement vector, $\varepsilon_{ij}(\boldsymbol{v})$ is the solid strain tensor:

$$\varepsilon_{ij}(\boldsymbol{v}) = \frac{1}{2}\left(\frac{\partial v_i}{\partial x_j} + \frac{\partial v_j}{\partial x_i}\right), \tag{6.1.10}$$

and $C_{ijk\ell}$ is the tensor of elasticity coefficients having the following properties of symmetry,

$$C_{ijk\ell} = C_{jik\ell} = C_{ij\ell k} = C_{k\ell ij}, \tag{6.1.11}$$

and positivity,

$$C_{ijk\ell}\,\xi_{ij}\xi_{k\ell} > 0, \quad \forall \xi_{ij}. \tag{6.1.12}$$

For isotropic elastic solids, there are only two independent coefficients,

$$C_{ijk\ell} = \lambda\delta_{ij}\delta_{k\ell} + G(\delta_{ik}\delta_{j\ell} + \delta_{i\ell}\delta_{jk}), \tag{6.1.13}$$

[1] Only Newtonian fluids are considered here.

hence

$$\sigma_{ij} = \lambda \frac{\partial v_k}{\partial x_k}\delta_{ij} + G\left(\frac{\partial v_i}{\partial x_j} + \frac{\partial v_j}{\partial x_i}\right) = \lambda \frac{\partial v_k}{\partial x_k}\delta_{ij} + 2G\varepsilon_{ij}(\boldsymbol{v}), \tag{6.1.14}$$

where G and λ are the Lamé constants, related to Young's modulus E and Poisson's ratio ν by,

$$\lambda = \frac{\nu E}{(1+\nu)(1-2\nu)}, \quad G = \frac{E}{2(1+\nu)}. \tag{6.1.15}$$

Let $\dot{\boldsymbol{v}} = \partial \boldsymbol{v}/\partial t = \{\dot{v}_i\}$ denote the solid velocity. On the solid/fluid interface Γ, the velocity components are continuous

$$u_i = \dot{v}_i \equiv \frac{\partial v_i}{\partial t}, \quad \boldsymbol{x} \in \Gamma, \tag{6.1.16}$$

as are the stresses

$$\Sigma_{ij} n_j = \sigma_{ij} n_j, \quad \boldsymbol{x} \in \Gamma, \tag{6.1.17}$$

where n_i is the unit normal vector pointing from the solid to the fluid. These requirements are, respectively, the kinematic and dynamic boundary conditions.

We now discuss separately the scaling for quasi-static and dynamic problems, which are distinguished mainly by very different time scales.

6.2. Scale Estimates

6.2.1. *Quasi-Static Poroelasticity*

As will be justified later, in slow processes such as soil consolidation under surface weight, inertia in both the pore fluid and the solid matrix can be ignored, i.e.,

$$\frac{\partial \Sigma_{ij}}{\partial x_j} = 0, \quad \boldsymbol{x} \in \Omega_f, \tag{6.2.1}$$

$$\frac{\partial \sigma_{ij}}{\partial x_j} = 0, \quad \boldsymbol{x} \in \Omega_s. \tag{6.2.2}$$

In order to examine the relative magnitude of various terms in the governing equations, let us make an *a priori* estimate of the time scale of the macroscale problem. Using the result for a rigid porous medium as a guide, we have the following relation between the scales of fluid velocity and pore pressure

$$[u] = \frac{[p]\ell^2}{\mu \ell'} = \epsilon \frac{[p]\ell}{\mu}, \quad \text{with } \epsilon = \frac{\ell}{\ell'} \ll 1, \tag{6.2.3}$$

where $[f]$ denotes the scale of f. From the continuity of velocity at the interface

$$u_i = \frac{\partial v_i}{\partial t}, \quad \boldsymbol{x} \in \Gamma, \tag{6.2.4}$$

we get

$$\frac{[v]}{[t]} = [u], \quad \text{hence } [v] = \frac{[p]\ell^2}{\mu\ell'}[t]. \tag{6.2.5}$$

From Sec. 3.2 we know that the fluid stress is dominated by the pore pressure. To be confirmed shortly, we anticipate the solid stress to be dominated by the strain on the macroscale, i.e.,

$$\varepsilon'_{ij}(\boldsymbol{v}) = \mathcal{O}\left(\frac{[v]}{\ell'}\right) = \mathcal{O}\left(\frac{[p]\ell^2[t]}{\mu\ell'^2}\right). \tag{6.2.6}$$

Since the solid and fluid stresses must be equal on the interface, we must have

$$E\left(\frac{[p][t]\ell^2}{\mu\ell'^2}\right) = \mathcal{O}([p]). \tag{6.2.7}$$

Therefore, the time scale should be

$$[t] = \frac{\mu\ell'^2}{\ell^2 E} = \mathcal{O}\left(\frac{\ell'^2}{KE}\right), \tag{6.2.8}$$

where $K = \mu/\ell^2$ is the scale of the hydraulic conductivity. This time scale characterizes the physics of soil consolidation first obtained in the pioneering work of Terzaghi. The product KE is essentially the consolidation coefficient (Terzaghi, 1943). Now we can assess the error in omitting inertia for soil consolidation problems. The ratio of fluid inertia to viscous force is

$$\frac{\rho(\partial u_i/\partial t)}{\mu\nabla^2 u_i} \sim \frac{\ell^2}{\ell'^2}\frac{KE}{\mu/\rho_f}.$$

For common soils $KE = \mathcal{O}(10^{-3} - 10^{-4})\,\text{cm}^2/\text{s}$ and $\mu/\rho_f \sim 10^{-2}\,\text{cm}^{-2}/\text{s}$. The above ratio is certainly very small, hence inertia is negligible.

Let the following dimensionless variables be introduced

$$\boldsymbol{x} = \ell\boldsymbol{x}^\dagger, \quad t = [t]t^\dagger, \quad (u_i, v_i) = [u](u_i^\dagger, v_i^\dagger)V = [u][t]V^\dagger$$

$$p = [p]p^\dagger, \quad \Sigma_{ij} = [p]\Sigma_{ij}^\dagger, \quad \sigma_{ij} = \frac{E[u][t]}{\ell}\sigma_{ij}^\dagger, \tag{6.2.9}$$

with $[p]$ and $[u]$ related by Eq. (6.2.3). The normalized equation for fluid mass conservation is

$$\epsilon\frac{G}{E_f}\frac{\partial p^\dagger}{\partial t^\dagger} + \frac{\partial u_i^\dagger}{\partial x_i^\dagger} = 0. \tag{6.2.10}$$

The momentum equations for the two phases (Eqs. (6.1.5) and (6.1.8)) are unchanged in form, but the stress–strain rate relation for the pore fluid is now

$$\Sigma_{ij}^{\dagger} = -p^{\dagger}\delta_{ij} + 2\epsilon\varepsilon_{ij}^{\dagger}(\boldsymbol{u}^{\dagger})\,. \tag{6.2.11}$$

On the other hand, Hooke's law becomes simply

$$\sigma_{ij}^{\dagger} = \frac{C_{ijk\ell}}{E}\varepsilon_{k\ell}^{\dagger}(\boldsymbol{v}^{\dagger})\,. \tag{6.2.12}$$

On the interface Γ, the kinematic condition of velocity continuity remains the same in form, but the condition for stress continuity becomes,

$$\sigma_{ij}^{\dagger}n_j = \epsilon\Sigma_{ij}^{\dagger}n_j\,. \tag{6.2.13}$$

Following our custom let us return to the governing equations in physical variables but rewrite Eqs. (6.1.4), (6.1.6), and (6.1.17) with the ordering parameter ϵ to indicate the relative importance of different terms, i.e.,

$$\frac{\epsilon}{C^2}\frac{\partial p}{\partial t} + \frac{\partial u_i}{\partial x_i} = 0\,, \tag{6.2.14}$$

$$\Sigma_{ij} = -p\delta_{ij} + 2\epsilon\mu\varepsilon_{ij}(\boldsymbol{u})\,, \tag{6.2.15}$$

$$\sigma_{ij}n_j = \epsilon\Sigma_{ij}n_j \quad \text{on } \Gamma\,. \tag{6.2.16}$$

The force balance equations (6.2.1) and (6.2.2) and the kinematic condition on the interface (Eq. (6.2.4)) are unchanged.

6.2.2. *Dynamic Poroelasticity*

As a basis for wave problems, we now return to the fuller momentum equations with inertia, Eqs. (6.1.5) and (6.1.8). All other equations and boundary conditions in Sec. 5.2 remain unchanged.

In wave propagation, the natural time scale is the characteristic wave period $T/2\pi = 1/\omega$, where ω is the characteristic frequency. Allowing the solid inertia to be comparable to the solid stress variation, we must have

$$\rho_s[u]\omega = G\frac{[u]}{\omega\ell'^2}\,. \tag{6.2.17}$$

Use has been made of $[u] = \omega[V]$ by continuity of velocity on the interface. It follows that

$$\ell' = \frac{1}{\omega}\sqrt{\frac{E}{\rho_s}}\,. \tag{6.2.18}$$

Thus the macroscale length is the classical elastic wavelength.

For sufficiently high frequency, the fluid inertia can also be as important as the gradients of pressure and the viscous stress, i.e.,

$$\rho_f[u]\omega = \frac{[p]}{\ell'} = \frac{\mu[u]}{\ell^2}. \tag{6.2.19}$$

It follows that

$$\ell = \delta \equiv \sqrt{\frac{\mu}{\rho_f \omega}}. \tag{6.2.20}$$

Thus inertia, hence dynamics, is important if the frequency is so high that Stokes' boundary-layer thickness δ is comparable to the microlength scale (pore or grain size). In water $\mu/\rho_f = 10^{-2}\,\mathrm{cm}^2/\mathrm{s}$. For gravity waves in the sea, $\omega = \mathcal{O}(1)\,\mathrm{rad/s}$, the Stokes layer thickness is $\mathcal{O}(1)\,\mathrm{mm}$ which is comparable to coarse sand but much greater than fine sand or silt. For sound waves the frequencies can be much higher ($\omega =$ 100–10,000 rad/s); the Stokes layer thickness is $\mathcal{O}(0.1\text{–}0.01)$ mm or $\mathcal{O}(100\text{–}10)\mu$m.

We introduce the dimensionless variables as follows,

$$\boldsymbol{x} = \ell \boldsymbol{x}^\dagger, \quad t = \frac{t^\dagger}{\omega}, \quad p = [p]p^\dagger, \quad u_i = [u]u_i^\dagger, \quad v_i = \frac{[u]v_i^\dagger}{\omega}. \tag{6.2.21}$$

In dimensionless variables, the equation of fluid momentum reads:

$$\epsilon \frac{\partial u_i^\dagger}{\partial t^\dagger} = -\frac{\partial p^\dagger}{\partial x_i^\dagger} + \epsilon \frac{\partial^2 u_i^\dagger}{\partial x_k^\dagger \partial x_k^\dagger}. \tag{6.2.22}$$

The equation of solid momentum is now,

$$\epsilon^2 \frac{\partial^2 v_i^\dagger}{\partial {t^\dagger}^2} = \frac{\partial \sigma_{ij}^\dagger}{\partial x_j^\dagger}. \tag{6.2.23}$$

The two constitutive equations are still

$$\Sigma_{ij}^\dagger = -p^\dagger \delta_{ij} + 2\epsilon \varepsilon_{ij}^\dagger(\boldsymbol{u}^\dagger), \tag{6.2.24}$$

and

$$\sigma_{ij}^\dagger = \frac{C_{ijk\ell}}{G} \varepsilon_{k\ell}^\dagger(\boldsymbol{v}^\dagger). \tag{6.2.25}$$

On the interface Γ, the kinematic condition of velocity continuity (Eq. (6.2.4)) and the dynamic condition for stress continuity (Eq. (6.2.13)) remain the same as in the quasi-static case.

Returning to dimensional forms with the order symbol $\epsilon = \ell/\ell' \ll 1$, the conservation law of fluid mass (Eq. (6.2.14)), the stress–strain relations (Eqs. (6.2.15) and (6.1.9)), and the interface conditions (Eqs. (6.2.4) and (6.2.16)) remain the same. Only the two momentum equations must be changed to include inertia:

$$\epsilon\rho_f \frac{\partial u_i}{\partial t} = \frac{\partial \Sigma_{ij}}{\partial x_j} = -\frac{\partial p}{\partial x_i} + 2\epsilon\mu \frac{\partial \varepsilon_{ij}(\boldsymbol{u})}{\partial x_j}, \tag{6.2.26}$$

and

$$\epsilon^2 \rho_s \frac{\partial^2 v_i}{\partial t^2} = \frac{\partial \sigma_{ij}}{\partial x_j}. \tag{6.2.27}$$

6.3. Multiple-Scale Expansions

Since the dynamic problem differs from the static one only by the presence of the inertia term, we shall treat both together. Let multiple-scale expansions be introduced for all the unknowns, i.e.,

$$f = f^{(0)} + \epsilon f^{(1)} + \epsilon^2 f^{(2)} + \cdots, \quad \text{where } f = (u_i, v_i, p, \Sigma_{ij}), \tag{6.3.1}$$

and

$$\sigma_{ij} = \frac{1}{\epsilon}\sigma_{ij}^{(-1)} + \sigma_{ij}^{(0)} + \epsilon\sigma_{ij}^{(1)} + \cdots. \tag{6.3.2}$$

Each of the perturbation terms is a function of $\boldsymbol{x}, \boldsymbol{x}'$ and t. Note that with fast and slow coordinates $\varepsilon_{ij}(\boldsymbol{u})$ is transformed to

$$\varepsilon_{ij}(\boldsymbol{u}) \to \varepsilon_{ij}(\boldsymbol{u}) + \epsilon\varepsilon'_{ij}(\boldsymbol{u}). \tag{6.3.3}$$

The first ($\varepsilon_{ij}(\boldsymbol{u})$) and second ($\varepsilon'_{ij}(\boldsymbol{u})$) parts represent the rate of strain on the micro- and macroscales, respectively, where

$$\varepsilon'_{ij}(\boldsymbol{u}) \equiv \frac{1}{2}\left(\frac{\partial u_i}{\partial x'_j} + \frac{\partial u_j}{\partial x'_i}\right). \tag{6.3.4}$$

Again ε_{ij} and ε'_{ij} are regarded as differential operators on the micro- and macroscales, respectively.

For the fluid phase the law of mass conservation gives, at orders $\mathcal{O}(\epsilon^0)$ and $\mathcal{O}(\epsilon)$,

$$\frac{\partial u_i^{(0)}}{\partial x_i} = 0, \tag{6.3.5}$$

$$\frac{1}{\rho_f E_f}\frac{\partial p^{(0)}}{\partial t} + \frac{\partial u_i^{(0)}}{\partial x'_i} + \frac{\partial u_i^{(1)}}{\partial x_i} = 0, \tag{6.3.6}$$

$$\vdots$$

From Eq. (6.1.6), we get

$$\Sigma_{ij}^{(0)} = -p^{(0)}\delta_{ij}\,, \tag{6.3.7}$$

$$\Sigma_{ij}^{(1)} = -p^{(1)}\delta_{ij} + 2\mu\varepsilon_{ij}(\boldsymbol{u}^{(0)}), \tag{6.3.8}$$

$$\vdots$$

it follows from Eqs. (6.2.1) and (6.2.26) that

$$\frac{\partial \Sigma_{ij}^{(0)}}{\partial x_j} = -\frac{\partial p^{(0)}}{\partial x_i} = 0\,, \tag{6.3.9}$$

$$\frac{\partial \Sigma_{ij}^{(1)}}{\partial x_j} + \frac{\partial \Sigma_{ij}^{(0)}}{\partial x_j'} = -\frac{\partial p^{(0)}}{\partial x_i'} - \frac{\partial p^{(1)}}{\partial x_i} + \mu\frac{\partial^2 u_i^{(0)}}{\partial x_k \partial x_k} = \underline{\rho_f \frac{\partial u_i^{(0)}}{\partial t}}\,, \tag{6.3.10}$$

$$\vdots$$

The perturbation equations for the solid momentum are obtained similarly from Eqs. (6.2.2) and (6.2.27),

$$\frac{\partial \sigma_{ij}^{(-1)}}{\partial x_j} = 0\,, \tag{6.3.11}$$

$$\frac{\partial \sigma_{ij}^{(0)}}{\partial x_j} + \frac{\partial \sigma_{ij}^{(-1)}}{\partial x_j'} = 0\,, \tag{6.3.12}$$

$$\frac{\partial \sigma_{ij}^{(1)}}{\partial x_j} + \frac{\partial \sigma_{ij}^{(0)}}{\partial x_j'} = \underline{\rho_s \frac{\partial^2 v_i^{(0)}}{\partial t^2}}\,, \tag{6.3.13}$$

$$\vdots$$

In Eqs. (6.3.10) and (6.3.13), the inertia terms (underlined) are present for the dynamical case only. In quasi-statics, they should be neglected.

Expansion of Hooke's law gives

$$\sigma_{ij}^{(-1)} = C_{ijk\ell}\varepsilon_{k\ell}(\boldsymbol{v}^{(0)})\,, \tag{6.3.14}$$

$$\sigma_{ij}^{(0)} = C_{ijk\ell}\varepsilon_{k\ell}(\boldsymbol{v}^{(1)}) + C_{ijk\ell}\varepsilon_{k\ell}'(\boldsymbol{v}^{(0)})\,, \tag{6.3.15}$$

$$\vdots$$

On the interface Γ, we get from the kinematic boundary condition (Eq. (6.2.4))

$$u_i^{(0)} = \frac{\partial v_i^{(0)}}{\partial t}, \tag{6.3.16}$$

$$u_i^{(1)} = \frac{\partial v_i^{(1)}}{\partial t}, \tag{6.3.17}$$

$$\vdots$$

and from the dynamic boundary condition (Eq. (6.2.16))

$$\sigma_{ij}^{(-1)} n_j = 0, \tag{6.3.18}$$

$$\sigma_{ij}^{(0)} n_j = -p^{(0)} \delta_{ij} n_j, \tag{6.3.19}$$

$$\sigma_{ij}^{(1)} n_j = \Sigma_{ij}^{(1)} n_j = [-p^{(1)} \delta_{ij} + 2\mu \varepsilon_{ij}(\boldsymbol{u}^{(0)})] n_j, \tag{6.3.20}$$

$$\vdots$$

We now examine the perturbation problems order by order. At the leading order $\mathcal{O}(\epsilon^{-1})$, we obtain from Eq. (6.3.9)

$$p^{(0)} = p^{(0)}(\boldsymbol{x}', t), \tag{6.3.21}$$

as in a rigid porous medium. From Eq. (6.3.11) and the boundary condition (Eq. (6.3.18)) it is clear that

$$\sigma_{ij}^{(-1)} \equiv 0, \quad \varepsilon_{ij}(\boldsymbol{v}^{(0)}) \equiv 0; \quad \boldsymbol{x} \in \Omega_s. \tag{6.3.22}$$

Consequently $\boldsymbol{v}^{(0)}$ is independent of the fast variable,

$$\boldsymbol{v}^{(0)} = \boldsymbol{v}^{(0)}(\boldsymbol{x}', t). \tag{6.3.23}$$

6.4. Averaged Total Momentum of the Composite

At the order $\mathcal{O}(\epsilon^0)$, $\boldsymbol{v}^{(1)}$ is governed by

$$\frac{\partial \sigma_{ij}^{(0)}}{\partial x_j} = \frac{\partial}{\partial x_j}\{C_{ijk\ell}[\varepsilon_{k\ell}(\boldsymbol{v}^{(1)}) + \varepsilon'_{k\ell}(\boldsymbol{v}^{(0)})]\} = 0, \quad \boldsymbol{x} \in \Omega_s \tag{6.4.1}$$

subject to the boundary condition on the interface

$$\{C_{ijk\ell}[\varepsilon_{k\ell}(\boldsymbol{v}^{(1)}) + \varepsilon'_{k\ell}(\boldsymbol{v}^{(0)})]\} n_j = -p^{(0)} n_i \quad \boldsymbol{x} \in \Gamma, \tag{6.4.2}$$

and the condition of Ω-periodicity.

Let us formally express $v_i^{(1)}$ in terms of $v_i^{(0)}$ and $p^{(0)}$:

$$v_i^{(1)}(\boldsymbol{x}, \boldsymbol{x}', t) = \chi_i^{pq} \varepsilon'_{pq}(\boldsymbol{v}^{(0)}) - \eta_i p^{(0)}, \quad \boldsymbol{x} \in \Omega_s, \tag{6.4.3}$$

where $\boldsymbol{\eta} = \{\eta_i\}$ is a vector and $\boldsymbol{\chi}^{pq} = \{\chi_i^{pq}\}$ a tensor of rank 3. Both of them are yet known functions of $\boldsymbol{x}$. We now choose a unit cell Ω of which the parts taken by fluid and solid phases are denoted by Ω_s and Ω_f, respectively. It is easy to check that $\boldsymbol{\chi}^{pq} = \{\chi_i^{pq}\}$ is governed by the following inhomogeneous partial differential equation in a Ω-cell:

$$\boxed{\frac{\partial}{\partial x_j}\left\{C_{ijk\ell}\left[\varepsilon_{k\ell}(\boldsymbol{\chi}^{pq}) + \delta_{kp}\delta_{\ell q}\right]\right\} = 0, \quad \boldsymbol{x} \in \Omega_s}, \tag{6.4.4}$$

subject to the inhomogeneous interface condition:

$$\boxed{\left\{C_{ijk\ell}\left[\varepsilon_{k\ell}(\boldsymbol{\chi}^{pq}) + \delta_{kp}\delta_{\ell q}\right]\right\} n_j = 0, \quad \boldsymbol{x} \in \Gamma}, \tag{6.4.5}$$

where

$$\varepsilon_{k\ell}(\boldsymbol{\chi}^{pq}) = \frac{1}{2}\left(\frac{\partial \chi_\ell^{pq}}{\partial x_k} + \frac{\partial \chi_k^{pq}}{\partial x_\ell}\right) \tag{6.4.6}$$

is a tensor of rank 4. In addition $\boldsymbol{\chi}^{pq}$ must be Ω-periodic.

Let $\langle\cdot\rangle$ denote the cell-average of each phase:

$$\langle \text{solid} \rangle = \frac{1}{\Omega}\iiint_{\Omega_s} (\text{solid})\, d\Omega, \quad \langle \text{fluid} \rangle = \frac{1}{\Omega}\iiint_{\Omega_f} (\text{fluid})\, d\Omega. \tag{6.4.7}$$

For uniqueness we also require

$$\boxed{\langle \chi_i^{pq} \rangle = 0}, \quad \text{or} \quad \boxed{\langle \boldsymbol{\chi}^{pq} \rangle = 0}. \tag{6.4.8}$$

Physically $\boldsymbol{\chi}^{pq}$ is the static displacement field under the influence of a known body force and surface stress.

Similarly, the vector η_j is governed by

$$\boxed{\frac{\partial}{\partial x_j}\{C_{ijk\ell}\varepsilon_{k\ell}(\boldsymbol{\eta})\} = 0, \quad \boldsymbol{x} \in \Omega_s}, \tag{6.4.9}$$

subject to the inhomogeneous interface condition:

$$\boxed{\{C_{ijk\ell}\varepsilon_{k\ell}(\boldsymbol{\eta})\} n_j = n_i \quad \boldsymbol{x} \in \Gamma}. \tag{6.4.10}$$

The displacement vector $\boldsymbol{\eta}$ must be Ω-periodic in $\boldsymbol{x}$ and satisfying

$$\boxed{\langle \eta_j \rangle = 0}. \tag{6.4.11}$$

Note that the two cell problems for χ^{pq} and η represent static equilibrium under forcing, therefore the solutions are independent of time. Numerical results found by the existing method of finite elements will be presented in a later section. Note also that the periodicity conditions in both cell problems rule out the seemingly trivial solution of constant stress, to which the corresponding displacements χ^{pq} and η would be linear in x and not Ω-periodic. These cell problems must in general be solved numerically for a prescribed cell geometry.

Once χ^{pq} and η are solved, we can take the solid-phase average to get the solid stress

$$\begin{aligned}
\langle \sigma_{ij}^{(0)} \rangle &= \frac{1}{\Omega} \iiint_{\Omega_s} \sigma_{ij}^{(0)} d\Omega \\
&= \langle C_{ijk\ell} \varepsilon_{k\ell}(\boldsymbol{v}^{(1)}) \rangle + \langle C_{ijk\ell} \rangle \varepsilon'_{k\ell}(\boldsymbol{v}^{(0)}) \\
&= [\langle C_{ijk\ell} \varepsilon_{k\ell}(\boldsymbol{\chi}^{pq}) \rangle + \langle C_{ijk\ell} \rangle \delta_{kp} \delta_{\ell q}] \varepsilon'_{pq}(\boldsymbol{v}^{(0)}) - \langle C_{ijk\ell} \varepsilon_{k\ell}(\boldsymbol{\eta}) \rangle p^{(0)} \\
&= \mathcal{C}_{ijpq} \varepsilon'_{pq}(\boldsymbol{v}^{(0)}) - \langle C_{ijk\ell} \varepsilon_{k\ell}(\boldsymbol{\eta}) \rangle p^{(0)} , \qquad (6.4.12)
\end{aligned}$$

where

$$\mathcal{C}_{ijpq} \equiv \langle C_{ijk\ell} \varepsilon_{k\ell}(\boldsymbol{\chi}^{pq}) \rangle + \langle C_{ijpq} \rangle . \qquad (6.4.13)$$

Equation (6.4.12) is the macroscale Hooke's law in the solid phase and is affected by fluid pressure.

Let us consider the averaged momentum in the composite by defining the total stress tensor as follows

$$T_{ij} = \begin{cases} \Sigma_{ij} , & \boldsymbol{x} \in \Omega_f ; \\ \sigma_{ij} , & \boldsymbol{x} \in \Omega_s . \end{cases} \qquad (6.4.14)$$

The total stress tensor for the bulk composite is

$$\begin{aligned}
\langle T_{ij}^{(0)} \rangle &\equiv \frac{1}{\Omega} \left\{ \iiint_{\Omega_s} \sigma_{ij}^{(0)} d\Omega + \iiint_{\Omega_f} \Sigma_{ij}^{(0)} d\Omega \right\} \\
&= \langle \Sigma_{ij}^{(0)} \rangle - \theta p^{(0)} \delta_{ij} \\
&= [\langle C_{ijk\ell} \varepsilon_{k\ell}(\boldsymbol{\chi}^{pq}) \rangle + \langle C_{ijk\ell} \rangle \delta_{kp} \delta_{\ell q}] \varepsilon'_{pq}(\boldsymbol{v}^{(0)}) \\
&\quad - [\theta \delta_{ij} + \langle C_{ijk\ell} \varepsilon_{k\ell}(\boldsymbol{\eta}) \rangle] p^{(0)} \\
&= \mathcal{C}_{ijpq} \varepsilon'_{pq}(\boldsymbol{v}^{(0)}) - \alpha'_{ij} p^{(0)} , \qquad (6.4.15)
\end{aligned}$$

where

$$\alpha'_{ij} \equiv \theta \delta_{ij} + \langle C_{ijk\ell} \varepsilon_{k\ell}(\boldsymbol{\eta}) \rangle . \qquad (6.4.16)$$

Physically the total stress tensor of the composite consists of two parts: one part is linear in the effective strain in the solid phase, and can be called the *effective solid stress.*

$$\sigma'_{ij} \equiv C_{ijpq}\varepsilon'_{pq}(\boldsymbol{v}^{(0)}) . \tag{6.4.17}$$

The other part is linear in the pore pressure in the fluid phase, with α'_{ij} being the coefficient tensor of proportionality. Since they are obtained from the cell-averages of $\boldsymbol{\chi}^{pq}$ and $\boldsymbol{\eta}$, the coefficient tensors ε'_{ij} and α'_{ij} are independent of time, namely their values are the same for quasi-static and dynamic problems.

We remark that in soil mechanics, the defining relation between the total and effective stresses is $\langle T_{ij}^{(0)}\rangle \equiv \Sigma_{ij}^e - p^{(0)}\delta_{ij}$, i.e., $\alpha'_{ij} = \delta_{ij}$. This relation is known as Terzaghi's effective stress principle derived on heuristic ground. This difference from Eq. (6.4.15) will be reconciled after Eq. (6.8.18).

Adding the phase averages of Eqs. (6.3.10) and (6.3.13) and using Gauss's theorem and the interface condition (Eq. (6.3.20)), we get finally the effective momentum equation for the composite,

$$\underline{\rho_f \frac{\partial \langle u_i^{(0)}\rangle}{\partial t} + \rho_s(1-\theta)\frac{\partial^2 v_i^{(0)}}{\partial t^2}} = \frac{\partial \langle T_{ij}^{(0)}\rangle}{\partial x'_j} = \frac{\partial}{\partial x'_j}[C_{ijpq}\varepsilon'_{pq}(\boldsymbol{v}^{(0)}) - \alpha'_{ij}p^{(0)}] . \tag{6.4.18}$$

6.5. Averaged Mass Conservation of Fluid Phase

Let us take the phase average of Eq. (6.3.6) for fluid mass conservation

$$\begin{aligned}\frac{\theta}{\rho_f E_f}\frac{\partial p^{(0)}}{\partial t} + \frac{\partial \langle u_i^{(0)}\rangle}{\partial x'_j} &= -\frac{1}{\Omega}\iiint_{\Omega_f}\frac{\partial u_i^{(1)}}{\partial x_i}d\Omega = \frac{1}{\Omega}\iiint_{\Gamma} u_i^{(1)} n_i dS \\ &= \frac{1}{\Omega}\iiint_{\Gamma}\dot{v}_i^{(1)} n_i dS = \frac{1}{\Omega}\iiint_{\Omega_s}\frac{\partial \dot{v}_i^{(1)}}{\partial x_i}d\Omega ,\end{aligned} \tag{6.5.1}$$

where use has been made of the interface condition of velocity continuity, Gauss' theorem and the fact that the unit normal is outward from the solid. Because of Eq. (6.4.3) the last term in Eq. (6.5.1) can be written as

$$\left\langle \frac{\partial \chi_i^{k\ell}}{\partial x_i}\right\rangle \varepsilon'_{k\ell}(\dot{\boldsymbol{v}}^{(0)}) - \left\langle \frac{\partial \eta_i}{\partial x_i}\right\rangle \frac{\partial p^{(0)}}{\partial t} .$$

It follows that

$$\left(\frac{\theta}{\rho_f E_f} + \left\langle \frac{\partial \eta_i}{\partial x_i}\right\rangle\right)\frac{\partial p^{(0)}}{\partial t} + \frac{\partial \langle u_i^{(0)}\rangle}{\partial x'_i} = \left\langle \frac{\partial \chi_i^{k\ell}}{\partial x_i}\right\rangle [\varepsilon'_{k\ell}(\dot{\boldsymbol{v}}^{(0)})] , \tag{6.5.2}$$

which again couples the pore pressure with the solid strain. This result is valid for both quasi-static and dynamic cases.

We now turn to the averaged fluid momentum and separate the quasi-static and dynamic cases.

6.6. Averaged Fluid Momentum

6.6.1. *Quasi-Static Case*

The microscale problem for $u_i^{(0)}$ and $p^{(1)}$ is governed by Eqs. (6.3.5) and (6.3.10) in Ω_f, to which the following formal solution can be assumed,

$$u_i^{(0)} - \dot{v}_i^{(0)} = -K_{ij}\frac{\partial p^{(0)}}{\partial x'_j}, \tag{6.6.1}$$

where use is made of the fact that $\dot{v}_i^{(0)} \equiv \partial v_i^{(0)}/\partial t$ is independent of $\boldsymbol{x}$ and

$$p^{(1)} = -A_j\delta_{ij}\frac{\partial p^{(0)}}{\partial x'_j}, \quad \langle A_j\rangle = 0\,. \tag{6.6.2}$$

Clearly K_{ij}, A_j are governed by the same Stokes equations for the rigid matrix problem in Sec. 3.2. Taking the cell-average of Eq. (6.6.1) over the fluid phase gives Darcy's law for a deformable medium,

$$\langle u_i^{(0)}\rangle - n\dot{v}_i^{(0)} = -\mathcal{K}_{ij}\frac{\partial p^{(0)}}{\partial x'_j}, \tag{6.6.3}$$

where

$$\mathcal{K}_{ij} \equiv \langle K_{ij}\rangle \tag{6.6.4}$$

is the time-independent permeability for a rigid matrix. Equation (6.6.3) is the momentum equation for the fluid phase.

In summary, Eq. (6.4.18) with zero on the left-hand side, Eqs. (6.5.2) and (6.6.3) constitute seven scalar equations for seven unknowns: $u_i^{(0)}, v_i^{(0)}$ and $p^{(0)}$.

Equation (6.6.3) can also be combined with Eq. (6.5.2) to give

$$\frac{\partial}{\partial x'_i}\left(\mathcal{K}_{ij}\frac{\partial p^{(0)}}{\partial x'_j}\right) = \frac{\partial}{\partial x'_i}(\theta\dot{v}_i^{(0)}) + \left[\theta\delta_{k\ell} - \left\langle\frac{\partial\chi_i^{k\ell}}{\partial x_i}\right\rangle\right][\varepsilon'_{k\ell}(\dot{\boldsymbol{v}}^{(0)})] + \left(\frac{\theta}{\rho_f E_f} + \left\langle\frac{\partial\eta_i}{\partial x_i}\right\rangle\right)\frac{\partial p^{(0)}}{\partial t}, \tag{6.6.5}$$

or

$$\frac{\partial}{\partial x_i'}\left(\mathcal{K}_{ij}\frac{\partial p^{(0)}}{\partial x_j'}\right) = \frac{\partial}{\partial x_i'}(\theta \dot{v}_i^{(0)}) + \gamma_{k\ell}'[\varepsilon_{k\ell}'(\dot{\boldsymbol{v}}^{(0)})] + \left(\frac{\theta}{\rho_f E_f} + \beta'\right)\frac{\partial p^{(0)}}{\partial t}, \tag{6.6.6}$$

where

$$\gamma_{k\ell}' = \theta\delta_{k\ell} - \left\langle \frac{\partial \chi_i^{k\ell}}{\partial x_i}\right\rangle = \theta\delta_{k\ell} - \langle \varepsilon_{ij}(\boldsymbol{\chi}^{k\ell})\rangle\delta_{ij}, \tag{6.6.7}$$

which can be computed from the solution of the cell problem for $\boldsymbol{\chi}$, and

$$\beta' = \left\langle \frac{\partial \eta_i}{\partial x_i}\right\rangle, \tag{6.6.8}$$

which follows from the solution of the similar cell problem for $\boldsymbol{\eta}$. Physically the first two terms on the right-hand side of Eq. (6.6.7) represent an apparent source of mass to the fluid phase, and β' is the modification of fluid compressibility, due to the solid strain. It will be shown shortly that $\gamma_{k\ell}'$ is equal to $\alpha_{k\ell}'$ defined in Eq. (6.4.16).

Thus Eq. (6.4.18) with zero on the left-hand side and Eq. (6.6.6) give four scalar equations for the unknowns: $v_i^{(0)}$ and $p^{(0)}$. The seepage velocity follows from Eq. (6.6.3).

6.6.2. *Dynamic Case*

In view of the kinematic boundary condition (Eq. (6.3.16)) on the interface and Eq. (6.3.23), we rewrite Eqs. (6.3.5) and (6.3.10) as

$$\frac{\partial}{\partial x_i}\left(u_i^{(0)} - \dot{v}_i^{(0)}\right) = 0, \tag{6.6.9}$$

and

$$\rho_f \frac{\partial(u_i^{(0)} - \dot{v}_i^{(0)})}{\partial t} = -\left(\rho_f\frac{\partial \dot{v}_i^{(0)}}{\partial t} + \frac{\partial p^{(0)}}{\partial x_i'}\right) - \frac{\partial p^{(1)}}{\partial x_i} + \mu\frac{\partial^2(u_i^{(0)} - \dot{v}_i^{(0)})}{\partial x_k \partial x_k}. \tag{6.6.10}$$

For the transient case in general, we assume that the initial values of $p^{(0)}$ and $v_i^{(0)}$ are all zero, and formally introduce $K_{ij}(\boldsymbol{x},t), A_j(\boldsymbol{x},t)$ via the following convolution integrals,

$$u_i^{(0)} - \dot{v}_i^{(0)} = -\int_0^t K_{ij}(\boldsymbol{x}, t-t')\left(\rho_f \ddot{v}_j^{(0)}(\boldsymbol{x}',t') + \frac{\partial p^{(0)}(\boldsymbol{x}',t')}{\partial x_j'}\right)dt', \tag{6.6.11}$$

and

$$p^{(1)} = -\int_0^t A_j(\boldsymbol{x}, t - t') \left(\rho_f \ddot{v}_j^{(0)}(\boldsymbol{x}', t') + \frac{\partial p^{(0)}(\boldsymbol{x}', t')}{\partial x'_j} \right) dt' , \quad (6.6.12)$$

with the constraint $\langle A_i \rangle = 0$. For brevity, dependences on $\boldsymbol{x}$ and $\boldsymbol{x}'$ are suppressed but implied. Equations (6.6.9) and (6.6.10) are satisfied if K_{ij} and A_j are the solutions of the following initial-boundary-value problem in a cell,

$$\frac{\partial K_{ij}}{\partial x_i} = 0 , \quad (6.6.13)$$

and

$$\rho_f \frac{\partial K_{ij}}{\partial t} = -\frac{\partial A_j}{\partial x_i} + \mu \frac{\partial^2 K_{ij}}{\partial x_k \partial x_k} \quad (6.6.14)$$

subject to the initial conditions

$$K_{ij}(\boldsymbol{x}, 0) = \delta_{ij} , \quad A_i(\boldsymbol{x}, 0) = 0 , \quad \boldsymbol{x} \in \Omega_f , \quad t = 0 , \quad (6.6.15)$$

the boundary condition

$$K_{ij}(\boldsymbol{x}, t) = 0 , \quad \boldsymbol{x} \in \Gamma , \quad t > 0 , \quad (6.6.16)$$

and the requirement of Ω-periodicity. Once the transient problem for K_{ij} is solved, the cell average of Eq. (6.6.11) over the fluid phase gives the dynamic Darcy's law,

$$\langle u_i^{(0)} \rangle - \theta \dot{v}_i^{(0)} = -\int_0^t \mathcal{K}_{ij}(\boldsymbol{x}, t - t') \left(\rho_f \ddot{v}_j^{(0)}(\boldsymbol{x}', t') + \frac{\partial p^{(0)}(\boldsymbol{x}', t')}{\partial x'_i} \right) dt' , \quad (6.6.17)$$

where $\mathcal{K}_{ij}(\boldsymbol{x}, t) \equiv \langle K_{ij}(\boldsymbol{x}, t) \rangle$ is the dynamic permeability.

In summary, Eqs. (6.4.18), (6.5.2), and (6.6.17) give seven equations for seven unknowns $p^{(0)}$, $\langle u_i^{(0)} \rangle$, and $v_i^{(0)}$.

6.7. Time-Harmonic Motion

In the important special case where the motion is simple harmonic in time, $u_i^{(0)}$, $v_i^{(0)}$, and $p^{(0)}$ are all proportional to $e^{-i\omega t}$, i.e.,

$$\{u_i^{(0)}, v_i^{(0)}, p^{(0)}\} = \mathrm{Re}\left[\{\widetilde{u}_i^{(0)}, \widetilde{v}_i^{(0)}, \widetilde{p}^{(0)}\} e^{-i\omega t}\right] . \quad (6.7.1)$$

The amplitude functions are governed on the microscale by

$$\frac{\partial}{\partial x_i}(\widetilde{u}_i^{(0)} + i\omega \widetilde{v}_i^{(0)}) = 0 , \quad (6.7.2)$$

and

$$-i\omega\rho_f\left(\widetilde{u}_i^{(0)} + i\omega\widetilde{v}_i^{(0)}\right) = -\left(-\rho_f\omega^2\widetilde{v}_i^{(0)} + \frac{\partial\widetilde{p}^{(0)}}{\partial x_i'}\right) - \frac{\partial\widetilde{p}^{(1)}}{\partial x_i} + \mu\frac{\partial^2\left(\widetilde{u}_i^{(0)} + i\omega\widetilde{v}_i^{(0)}\right)}{\partial x_k \partial x_k}. \tag{6.7.3}$$

Now we define $\widetilde{K}_{ij}(\boldsymbol{x},\omega)$, $\widetilde{A}_j(\boldsymbol{x},\omega)$ by

$$\widetilde{u}_i^{(0)} + i\omega\widetilde{v}_i^{(0)} = -\widetilde{K}_{ij}\left(-\rho_f\omega^2\widetilde{v}_j^{(0)} + \frac{\partial\widetilde{p}^{(0)}}{\partial x_j'}\right), \tag{6.7.4}$$

and

$$\widetilde{p}^{(1)} = -\widetilde{A}_j\left(-\rho_f\omega^2\widetilde{v}_j^{(0)} + \frac{\partial\widetilde{p}^{(0)}}{\partial x_j'}\right), \tag{6.7.5}$$

with the constraint $\langle\widetilde{A}_i\rangle = 0$. The cell problem for $\widetilde{K}_{ij}(\boldsymbol{x},\omega)$, $\widetilde{A}_j(\boldsymbol{x},\omega)$ is defined by

$$\frac{\partial\widetilde{K}_{ij}}{\partial x_i} = 0\,, \tag{6.7.6}$$

and

$$-i\omega\rho_f\widetilde{K}_{ij} = \delta_{ij} - \frac{\partial\widetilde{A}_j}{\partial x_i} + \mu\frac{\partial^2\widetilde{K}_{ij}}{\partial x_k\partial x_k}\,, \tag{6.7.7}$$

subject to the boundary condition

$$\widetilde{K}_{ij}(\boldsymbol{x},\omega) = 0\,, \quad \boldsymbol{x}\in\Gamma\,, \tag{6.7.8}$$

and Ω-periodicity. After its solution, the dynamic Darcy's law follows by cell averaging,

$$\langle\widetilde{u}_i^{(0)}\rangle + i\omega\theta\widetilde{v}_i^{(0)} = -\widetilde{\mathcal{K}}_{ij}(\omega)\left(-\rho_f\omega^2\widetilde{v}_j^{(0)} + \frac{\partial\widetilde{p}^{(0)}}{\partial x_i'}\right), \tag{6.7.9}$$

where the dynamic permeability is given by

$$\widetilde{\mathcal{K}}_{ij}(\omega) = \langle\widetilde{K}_{ij}(\boldsymbol{x},\omega)\rangle\,. \tag{6.7.10}$$

We stress that $\widetilde{\mathcal{K}}_{ij}$ is a complex function of ω. In views of Eq. (6.7.7), the importance of frequency dependence is measured by the ratio

$$\frac{\omega\ell^2}{\mu/\rho_f} = \frac{\ell^2}{\delta^2}\,, \tag{6.7.11}$$

where $\delta = \sqrt{2\mu/\rho_f\omega}$ is the thickness of Stokes boundary layer. If this ratio is small (i.e., low frequency), the permeability is independent of frequency and dominated by viscous stress. On the other hand, if the ratio is very large, viscosity is not as important as fluid inertia, and permeability depends strongly on frequency.

We point out that the dynamic permeability defined by the cell problem does not depend on the deformation of the solid elasticity, hence holds for an oscillatory flow through a rigid solid matrix. For the idealized case of one-dimensional pores modeled as straight and long tubes or channels, the permeability can be calculated analytically (Biot, 1956; Auriault *et al.*, 1985; Sheng and Zhou, 1988; Zhou and Sheng, 1989). For a rigid solid matrix with three-dimensional grains of uniform size in cubic packing, Zhou and Sheng (1989) have applied the homogenization technique and solved the cell problem by finite elements for the complex dynamic permeability for three microstructures. In particular, they have found a surprising result that the ratio of dynamic to static permeabilities K/K_0 is a universal function of the frequency ratio ω/ω_0, i.e., $K/K_0 = f(\omega/\omega_0)$ where ω_0 is a characteristic frequency of the medium and f is independent of the microstructure.

In the special case of an isotropic and homogeneous composite, the various coefficients take on simpler forms,

$$\mathcal{C}_{ijk\ell} = \lambda'\delta_{ij}\delta_{k\ell} + G'(\delta_{ik}\delta_{j\ell} + \delta_{im}\delta_{jk}), \quad \gamma'_{ij} = \alpha'_{ij} = \alpha'\delta_{ij}, \quad \mathcal{K}_{ij} = \mathcal{K}\delta_{ij}, \tag{6.7.12}$$

where

$$\lambda' = \frac{E'\nu'}{(1+\nu')(1-2\nu')}, \quad G' = \frac{E'}{2(1+\nu')}, \tag{6.7.13}$$

as well as α' and $\mathcal{K}$ are constants. In terms of which the effective solid stress is

$$\Sigma^e_{ij} = \lambda'\frac{\partial v_k^{(0)}}{\partial x'_k}\delta_{ij} + G'\left(\frac{\partial v_i^{(0)}}{\partial x'_j} + \frac{\partial v_j^{(0)}}{\partial x'_i}\right). \tag{6.7.14}$$

Let us consider the dynamic case with simple harmonic dependence on time. The momentum equation for the bulk composite (Eq. (6.4.18)) becomes,

$$\begin{aligned} &-i\omega\rho_f\langle\tilde{u}_i^{(0)}\rangle - \omega^2\rho_s(1-\theta)\tilde{v}_i^{(0)} \\ &= -\alpha'\frac{\partial\tilde{p}^{(0)}}{\partial x'_i} + G'\left(\frac{\partial^2\tilde{v}_i^{(0)}}{\partial x'_k\partial x'_k} + \frac{1}{1-2\nu}\frac{\partial}{\partial x'_i}\frac{\partial\tilde{v}_k^{(0)}}{\partial x'_k}\right). \end{aligned} \tag{6.7.15}$$

The fluid mass conservation equation (6.5.2) reduces to

$$-i\omega\frac{n}{B_f}\tilde{p}^{(0)} + \frac{\partial\langle\tilde{u}_j^{(0)}\rangle}{\partial x'_j} - i\omega(\alpha'-\theta)\frac{\partial\tilde{v}_j^{(0)}}{\partial x'_j} = 0, \tag{6.7.16}$$

where B_f denotes the effective bulk modulus of the pore fluid defined by

$$\frac{\theta}{B_f} = \frac{\theta}{\rho_f E_f} + \beta' . \tag{6.7.17}$$

Darcy's law (Eq. (6.6.17)) becomes,

$$\langle \widetilde{u}_i^{(0)} \rangle + i\omega\theta\widetilde{v}_i^{(0)} = -\widetilde{\mathcal{K}}\left(-\omega^2 \rho_f \widetilde{v}_i^{(0)} + \frac{\partial \widetilde{p}^{(0)}}{\partial x_i'} \right) . \tag{6.7.18}$$

Equations (6.7.15), (6.7.16) and (6.7.18) govern the seven scalar unknowns $\langle \widetilde{u}_i^{(0)} \rangle$, $\widetilde{v}_i^{(0)}$ and $\widetilde{p}^{(0)}$.

6.8. Properties of the Effective Coefficients

As in the case of rigid porous media, several fundamental properties of the constitutive coefficients can be derived from the defining cell boundary problems. We first derive some results for general media, then for homogeneous and isotropic media, following Auriault and Sanchez-Palencia (1977).

6.8.1. *Three Identities for General Media*

(i) We first show the following symmetry of the effective elasticity tensor defined in Eq. (6.4.13)

$$\boxed{\mathcal{C}_{ijk\ell} = \mathcal{C}_{jik\ell} = \mathcal{C}_{ij\ell k} = \mathcal{C}_{k\ell ij}} . \tag{6.8.1}$$

Recall first the known symmetry of the material elasticity tensor

$$C_{ijk\ell} = C_{jik\ell} = C_{ij\ell k} = C_{k\ell ij} . \tag{6.8.2}$$

The first two equalities in Eq. (6.8.1) follow by definition and by Eq. (6.8.2). For the last, it is only necessary to show that

$$\langle C_{rsk\ell}\varepsilon_{k\ell}(\boldsymbol{\chi}^{pq}) \rangle = \langle C_{pqk\ell}\varepsilon_{k\ell}(\boldsymbol{\chi}^{rs}) \rangle . \tag{6.8.3}$$

Scalar-multiplying Eq. (6.4.4) by χ_i^{rs} and integrating over Ω_s we get after using Gauss' theorem and the boundary condition (Eq. (6.4.5)),

$$\iiint_{\Omega_s} \frac{\partial \chi_i^{rs}}{\partial x_j} (C_{ijk\ell}[\varepsilon_{k\ell}(\boldsymbol{\chi}^{pq}) + \delta_{kp}\delta_{\ell q}]) d\Omega = 0 .$$

This result can be rewritten, on account of the symmetry $C_{ijk\ell} = C_{jik\ell}$, as

$$\iiint_{\Omega_s} \varepsilon_{ij}(\boldsymbol{\chi}^{rs})(C_{ijk\ell}[\varepsilon_{k\ell}(\boldsymbol{\chi}^{pq}) + \delta_{kp}\delta_{\ell q}])d\Omega = 0\,, \tag{6.8.4}$$

therefore,

$$\begin{aligned}
&\iiint_{\Omega_s} \varepsilon_{ij}(\boldsymbol{\chi}^{rs})C_{ijk\ell}\varepsilon_{k\ell}(\boldsymbol{\chi}^{pq})d\Omega \\
&\quad = -\iiint_{\Omega_s} \varepsilon_{ij}(\boldsymbol{\chi}^{rs})C_{ijpq}d\Omega = -\iiint_{\Omega_s} C_{pqij}\varepsilon_{ij}(\boldsymbol{\chi}^{rs})d\Omega\,.
\end{aligned} \tag{6.8.5}$$

Let us rewrite the preceding equality by interchanging r and s, p and q, and then invoke the symmetry $C_{ijk\ell} = C_{k\ell ij}$,

$$\iiint_{\Omega_s} \varepsilon_{ij}(\boldsymbol{\chi}^{pq})C_{k\ell ij}\varepsilon_{k\ell}(\boldsymbol{\chi}^{rs})d\Omega = -\iiint_{\Omega_s} C_{rsij}\varepsilon_{ij}(\boldsymbol{\chi}^{pq})d\Omega\,. \tag{6.8.6}$$

The left-hand sides of Eqs. (6.8.5) and (6.8.6) are the same, so must be the right-hand sides, hence Eq. (6.8.3) is proven.

(ii) We next show that

$$\boxed{\gamma'_{ij} = \alpha'_{ij}}\,. \tag{6.8.7}$$

Although γ'_{ij} and α'_{ij} are defined by two different cell problems for $\boldsymbol{\chi}$ and $\boldsymbol{\eta}$, respectively, by this equality the numerical task for $\boldsymbol{\eta}$ can be avoided.

Integrating the scalar product of Eq. (6.4.4) with η_i over Ω_s and using Gauss' theorem and the boundary conditions, we get

$$\begin{aligned}
0 &= \iiint_{\Omega_s} \eta_i \frac{\partial}{\partial x_j}\{C_{ijk\ell}[\varepsilon_{k\ell}(\boldsymbol{\chi}^{pq}) + \delta_{kp}\delta_{\ell q}]\}dV \\
&= -\iiint_{\Omega_s} \frac{\partial \eta_i}{\partial x_j}\{C_{ijk\ell}[\varepsilon_{k\ell}(\boldsymbol{\chi}^{pq}) + \delta_{kp}\delta_{\ell q}]\}d\Omega \\
&= -\iiint_{\Omega_s} \frac{1}{2}\left(\frac{\partial \eta_i}{\partial x_j} + \frac{\partial \eta_i}{\partial x_j}\right)\{C_{ijk\ell}[\varepsilon_{k\ell}(\boldsymbol{\chi}^{pq}) + \delta_{kp}\delta_{\ell q}]\}d\Omega \\
&= -\iiint_{\Omega_s} \varepsilon_{ij}(\boldsymbol{\eta})\{C_{ijk\ell}[\varepsilon_{k\ell}(\boldsymbol{\chi}^{pq}) + \delta_{kp}\delta_{\ell q}]\}d\Omega \\
&= -\iiint_{\Omega_s} \varepsilon_{ij}(\boldsymbol{\eta})C_{ijk\ell}\varepsilon_{k\ell}(\boldsymbol{\chi}^{pq})d\Omega - \iiint_{\Omega_s} C_{pqij}\varepsilon_{ij}(\boldsymbol{\eta})d\Omega\,,
\end{aligned}$$

where the symmetry of $C_{ijpq} = C_{pqij}$ has been used, hence

$$\langle \varepsilon_{ij}(\boldsymbol{\eta})C_{ijk\ell}\varepsilon_{k\ell}(\boldsymbol{\chi}^{pq})\rangle = -\langle C_{pqij}\varepsilon_{ij}(\boldsymbol{\eta})\rangle\,. \tag{6.8.8}$$

Let us multiply Eq. (6.4.9) by χ_i^{pq}, use Gauss' theorem and then Eq. (6.4.10) to get

$$
\begin{aligned}
0 &= \iiint_{\Omega_s} \chi_i^{pq} \frac{\partial}{\partial x_j} C_{ijk\ell}\varepsilon_{k\ell}(\boldsymbol{\eta}) d\Omega \\
&= \iint_{\partial\Omega_s} \chi_i^{pq} n_i dS - \iiint_{\Omega_s} \frac{1}{2}\left(\frac{\partial \chi_{ipq}}{\partial x_j} + \frac{\partial \chi_j^{pq}}{\partial x_i}\right) C_{ijk\ell}\varepsilon_{k\ell}(\boldsymbol{\eta}) d\Omega \\
&= \iiint_{\Omega_s} \frac{\partial \chi_i^{pq}}{\partial x_i} d\Omega - \iiint_{\Omega_s} \varepsilon_{ij}(\boldsymbol{\chi}^{pq}) C_{ijk\ell}\varepsilon_{k\ell}(\boldsymbol{\eta}) d\Omega ,
\end{aligned}
$$

hence

$$
\left\langle \frac{\partial \chi_i^{pq}}{\partial x_i} \right\rangle = \langle \varepsilon_{ij}(\boldsymbol{\chi}^{pq})\rangle \delta_{ij} = \langle \varepsilon_{ij}(\boldsymbol{\chi}^{pq}) C_{k\ell ij}\varepsilon_{k\ell}(\boldsymbol{\eta})\rangle = \langle \varepsilon_{ij}(\boldsymbol{\eta}) C_{ijk\ell}\varepsilon_{k\ell}(\boldsymbol{\chi}^{pq})\rangle , \tag{6.8.9}
$$

after invoking the symmetry of $C_{ijk\ell}$. Using Eqs. (6.8.8) and (6.8.9) in the definitions (Eqs. (6.4.16) and (6.6.7)), Eq. (6.8.7) is verified.

(iii) Finally we show that

$$
\boxed{\beta' > 0} \,. \tag{6.8.10}
$$

Let us multiply Eq. (6.4.9) by η_j and integrate

$$
\begin{aligned}
0 &= \iiint_{\Omega_s} \eta_i \frac{\partial}{\partial x_j} C_{ijk\ell}\varepsilon_{k\ell}(\boldsymbol{\eta}) d\Omega \\
&= \iint_{\partial\Omega_s} \eta_i n_i dS - \iiint_{\Omega_s} \frac{1}{2}\left(\frac{\partial \eta_i}{\partial x_j} + \frac{\partial \eta_j}{\partial x_i}\right) C_{ijk\ell}\varepsilon_{k\ell}(\boldsymbol{\eta}) d\Omega \\
&= \iiint_{\Omega_s} \frac{\partial \eta_i}{\partial x_i} d\Omega - \iiint_{\Omega_s} \varepsilon_{ij}(\boldsymbol{\eta}) C_{ijk\ell}\varepsilon_{k\ell}(\boldsymbol{\eta}) d\Omega .
\end{aligned}
$$

Since the second integral above is positive, so is the first; Eq. (6.8.10) follows.

6.8.2. Homogeneous and Isotropic Grains

Aside from the results above, Auriault and Sanchez-Palencia (1977) have derived two other explicit relations among the constitutive coefficients for grains that are made up of a homogeneous and isotropic solid.

Let us define the compliance tensor $B_{\ell mij}$, i.e., the reciprocal of the elasticity tensor, by

$$\varepsilon_{\ell m}(\boldsymbol{v}) = B_{\ell mij}\sigma_{ij}\,. \tag{6.8.11}$$

We also cite the following known properties from the theory of elasticity (Fung, 1965)

$$C_{ijkh}B_{\alpha\beta kh} = \frac{1}{2}(\delta_{\alpha i}\delta_{\beta j} + \delta_{\alpha j}\delta_{\beta i})\,, \tag{6.8.12}$$

$$B_{\alpha\beta ij}C_{ij\ell m} = \frac{1}{2}(\delta_{\alpha m}\delta_{\beta \ell} + \delta_{\alpha \ell}\delta_{\beta m})\,, \tag{6.8.13}$$

and the special property of an isotropic solid,

$$B_{\alpha\beta rr} = \frac{\delta_{\alpha\beta}}{3\lambda + 2G}\,. \tag{6.8.14}$$

For convenience, a proof of the preceding formula is reproduced in Appendix 6A (Eq. (A.12)).

We first show that, for materials homogeneous on the microscale,

$$\gamma'_{ij} = \alpha'_{ij} = \delta_{ij} - C_{pqij}B_{pqrr}\,. \tag{6.8.15}$$

If in addition, the solid is isotropic on the microscale, then

$$\boxed{\gamma'_{ij} = \alpha'_{ij} = \delta_{ij} - \frac{C_{pqij}\delta_{pq}}{3\lambda + 2G}}\,. \tag{6.8.16}$$

To prove these results, we start from Eq. (6.4.13),

$$C_{ijpq} = \langle C_{ijk\ell}\varepsilon_{k\ell}(\boldsymbol{\chi}^{pq})\rangle + \langle C_{ijpq}\rangle\,.$$

For a homogeneous material, C_{ijpq} is independent of $\boldsymbol{x}$, so that

$$C_{ijpq} = C_{ijk\ell}\langle\varepsilon_{k\ell}(\boldsymbol{\chi}^{pq})\rangle + (1-\theta)C_{ijpq}\,.$$

Scalar-multiplying both sides by the compliance tensor,

$$\begin{aligned} B_{ijpq}C_{ijk\ell} &= B_{ijpq}C_{ij\alpha\beta}\langle\varepsilon_{\alpha\beta}(\boldsymbol{\chi}^{k\ell})\rangle + (1-\theta)B_{ijpq}C_{ijk\ell} \\ &= \delta_{p\alpha}\delta_{q\beta}\langle\varepsilon_{\alpha\beta k\ell}(\boldsymbol{\chi}^{k\ell})\rangle + (1-\theta)\delta_{pk}\delta_{q\ell} \\ &= \langle\varepsilon_{pq}(\boldsymbol{\chi}^{k\ell})\rangle + (1-\theta)\delta_{pk}\delta_{q\ell}\,. \end{aligned}$$

Use has been made of Eq. (6.8.13). Contracting by setting $p = q$, the preceding formula becomes

$$B_{ijpp}C_{ijk\ell} = \langle\varepsilon_{pp}(\boldsymbol{\chi}^{k\ell})\rangle + (1-\theta)\delta_{k\ell}\,,$$

which can be used in the definition of γ'_{ij} and α'_{ij}, to give

$$\begin{aligned}\gamma'_{ij} = \alpha'_{ij} &= \theta\delta_{ij} - \langle\varepsilon_{pp}(\boldsymbol{\chi}^{ij})\rangle \\ &= \theta\delta_{ij} - B_{k\ell pp}C_{k\ell ij} + (1-\theta)\delta_{ij} \\ &= \delta_{ij} - B_{k\ell pp}C_{k\ell ij}\,.\end{aligned} \tag{6.8.17}$$

Hence Eq. (6.8.15) follows.

If the solid is isotropic on the microscale, we use the identity (Eq. (6.8.14)) in (Eq. (6.8.15)) to give Eq. (6.8.16).

If furthermore the composite is also isotropic on the macroscale,

$$\mathcal{C}_{k\ell ij} = \lambda'\delta_{k\ell}\delta_{ij} + G'(\delta_{ki}\delta_{\ell j} + \delta_{\ell i}\delta_{kj})\,,$$

then

$$\mathcal{C}_{k\ell ij}\delta_{k\ell} = (3\lambda' + 2G')\delta_{ij}\,,$$

hence $\gamma'_{ij} = \alpha'_{ij} = \gamma'\delta_{ij} = \alpha'\delta_{ij}$ are diagonal with

$$\boxed{\gamma' = \alpha' = \left(1 - \frac{3\lambda' + 2G'}{3\lambda + 2G}\right)}\,. \tag{6.8.18}$$

Recall from Sec. 6.4 that Terzaghi reasoned in his effective stress principle that $\alpha'_{ij} = \delta_{ij}$. For many natural soils the macroscale coefficients from soil testing are often very small, $((\lambda', G') = \mathcal{O}(10^7\text{–}10^8)\,\text{N/m}^2)$ (see Lamb and Whitman, 1969, p. 159) compared to the microscale values of quartz grains $((\lambda, G) = \mathcal{O}(10^{10}\text{–}10^{11})\,\text{N/m}^2)$ due possibly to the slippage among grains. Therefore Terzaghi's assumption is a good approximation for soils.

The second relation is concerned with β'. We show for solids homogeneous on the microscale that

$$\boxed{\beta' = B_{\ell\ell ij}[(1-\theta)\delta_{ij} - \mathcal{C}_{pqij}B_{pqrr}]}\,. \tag{6.8.19}$$

Again β' can be obtained without solving the cell problem for $\boldsymbol{\eta}$. If the solid is also isotropic we further have

$$\boxed{\beta' = \frac{1}{3\lambda + 2G}\left[3(1-\theta) - \frac{\mathcal{C}_{ppjj}}{3\lambda + 2G}\right]}\,. \tag{6.8.20}$$

From the definition of α'_{ij}, we have for homogeneous solids,

$$\alpha'_{ij} = \theta\delta_{ij} + \langle C_{ijk\ell}\varepsilon_{k\ell}(\boldsymbol{\eta})\rangle = \theta\delta_{ij} + C_{ijk\ell}\langle\varepsilon_{k\ell}(\boldsymbol{\eta})\rangle\,.$$

Making use of Eq. (6.8.17), we get

$$C_{ijk\ell}\langle\varepsilon_{k\ell}(\boldsymbol{\eta})\rangle = (1-\theta)\delta_{ij} - B_{k\ell pp}\mathcal{C}_{k\ell ij}\,.$$

Multiplying both sides by $B_{\alpha\beta ij}$, we get from the left-hand side

$$B_{\alpha\beta ij}C_{ijk\ell}\langle\varepsilon_{k\ell}(\boldsymbol{\eta})\rangle = \delta_{\alpha k}\delta_{\beta\ell}\langle\varepsilon_{k\ell}(\boldsymbol{\eta})\rangle = \langle\varepsilon_{\alpha\beta}(\boldsymbol{\eta})\rangle\,,$$

and from the right-hand side

$$B_{\alpha\beta ij}(1-\theta)\delta_{ij} - B_{\alpha\beta ij}C_{k\ell ij}B_{k\ell pp}\,.$$

It follows after contraction,

$$\beta' = \langle\varepsilon_{\alpha\alpha}(\boldsymbol{\eta})\rangle = B_{\alpha\beta ij}(1-\theta)\delta_{ij} - B_{\alpha\beta ij}C_{k\ell ij}B_{k\ell pp}\,, \tag{6.8.21}$$

which proves Eq. (6.8.19).

If the solid material is further isotropic, Eq. (6.8.14) applies, so that

$$\begin{aligned}\beta' = \langle\varepsilon_{\alpha\alpha}(\boldsymbol{\eta})\rangle &= \frac{\delta_{ij}}{3\lambda+2G}\left[(1-\theta)\delta_{ij} - C_{k\ell ij}\frac{\delta_{k\ell}}{3\lambda+2G}\right] \\ &= \frac{1}{3\lambda+2G}\left[3(1-\theta) - \frac{C_{kkii}}{3\lambda+2G}\right], \end{aligned} \tag{6.8.22}$$

which proves Eq. (6.8.20). If the composite is also isotropic on the macroscale, then

$$C_{kkii} = \lambda'\delta_{kk}\delta_{ii} + 2G'\delta_{kk}\delta_{ii} = 3(3\lambda'+2G')\,,$$

so that

$$\beta' = \frac{3}{3\lambda+2G}\left[(1-\theta) - \frac{3\lambda'+2G'}{3\lambda+2G}\right]. \tag{6.8.23}$$

The relations in this subsection show that once the elastic coefficients $C_{ijk\ell}$ are computed by some numerical scheme, the coefficients α'_{ij}, β_{ij} and γ'_{ij} can be obtained simply.

6.9. Computed Elastic Coefficients

The main numerical task is to solve the cell problem for the 3-tensor $\chi^{k\ell}$. The complexity of computational task depends of course on the microscale geometry. The elastostatic problem can be solved by the well-known method of finite elements. Some numerical results have been reported by Lee and Mei (1997.II) for the simple case of identical and homogeneous Wigner–Seitz grains in cubic packing. Thus each cell Ω is a cube containing a single grain. With this symmetry it can be shown (Lee, 1994) that among the 21 coefficients C_{ijmn}, only three are independent:

$$
\begin{aligned}
C_I &= C_{xxxx} = C_{yyyy} = C_{zzzz}\,, \\
C_{II} &= C_{xxyy} = C_{yyzz} = C_{zzxx}\,, \\
C_{III} &= C_{xyxy} = C_{yzyz} = C_{zxzx}\,.
\end{aligned}
\tag{6.9.1}
$$

A material with this property is said to have cubic symmetry (Love, 1944; Lekhnitskii, 1963). If further, $C_I - C_{II} = 2C_{III}$, i.e.,

$$
C_I = \lambda' + 2\mu'\,, \quad C_{II} = \lambda'\,, \quad C_{III} = \mu'\,,
\tag{6.9.2}
$$

the material is isotropic. For cubically symmetric materials the anisotropy factor is defined to be

$$
A_N = \frac{2C_{III}}{C_I - C_{II}}\,,
\tag{6.9.3}
$$

which is unity if isotropic.

Lee and Mei (1997.II) first converted the cell boundary-value problem for $\boldsymbol{\chi}^{mn} = \{\chi_j^{mn}\}$ to the extremization of the following functional,

$$
\mathcal{F} = \iint_{\Omega_s} \frac{1}{2} C_{ijpq}\varepsilon_{ij}(\boldsymbol{\chi}^{mn})\varepsilon_{pq}(\boldsymbol{\chi}^{mn})d\Omega - \iint_\Gamma C_{ijmn}\chi_j^{mn} n_i dS
\tag{6.9.4}
$$

under the constraint that

$$
\langle \chi_j^{mn} \rangle = 0\,,
\tag{6.9.5}
$$

as shown in Appendix B. The grains are then discretized into finite elements. Figure 6.1 shows some sample results of the normalized elastic coefficients $(C_I, C_{II}, C_{III})/E$.

6.10. Boundary-Layer Approximation for Macroscale Problems

Despite linearity, the macroscale equations of dynamic poroelasticity coupling two phases are quite complex. Analytical solutions are not easy and are available only for simple geometries (see Wang, 2000). For waves of low frequency, Biot (1956) found two types of dilatational waves. Waves of the first kind are close to those in a pure elastic solid; the pore fluid and the solid matrix essentially move in phase. Waves of the second kind is strongly diffusive and rapidly attenuating in space; the pore fluid and the solid matrix move in opposite directions. The second wave is obviously

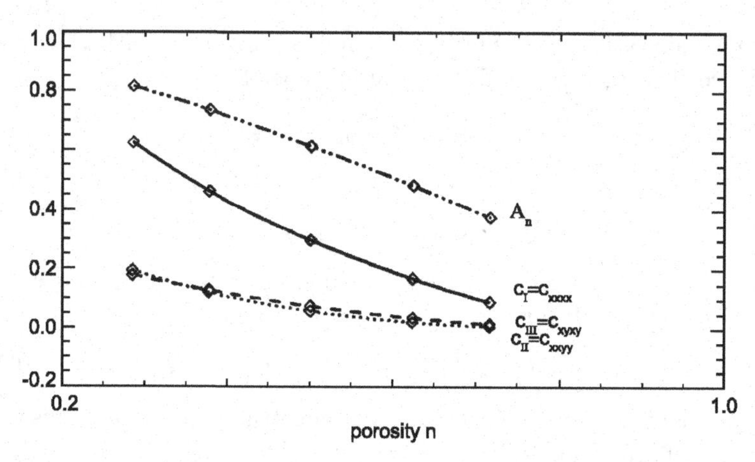

Fig. 6.1. Normalized elastic coefficients C_I/E, C_{II}/E, C_{III}/E, computed for uniform Wigner–Seitz grains in cubic packing for fixed Poisson ratio $\nu = 0.3$. From Lee and Mei (1997.II), *Adv. Water Resour.*

of importance primarily near an unsealed ground surface or the mudline (top of the porous seabed). This suggests a boundary-layer approximation which was formalized by Mei and Foda (1981) to expedite the analysis of several geophysical problems.

The idea is to deduce first an outer approximation valid away from the unsealed boundary, and then to add a correction within the boundary layer. For brevity we shall only treat the case where the bulk composite is isotropic and homogeneous, and all primes will be omitted from macroscale quantities. The amplitudes of fluid and solid velocities, the solid displacement, and the pore pressure will simply be denoted, respectively, by u_i, v_i, and p without the superscripts and without the cell-average symbol $\langle\cdot\rangle$. We shall also denote the solid velocity amplitude by $\mathcal{V}_i$ so that $\mathcal{V}_i = -i\omega\widetilde{v}_i^{(0)}$.

6.10.1. *The Outer Approximation*

Let us designate the region far away from the unsealed boundary as the outer region, where the characteristic length L on the macrolevel is chosen to be the typical elastic wavelength $L = \omega^{-1}\sqrt{G/\rho_s}$. Let the time scale be $1/\omega$ and stress scale be P_0. Then the independent variables are normalized by

$$x_i = Lx_i^\dagger, \quad \text{with } L = \frac{1}{\omega}\sqrt{\frac{G}{\rho_s}}. \tag{6.10.1}$$

The outer pore pressure and stresses are normalized as follows,

$$\left(p^{(o)}, T_{ij}^{(o)}\right) = P_0\left(p^\dagger, T_{ij}^\dagger\right), \tag{6.10.2}$$

where the superscript $(\cdot)^{(o)}$ denotes outer variables. In accordance with Hooke's law, the outer velocities and displacements are normalized by

$$\left(u_i^{(o)}, \mathcal{V}_i^{(o)}\right) = \frac{P_0 \omega L}{G}\left(u_i^\dagger, \mathcal{V}_i^\dagger\right), \quad v_i^{(o)} = \frac{P_0}{G} v_i^\dagger . \tag{6.10.3}$$

The normalized fluid momentum equation becomes

$$-i\frac{\rho_f}{\rho_s}\left(u_i^\dagger - \left(u_i^\dagger - \theta \mathcal{V}_i^\dagger\right)\right) = -\theta \frac{\partial p^\dagger}{\partial x_i^\dagger} - \frac{nL^2\omega}{GK}\left(u_i^\dagger - \theta \mathcal{V}_i^\dagger\right), \tag{6.10.4}$$

while the solid momentum equation becomes

$$-i(1-\theta)v_i^\dagger - i\frac{\rho_f}{\rho_s}\left(u_i^\dagger - \theta \mathcal{V}_i^\dagger\right) = -(\alpha - \theta)\frac{\partial p^\dagger}{\partial x_i^\dagger} + \frac{\partial \sigma_{ij}^{\dagger e}}{\partial x_j^\dagger} + \frac{\theta L^2 \omega}{GK}\left(u_i^\dagger - \theta \mathcal{V}_i^\dagger\right), \tag{6.10.5}$$

where

$$\sigma_{ij}^{\dagger e} = \frac{\partial v_i^\dagger}{\partial x_j^\dagger} + \frac{\partial v_j^\dagger}{\partial x_i^\dagger} + \frac{2\nu}{1-2\nu}\frac{\partial v_k^\dagger}{\partial x_k^\dagger}\delta_{ij} . \tag{6.10.6}$$

The dimensionless parameter

$$\frac{\theta L^2 \omega}{GK} \sim \frac{\theta}{\rho_s \omega K} \tag{6.10.7}$$

represents the importance of the Darcy friction term relative to the pressure and stress gradients. Typical permeabilities are $K = 10^{-8}\,\mathrm{m^3 s/kg}$ for fine sand and $K = 10^{-6}\,\mathrm{m^3 s/kg}$ for coarse sand. Taking $\theta = 0.3$, $\rho_s = 2.5 \times 10^3\,\mathrm{kg/m^3}$, and frequencies typical of earthquake $\omega = 10\pi\,\mathrm{rad/s}$ ($f \equiv \omega/2\pi = 5\,\mathrm{Hz}$) and ocean surface waves $\omega = 2\pi/10\,\mathrm{rad/s}$ ($f \equiv \omega/2\pi = 0.1\,\mathrm{Hz}$), the ratio above is very large as estimated in Table 6.1. Consequently we have, to the leading order of $\rho_s\omega K/\theta$,

$$u_i^\dagger - \theta \mathcal{V}_i^\dagger = u_i^\dagger + i\omega\theta v_i^\dagger \approx 0 , \tag{6.10.8}$$

implying that the pore fluid and the solid matrix move together as one phase, since the averaged pore fluid velocity is $u^\dagger/\theta$.

Table 6.1. Ratio $\theta/\rho_s\omega K$ for frequencies typical of earthquakes and sea waves.

	$K = 10^{-8}\,\mathrm{m^3\,s/kg}$	$K = 10^{-6}\,\mathrm{m^3\,s/kg}$
$f = 5\,\mathrm{Hz}$	2.4×10^4	2.4×10^2
$f = 0.1\,\mathrm{Hz}$	1.2×10^5	1.2×10^3

Using this result in the normalized equation of bulk momentum (Eq. (6.7.15)) we get

$$-\frac{\rho_e}{\rho_f}v_i^\dagger = -\frac{\partial p^\dagger}{\partial x_i^\dagger} + \frac{\partial}{\partial x_j^\dagger}\left(\frac{\partial v_i^\dagger}{\partial x_j^\dagger} + \frac{\partial v_j^\dagger}{\partial x_i^\dagger}\right) + \frac{2\nu}{1-2\nu}\frac{\partial^2 v_k^\dagger}{\partial x_i^\dagger \partial x_k^\dagger} = \frac{\partial T_{ij}^\dagger}{\partial x_j^\dagger}, \tag{6.10.9}$$

where $\rho_e \equiv [\theta\rho_f + (1-\theta)\rho_s]$ is the effective density of the bulk. We have taken $\alpha = 1$ as in Terzaghi's effective stress principle. The fluid mass conservation equation (6.7.16) becomes

$$p^\dagger = -\frac{B_f}{\theta G}\frac{\partial v_i^\dagger}{\partial x_i^\dagger}. \tag{6.10.10}$$

Thus the outer pore pressure is directly related to the solid dilatation in the outer region. The bulk momentum equation can be further reduced to

$$-\frac{\rho_e}{\rho_f}v_i^\dagger = \frac{\partial T_{ij}^\dagger}{\partial x_j^\dagger}, \tag{6.10.11}$$

where

$$T_{ij}^\dagger = \left(\frac{2\nu}{1-2\nu} + \frac{B_f}{\theta G}\right)\frac{\partial v_k^\dagger}{\partial x_k^\dagger}\delta_{ij} + \frac{\partial v_i^\dagger}{\partial x_j^\dagger} + \frac{\partial v_j^\dagger}{\partial x_i^\dagger}. \tag{6.10.12}$$

Returning to physical variable the outer momentum equation is:

$$-\omega^2\rho_e v_i^{(o)} = \frac{\partial T_{ij}^{(o)}}{\partial x_j}, \tag{6.10.13}$$

where the effective Hooke's law reads,

$$T_{ij} = G\left(\frac{2\nu}{1-2\nu} + \frac{\alpha^2 B_f}{\theta G}\right)\frac{\partial v_k^{(o)}}{\partial x_k}\delta_{ij} + G\left(\frac{\partial v_i^{(o)}}{\partial x_j} + \frac{\partial v_j^{(o)}}{\partial x_i}\right). \tag{6.10.14}$$

These are the usual elastodynamic equations for a single-phase medium. We can define the effective Lamé constant by

$$\lambda_e = \frac{2\nu G}{1-2\nu} + \frac{B_f}{\theta}, \tag{6.10.15}$$

and the effective Poisson ratio by

$$\nu_e = \frac{\lambda_e/G}{2\left(1 + (\lambda_e/G)\right)} = \frac{(2\nu/(1-2\nu)) + (B_f/\theta G)}{2\left((1/(1-2\nu)) + (B_f/\theta G)\right)}. \tag{6.10.16}$$

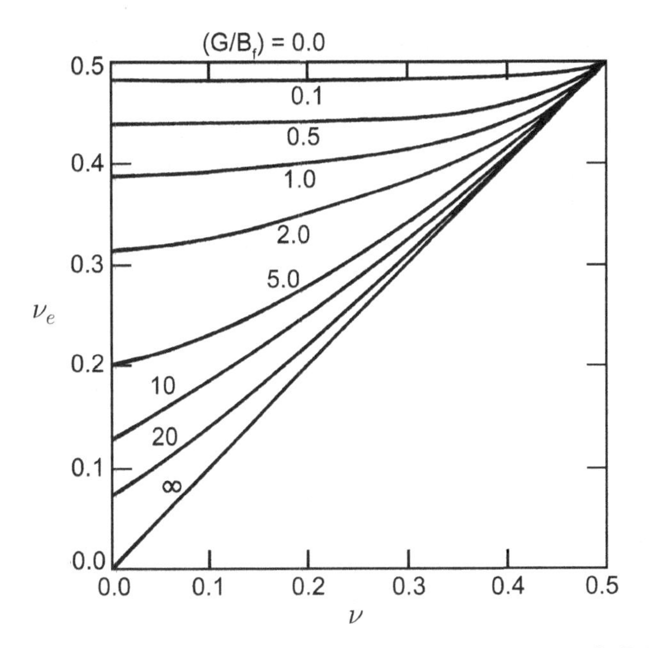

Fig. 6.2. Relation between equivalent Poisson ratio and the macroscale Poisson ratio for a water-saturated granular composite, for various values of nG/B_f. From Mei and Foda (1981), *Geophys. J. R. Soc.*

Equation (6.10.13) can be written as

$$G\nabla^2 v_i^{(o)} + \frac{G}{1-2\nu_e}\frac{\partial^2 v_j^{(o)}}{\partial x_j \partial x_i} = -\rho_e \omega^2 v_i^{(o)} . \qquad (6.10.17)$$

In Fig. 6.2, the relation between ν_e and ν is plotted for different $\theta G/\alpha^2 B_f$.

Once the one-phase elastodynamic problem is solved, the outer pore pressure is given by Eq. (6.10.10). However, in general the outer solution alone cannot satisfy all the boundary conditions on the unsealed surface; a boundary-layer correction is needed.

6.10.2. *Boundary-Layer Correction*

In physical variables let each unknown be the superposition of the outer approximation and a boundary-layer correction,

$$f = f^{(o)} + f^{(b)} .$$

Consider for simplicity that the porous composite is in the region $y<0$ and is bounded from above by an unsealed plane[2] at $y=0$. Let the unknown boundary-layer thickness be of the order $\delta \ll L$. In Eq. (6.7.15), the dominant stress term is $G\partial^2 v_i^{(b)}/\partial y^2$, relative to which the inertia terms are small since

$$\left(-\rho_s\omega^2 v_i^{(b)}\right)\left(G\frac{\partial^2 v_i^{(b)}}{\partial y^2}\right)^{-1} \sim \frac{\delta^2\rho_s\omega^2}{G} \sim \frac{\delta^2}{L^2} \ll 1\,.$$

Taking the curl of the remaining (quasi-static) part, we get, in vector form,

$$\nabla^2\nabla\times\boldsymbol{v}^{(b)} \approx \frac{\partial^2}{\partial y^2}\nabla\times\boldsymbol{v}^{(b)} \approx 0$$

throughout the boundary layer with a relative error of order $\mathcal{O}(\delta^2/L^2)$. Since the correction vanishes outside the boundary layer $-y \gg \delta$, $\nabla\times\boldsymbol{v}^{(b)} \approx 0$ everywhere inside, thus,

$$\frac{\partial v_i^{(b)}}{\partial y} - \frac{\partial v_2^{(b)}}{\partial x_i} \approx 0\,, \quad \text{with } i = 1,3\,.$$

This implies that

$$v_i^{(b)} \sim \frac{\delta}{L}v_2^{(b)} \ll v_2^{(b)}\,, \quad i=1,3\,. \tag{6.10.18}$$

Thus the longitudinal motion in the boundary layer is much smaller than the transverse motion by a factor of $\mathcal{O}(\delta/L)$. Using this fact the transverse (y) component of the momentum, Eq. (6.7.15), is dominated by

$$0 \approx -\frac{\partial p^{(b)}}{\partial y} + G\left(1+\frac{1}{1-2\nu}\right)\frac{\partial^2 v_2^{(b)}}{\partial y^2}\,,$$

which can be integrated to give

$$\frac{\partial v_2^{(b)}}{\partial y} = \frac{(1-2\nu)}{2(1-\nu)G}p^{(b)}\,. \tag{6.10.19}$$

Thus the pore pressure change causes the vertical dilatation of the solid matrix and vice versa. On the other hand the horizontal components of Eq. (6.7.15) is dominated by

$$G\left(\frac{\partial^2 v_j^{(b)}}{\partial y^2} + \frac{1}{1-2\nu}\frac{\partial^2 v_2^{(b)}}{\partial x\partial y}\right) - \frac{\partial p^{(b)}}{\partial x_j} = 0\,, \quad j=1,3\,. \tag{6.10.20}$$

[2]For other boundary geometries the approximation scheme can be extended, see Mei *et al.* (1984).

Inside the boundary layer the dynamic Darcy law in the transverse direction can be approximated by

$$u_2^{(b)} + i\omega v_2^{(b)} = -K\frac{\partial p^{(b)}}{\partial y}, \tag{6.10.21}$$

since

$$\frac{-\omega^2 \rho_f K v_2^{(b)}}{(i\omega v_2^{(b)})} \sim \rho_f K\omega \ll 1$$

in view of Eq. (6.10.7). The dominant part of the mass conservation equation (6.7.16) is

$$-i\omega\frac{\theta}{B}p^{(b)} + \frac{\partial u_2^{(b)}}{\partial y} - i\omega(1-\theta)\frac{\partial v_2^{(b)}}{\partial y} = 0. \tag{6.10.22}$$

Eliminating $u_2^{(b)}$ from Eqs. (6.10.21) and (6.10.22) we get

$$-i\omega\frac{\theta}{B_f}p^{(b)} - i\omega\frac{\partial v_2^{(b)}}{\partial y} = K\frac{\partial^2 p^{(b)}}{\partial y^2}. \tag{6.10.23}$$

Making use of Eq. (6.10.19) we finally obtain

$$-i\omega\hat{p} = \kappa_c\frac{\partial^2 \hat{p}}{\partial y^2}, \tag{6.10.24}$$

which is the time-harmonic version of Terzaghi's consolidation equation where

$$\kappa_c = \frac{K}{(\theta/B_f) + (1/G)((1-2\nu)/(2(1-\nu)))} \tag{6.10.25}$$

is the coefficient of consolidation. It is now evident that the boundary-layer thickness can be defined by

$$\delta = \sqrt{\frac{\kappa_c}{\omega}} = \frac{\sqrt{KG/\omega}}{\sqrt{(\theta G/B_f) + ((1-2\nu)/(2(1-\nu)))}}, \tag{6.10.26}$$

which is small for small permeability and/or high frequency. Its ratio to the outer length scale is

$$\frac{\delta}{L} = \sqrt{\frac{\kappa_c}{\omega}}\sqrt{\frac{\rho_s}{G}}\omega = \sqrt{\frac{2\rho_s K\omega}{(\theta G/B_f) + ((1-2\nu)/(2(1-\nu)))}}. \tag{6.10.27}$$

For very low permeability this ratio is very small, implying that relative motion between the fluid and solid phases is approximately one-dimensional and dissipative.

Let us introduce the following normalization for the boundary-layer variables,

$$(x', z') = L(X, Z), \quad y' = \delta Y, \quad t = \frac{T}{\omega}; \tag{6.10.28}$$

$$\left(p^{(b)}, \sigma_{ij}^{(b)}, T_{ij}^{(b)}\right) = P_0(\widehat{p}, \widehat{\sigma}_{ij}, \widehat{T}_{ij}). \tag{6.10.29}$$

In order that the solid stresses are of the same order as the fluid pressure, we choose the following normalization for velocities

$$\begin{aligned} u_j^{(b)} &= \frac{P_0 \omega L}{G} \frac{\delta^2}{L^2} \hat{u}_j, \quad v_j^{(b)} = \frac{P_0 L}{G} \frac{\delta^2}{L^2} \hat{v}_j; \; j = 1, 2, \\ u_2^{(b)} &= \frac{P_0 \omega L}{G} \frac{\delta}{L} \widehat{u}_j, \quad v_2^{(b)} = \frac{P_0 L}{G} \frac{\delta}{L} \widehat{v}_2. \end{aligned} \tag{6.10.30}$$

Use has been made of the difference in scaling between the transverse and longitudinal motions.

The normalized form of Eq. (6.10.24) is

$$-i\widehat{p} = \frac{\partial^2 \widehat{p}}{\partial Y^2}. \tag{6.10.31}$$

The pressure correction is clearly

$$\widehat{p} = F(X, Z) \exp\left(\frac{1-i}{\sqrt{2}} Y\right), \tag{6.10.32}$$

where $F(X, Z)$ is to be determined later by matching the total pressure $p^{(o)} + p^{(b)}$ to its value on the boundary at $y = 0$. From the normalized form of Eq. (6.10.19)

$$\frac{\partial \widehat{v}_2}{\partial Y} = \frac{(1-2\nu)}{2(1-\nu)} \widehat{p}, \tag{6.10.33}$$

we get by integration,

$$\widehat{v}_2 = \frac{1+i}{\sqrt{2}} \frac{1-2\nu}{2(1-\nu)} F \exp\left(\frac{1-i}{\sqrt{2}} Y\right). \tag{6.10.34}$$

Afterward the tangential components V_j can be obtained from the normalized form of Eq. (6.10.20),

$$\frac{\partial^2 \widehat{v}_j}{\partial Y^2} + \frac{1}{1-2\nu} \frac{\partial^2 \widehat{v}_2}{\partial X \partial Y} - \alpha \frac{\partial \widehat{p}}{\partial X_j} = 0, \quad j = 1, 3, \tag{6.10.35}$$

yielding

$$\widehat{v}_j = i\frac{1-2\nu}{2(1-\nu)}F\exp\left(\frac{1-i}{\sqrt{2}}Y\right), \quad j=1,3. \tag{6.10.36}$$

The corresponding corrections of effective solid stress in the boundary layer are dominated by $\partial\widehat{v}_2/\partial Y$,

$$\begin{bmatrix} \widehat{\sigma}^e_{11} = \widehat{\sigma}^e_{33} \\ \widehat{\sigma}^e_{22} \\ \widehat{\sigma}^e_{12} = \widehat{\sigma}^e_{13} \end{bmatrix} = \begin{bmatrix} \dfrac{\nu}{1-\nu}F + \mathcal{O}\left(\dfrac{\delta^2}{L^2}\right) \\ F + \mathcal{O}\left(\dfrac{\delta^2}{L^2}\right) \\ \mathcal{O}\left(\dfrac{\delta}{L}\right) \end{bmatrix} \exp\left(\frac{1-i}{\sqrt{2}}Y\right). \tag{6.10.37}$$

Consequently the total stress corrections are

$$\begin{bmatrix} \widehat{T}_{11} = \widehat{T}_{33} \\ \widehat{T}_{22} \\ \widehat{T}_{12} = \widehat{T}_{13} \end{bmatrix} = \begin{bmatrix} -\dfrac{1-2\nu}{1-\nu}F + \mathcal{O}\left(\dfrac{\delta^2}{L^2}\right) \\ \mathcal{O}\left(\dfrac{\delta^2}{L^2}\right) \\ \mathcal{O}\left(\dfrac{\delta}{L}\right) \end{bmatrix} \exp\left(\frac{1-i}{\sqrt{2}}Y\right). \tag{6.10.38}$$

It is remarkable that for the total normal stress $\hat{T}_{22}$ and the total tangential traction $(\hat{T}_{12}, \hat{T}_{32})$, no boundary-layer correction is needed, i.e., the outer solution is also valid inside the boundary layer, hence is uniformly valid everywhere. Thus, if on the unsealed boundary the traction and the fluid pressure are prescribed, the displacement field in the outer region can be solved first as a classical elastodynamic problem of one phase. The outer pore pressure as well as the solid effective stress then follows immediately. After adding the boundary-layer correction, the pore pressure is obtained inside the boundary layer. Afterward the solid stress in the boundary layer follows easily. Examples may be found in Mei and Foda (1981), Mei and McTigue (1984) for seabed responses under ocean waves, and Mei, Si and Cai (1984) for elastic waves scattered by a cylindrical cavity of circular cross-section.

As an illustration, we apply the boundary-layer approximation to the simple example of Rayleigh surface wave in a poroelastic half space (Foda and Mei, 1983). For the exact solution from the full Biot's equations, reference may be made to Deresiewicz and Rice (1962).

6.10.3. *Plane Rayleigh Wave in a Poroelastic Half Space*

The far field outside the boundary layer near the ground surface is a problem of single-phase elastodynamics. From standard treatises such as Fung (1965) the displacement components of the solid matrix can be found, in dimensional form

$$\begin{pmatrix} v_1^{(o)} \\ \bar{v}_2^{(o)} \end{pmatrix} = -\frac{A}{Gk}\begin{pmatrix} e^{qky} - \dfrac{2sq}{1+s^2}e^{sky} \\ iqe^{qky} - \dfrac{2iq}{1+s^2}e^{sky} \end{pmatrix} e^{ik(x-Ct)}, \tag{6.10.39}$$

where A is the arbitrary pressure amplitude, k is the wavenumber and $C = \omega/k$ the phase speed of the Rayleigh wave.

$$q^2 = 1 - \frac{C^2}{C_p^2}, \quad s^2 = 1 - \frac{C^2}{C_s^2}, \tag{6.10.40}$$

C_p and C_s are the phase speeds of the compressional (or longitudinal) and shear (or transverse) waves in an infinite space respectively,

$$C_p = \frac{2(1-\nu_e)}{1-2\nu_e}\frac{G}{\rho_e}, \quad C_s^2 = \frac{G}{\rho_e}. \tag{6.10.41}$$

The Rayleigh wave speed is the real positive root of

$$\frac{C^2}{C_s^2}\left[\frac{C^6}{C_s^6} - 8\frac{C^4}{C_s^4} + C^2\left(\frac{24}{C_s^2} - \frac{16}{C_p^2}\right) - 16\left(1 - \frac{C_s^2}{C_p^2}\right)\right] = 0 \tag{6.10.42}$$

in the range of $0 < C < C_s < C_p$. From Eq. (6.10.10) the outer pore pressure is calculated

$$p^{(o)} = \frac{iA\beta k}{Gn}(q^2 - 1)e^{qky}e^{ik(x-Ct)}. \tag{6.10.43}$$

For a pressure-free ground surface we must add the correction,

$$\bar{p}^{(b)} = -\frac{iA\beta k}{Gn}(q^2 - 1)\exp\left(\frac{(1-i)y}{\sqrt{2}\delta}\right)e^{ik(x-Ct)}, \tag{6.10.44}$$

so that $\bar{p}^{(o)} + \bar{p}^{(b)} = 0$ on $y = 0$. The corresponding correction of the effective stress in the solid can be derived from Eq. (6.10.37).

Figure 6.3 shows the depth profiles of the amplitude of the total pore pressure for sample fine and coarse sands. For the same shear modulus $G = 10^7\,\mathrm{N/m^2}$ two different bulk moduli of the pore fluid are

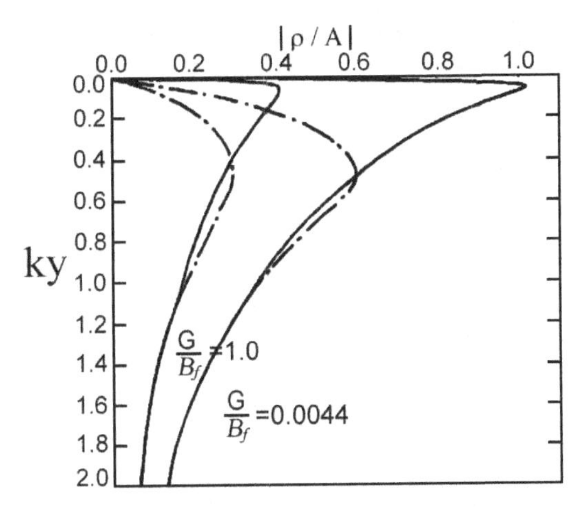

Fig. 6.3. Profiles of pore pressure amplitude for fine sand with $K = 10^{-8}\,\mathrm{m^3 s/kg}$ (solid), and coarse sand with $K = 10^{-6}\,\mathrm{m^3 s/kg}$ (dash-dot). From Foda and Mei (1983), *Soil Dyn. Earthq. Eng.*

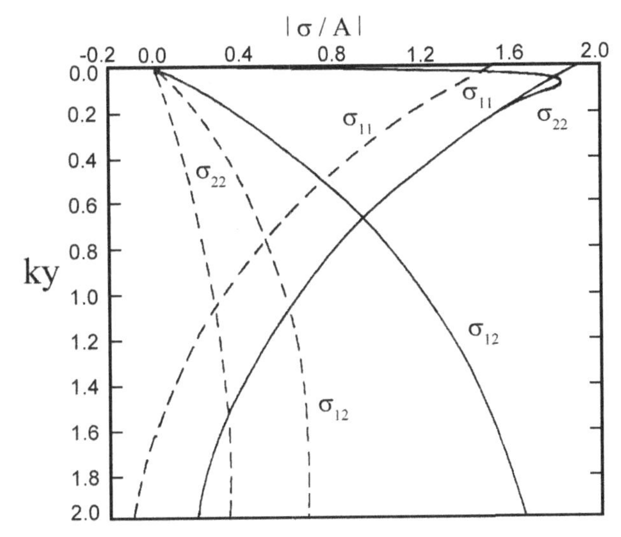

Fig. 6.4. Profiles of effective stress amplitudes for fine sand with $K = 10^{-8}\,\mathrm{m^3 s/kg}$ (solid), and coarse sand with $K = 10^{-6}\,\mathrm{m^3 s/kg}$ (dash-dot). From Foda and Mei (1983), *Soil Dyn. Earthq. Eng.*

compared: $B_f = 1.9 \times 10^9\,\mathrm{N/m^2}$ (fully saturated) and $B_f = 10^7\,\mathrm{N/m^2}$ (slightly unsaturated). It is seen that the total pore pressure is larger for the finer sand, implying greater trend toward failure. Figure 6.4 shows the profiles of the effective stress components.

Appendix 6A. Properties of the Compliance Tensor

Following Auriault and Sanchez-Palencia (1977), we write Hooke's law in two forms:

$$\sigma_{ij} = C_{ij\ell m}\varepsilon_{\ell m}\,, \tag{A.1}$$

and

$$\varepsilon_{\ell m} = B_{\ell mij}\sigma_{ij}\,. \tag{A.2}$$

Multiplying Eq. (A.1) by $B_{\alpha\beta ij}$, we get

$$B_{\alpha\beta ij}C_{ij\ell m}\varepsilon_{\ell m} = \varepsilon_{\alpha\beta}\,. \tag{A.3}$$

Hence

$$B_{\alpha\beta ij}C_{ij\ell m} = \delta_{\alpha\ell}\delta_{\beta m}\,. \tag{A.4}$$

Since $\varepsilon_{\alpha\beta} = \varepsilon_{\beta\alpha}$, Eq. (A.3) can be written as

$$B_{\alpha\beta ij}C_{ij\ell m}\varepsilon_{\ell m} = \varepsilon_{\beta\alpha}\,, \tag{A.5}$$

hence

$$B_{\alpha\beta ij}C_{ij\ell m} = \delta_{\alpha m}\delta_{\beta\ell}\,. \tag{A.6}$$

It follows that

$$B_{\alpha\beta ij}C_{ij\ell m} = \frac{1}{2}(\delta_{\alpha m}\delta_{\beta\ell} + \delta_{\alpha\ell}\delta_{\beta m})\,. \tag{A.7}$$

Invoking symmetry $C_{ij\ell m} = C_{\ell mij}$, we further get

$$C_{ijkh}B_{\alpha\beta kh} = \frac{1}{2}(\delta_{\alpha i}\delta_{\beta j} + \delta_{\alpha j}\delta_{\beta i})\,. \tag{A.8}$$

Relations (A.7) and (A.8) express the fact that the compliance tensor is the inverse of the elasticity tensor.

For isotropic solids,

$$C_{ij\ell m} = \lambda\delta_{ij}\delta_{\ell m} + G(\delta_{i\ell}\delta_{jm} + \delta_{im}\delta_{j\ell})\,, \tag{A.9}$$

therefore

$$C_{ijkh}B_{\alpha\beta kh} = \lambda\delta_{ij}B_{\alpha\beta rr} + G(B_{\alpha\beta ij} + B_{\alpha\beta ji})\,. \tag{A.10}$$

Combining Eqs. (A.1) and (A.2), we get

$$\lambda\delta_{ij}B_{\alpha\beta rr} + 2GB_{\alpha\beta ij} = \frac{1}{2}(\delta_{\alpha i}\delta_{\beta j} + \delta_{\alpha j}\delta_{\beta i})\,. \tag{A.11}$$

It follows by contraction with $i = j$ that

$$B_{\alpha\beta rr} = \frac{\delta_{\alpha\beta}}{3\lambda + 2G}\,. \tag{A.12}$$

Appendix 6B. Variational Principle for the Elastostatic Problem in a Cell

For homogeneous grains the elastostatic cell problem for $\boldsymbol{\chi}^{mn} = \{\chi_i^{mn}\}$ is governed by

$$\frac{\partial}{\partial x_i}\left[C_{ijpq}\varepsilon_{pq}(\boldsymbol{\chi}^{mn})\right] = 0, \quad \boldsymbol{x} \in \Omega_s \tag{B.1}$$

and the boundary conditions

$$C_{ijpq}\,\varepsilon_{pq}(\boldsymbol{\chi}^{mn})n_i = -C_{ijmn}n_i, \quad \boldsymbol{x} \in \Gamma, \tag{B.2}$$

$$\boldsymbol{\chi}^{mn} = \{\chi_i^{mn}\} \quad \text{is } \Omega\text{-periodic}, \tag{B.3}$$

$$\langle \chi_i^{mn} \rangle = 0\,. \tag{B.4}$$

We consider fixed m, n so that $\boldsymbol{\chi}^{mn} = \{\chi_i^{mn}\}$ is effectively a vector in the following manipulations and $\varepsilon_{pq}(\boldsymbol{\chi}^{mn})$ a second-rank tensor. No summation is carried out for repeated m, n.

As is standard in variational calculus, Eq. (B.1) is multiplied by $\delta\chi_j$, summed over the index j, and integrated over the volume of the solid grain Ω_s, we obtain,

$$\iiint_{\Omega_s} \frac{\partial}{\partial x_i}\left[C_{ijpq}\varepsilon_{pq}(\boldsymbol{\chi}^{mn})\delta\chi_j\right]\Omega - \iiint_{\Omega_s} C_{ijpq}\varepsilon_{pq}(\boldsymbol{\chi}^{mn})\frac{\partial\delta\chi_j}{\partial x_i} = 0\,. \tag{B.5}$$

Applying Gauss' theorem and the condition, Eq. (B.2), the first integral in Eq. (B.5) becomes

$$\begin{aligned}\iint_{\Gamma\cup\Gamma_s} C_{ijpq}\varepsilon_{pq}(\boldsymbol{\chi}^{mn})\delta\chi_j^{mn}n_i dS &= -\iint_{\Gamma} C_{ijmn}\delta\chi_j^{mn}n_i dS \\ &= -\delta\iint_{\Gamma} C_{ijmn}\chi_j^{mn}n_i dS\,. \end{aligned}\tag{B.6}$$

The surface integral over the solid part Γ_s of the cell boundary vanishes because of periodicity. Introducing the antisymmetric tensor

$$\omega_{ij}(\delta\boldsymbol{\chi}^{mn}) = \frac{1}{2}\left(\frac{\partial\delta\chi_j^{mn}}{\partial x_i} - \frac{\partial\delta\chi_i^{mn}}{\partial x_j}\right), \tag{B.7}$$

the second integral in Eq. (B.5) can be written as

$$\iiint_{\Omega_s} C_{ijpq}\left[\varepsilon_{ij}(\delta\boldsymbol{\chi}^{mn}) + \omega_{ij}(\delta\boldsymbol{\chi}^{mn})\right]\varepsilon_{pq}(\boldsymbol{\chi}^{mn})d\Omega\,. \tag{B.8}$$

In view of the properties that $C_{ijpq} = C_{jipq}, C_{ijpq} = C_{pqji}$, and $\omega_{ijmn} = -\omega_{jimn}$, the above integral reduces to

$$\iiint_{\Omega_s} C_{ijpq}\varepsilon_{ij}(\delta\boldsymbol{\chi}^{mn})\varepsilon_{pq}(\boldsymbol{\chi}^{mn})d\Omega = \delta \iiint_{\Omega_s} \frac{1}{2}C_{ijpq}\varepsilon_{ij}(\boldsymbol{\chi})\varepsilon_{pq}(\boldsymbol{\chi}^{mn})d\Omega\,. \tag{B.9}$$

It follows by substituting Eqs. (B.6) and (B.9) in Eq. (B.5) that

$$\delta\left\{\iint_{\Omega_s} \frac{1}{2}C_{ijpq}\varepsilon_{ij}(\boldsymbol{\chi}^{mn})\varepsilon_{pq}(\boldsymbol{\chi}^{mn})d\Omega - \iint_{\Gamma} C_{ijmn}\chi_j^{mn}n_i dS\right\} = 0. \tag{B.10}$$

Thus the following functional

$$\mathcal{F} = \iint_{\Omega_s} \frac{1}{2}C_{ijpq}\varepsilon_{ij}(\boldsymbol{\chi}^{mn})\varepsilon_{pq}(\boldsymbol{\chi}^{mn})d\Omega - \iint_{\Gamma} C_{ijmn}\chi_j^{mn}n_i dS \tag{B.11}$$

is extremum with the contraint that

$$\langle\chi_j^{mn}\rangle = 0\,. \tag{B.12}$$

References

Auriault, J. L. (1991). Dynamic behaviour of porous media. *NATO Advanced Study Institute on Transport in Porous Media*, eds. J. Bear and M. Y. Coraptoglu, Kluwer, pp. 471–459.

Auriault, J. L., L. Borne and R. Chambon (1985). Dynamics of porous saturated media-checking of the generalized law of Darcy. *J. Acoust. Soc. Amer.* **77**(5): 1641–1650.

Auriault, J. L. and E. Sanchez-Palencia (1977). Etude du comportment macroscopique d'un milieu poreux sature deformable. *J. Mech.* **16**(4): 575–603.

Biot, M. A. (1941). General theory of three-dimensional consolidation. *J. Appl. Phys.* **12**: 155–164.

Biot, M. A. (1956). Theory of propagation of elastic waves in a fluid-saturated porous solid I. Low-frequency range, II. High-frequency range. *J. Acout. Soc. Amer.* **28**(2): 168–190.

Biot, M. A. (1962a). Mechanics of deformation and acoustic propagation in porous media. *J. Appl. Phys.* **33**: 1482–1498.

Biot, M. A. (1962b). General theory of acoustic propagation porous dissipative media. *J. Acoust. Soc. Amer.* **34**: 1254–1264.

Burridge, R. and J. B. Keller (1981). Poroelasticity equations derived from microstructure. *J. Acoust. Soc. Amer.* **70**: 1140, 1148.

Cowan, S. C. and S. B. Doty (2007). *Tissue Mechanics*, Springer.

Deresiewicz, H. (1962). The effect of boundaries of wave propagation in a liquid-filled porous solid: IV. Surface waves in a half-space. *Bull. Seism. Soc. Am.* **52**: 627–638.

Deresiewicz, H. and J. T. Rice (1962). The effect of boundaries of wave propagation in a liquid-filled porous solid: III. Reflection of plane waves at a free surface boundary (General case). *Bull. Seism. Soc. Am.* **52**: 505–625.

Foda, M. A. and C. C. Mei (1983). A boundary layer theory for Rayleigh waves in a porous fluid-filled half space. *Soil Dyn. Earthq. Eng.* **2**: 62–65.

Fung, Y. C. (1965). *Foundations of Solid Mechanics*, Prentice-Hall, N.J.

Jaeger, J. C. and N. G. W. Cook (1969). *Fundamentals of Rock Mechanics*, Chapman and Hall, 593 pp.

Lamb, T. W. and R. V. Whitman (1969). *Soil Mechanics*, John Wiley, 553 pp.

Lee, C. K. (1994). *Thermoconsolidation and Thermal Dispersion in Deformable Porous Media.* PhD, thesis. Department of Civil and Environmental Engineering, Mass. Inst. Tech.

Lee, C. K. and C. C. Mei (1997.I). Thermal consolidation in porous media by homogenization theory-I. Derivation of macroscale equations. *Adv. Water Resour.* **20**: 127–144.

Lee, C. K. and C. C. Mei (1997.II). Thermal consolidation in porous media by homogenization theory-II. Calculation of effective coefficients. *Adv. Water Resour.* **20**: 145–156.

Lekhnitskii, S. G. (1963). *Theory of Elasticity of an Anisotropic Elastic Body*, Holden-Day San Francisco.

Levy, T. (1979). Propagation of waves in a fluid-saturated porous medium. *Int. J. Eng. Sci.* **17**: 1005–1014.

Love, A. E. H. (1944). *A Treatise on the Mathematical Theory of Elasticity*, Dover.

Mei, C. C. (1990). Wave–seabed interactions, *The Sea–Ocean Engineering Science*, eds. B. Lé Méhaute and D. M. Hines, **9B**: 889–918.

Mei, C. C. and M. A. Foda (1981). Wave-induced responses in a fluid-filled poroelastic sea-bed, *Geophys. J. R. Soc.* **66**: 597–637.

Mei, C. C. and D. McTigue (1984). Stresses in a submarine topography under ocean waves. *J. Energy Resour. Tech.* **106**: 311–318.

Mei, C. C., B. I. Si and D. Cai (1984). Scattering of simple harmonic waves by a circular cavity in a fluid-saturated poro-elastic medium. *Wave Motion* **6**: 265–278.

Rice, J. R. and M. P. Cleary (1976). Some basic stress-diffusion solutions for fluid-saturated elastic porous media with compressible constituents. *Rev. Geophys. Space Phys.* **14**: 227–241.

Santos, J. E., C. L. Ravazzoli and J. Geiser (2006). On the static and dynamic behavior of fluid-saturated porous solids: A homogenization approach. *Int. J. Solids Struc.* **43**: 1224–1238.

Sheng, P. and M. Y. Zhou (1988). Dynamic permeability of porous media. *Phys. Rev. Lett.* **61**: 1591–1594.

Terzaghi (1943). *Theoretical Soil Mechanics*, Wiley, 510 pp.

Veruijt, A. (1969). Elastic storage of aquifers. *Flow Through Porous Media*, ed. R. J. M. De Wiest, Academic.

Wang, H. F. (2000). *Theory of Linear Poroelasticity*, Princeton University Press.

Zhou, M. Y. and P. Sheng (1989). First-principle calculation of dynamic permeability of porous media. *Phys. Rev. B* **39**: 12027–12039.

Wave Propagation in Inhomogeneous Media

7

There is a vast literature of wave propagation where the process involves different scales in space or time, or both, as a result of inhomogeneity of the medium, or the complexity of external forcing, or inherent nonlinearity. Numerous examples can be found in the dynamics of ocean surface waves (Mei *et al.*, 2005) as well as waves of a variety of other physical origins (Whitham, 1974). A frequent objective is to predict the slow evolution of the envelope of a nearly sinusoidal wave train. The mathematical technique must yield a uniformly valid theory for time and space ranges much greater than the period and wavelength of the carrier wave. While the basic ideas are already present in the classical WKB method in quantum mechanics, the more recent techniques of multiple scales, i.e., homogenization, now provide a systematic formalism which can be applied to a wide variety of linear and nonlinear wave problems involving refraction and scattering. In Secs. 7.1–7.3 we first illustrate two physically different cases where the medium has a periodic microscale structure. In Sec. 7.4 we show that essentially the same analysis can be extended to weakly random media. Finally two examples involving weak disorder and weak nonlinearity are discussed in Secs. 7.5 and 7.6.

7.1. Long Wave Through a Compact Cylinder Array

If a periodic array of parallel cylinders of uniform radius is embedded in a medium, how does a monochromatic sound wave propagate through them? In general this is a problem of multiple scattering the analogs of which have long played a central role in crystallography. In the simplest case, there are three length scales affecting the physics: the incident wavelength $2\pi/k$, the size of the scatterers a, and their separation distance ℓ. Depending on the relative magnitudes of the two length ratios ka and $k\ell$, different physics may dominate and require different mathematical treatments. In this section we examine a special limit where a and ℓ are comparable but are much smaller than the wavelength, i.e., $ka, k\ell \ll 1$ and $a/\ell = \mathcal{O}(1)$. This is a problem of long waves propagating through a compact array of small objects. It will be shown that the main physical effect is a change of the index of refraction. A one-dimensional example with a different scale ratios ($ka \sim k\ell = \mathcal{O}(1)$) was treated in Chapter 1.

Consider two-dimensional propagation of acoustic waves through a periodic array of rigid cylinders. The medium between the cylinders is a compressible fluid such as air. Helmholtz equation governs the air pressure $\phi(x, y)e^{-i\omega t}$ so that

$$\nabla^2\phi + k^2\phi = 0\,, \quad \text{where } k = \frac{\omega}{C}\,, \tag{7.1.1}$$

and C is the sound speed in the unbounded fluid. For simplicity we assume the cylinders to be perfectly rigid so that

$$\frac{\partial\phi}{\partial r} = 0\,, \quad r = |\boldsymbol{x} - \boldsymbol{x}_N| = a\,, \tag{7.1.2}$$

where $\boldsymbol{x}_N$ denotes the center of the Nth cylinder.

A mathematically equivalent problem is the propagation of infinitesimal water waves in a sea of constant depth h, in which case the phase speed C is related to the wavelength and depth via the dispersion relation $C = \sqrt{(g/k)\tanh kh}$ (Mei *et al.*, 2005; Mei, 1989). For the special case of shallow sea ($kh \ll 1$) Hu and Chan (2005) derived an approximate theory for the effective refractive index for a square array of vertical cylinders. They employ the *Coherent Potential Approximation* (CPA) in theoretical physics by first positing that the inhomogeneous region populated by cylinders can be replaced by a homogeneous medium characterized by an effective wavenumber k_e and depth h_e. To find these two parameters they require the scattered wave from each cylinder to vanish at a certain radius R from the center of the cylinder, where R is chosen so that the area of the circle is the same as that of the square unit cell, i.e., $\ell^2 = \pi R^2$, or, $R = \ell/\sqrt{\pi}$. To get two conditions for two unknown parameters k_e and h_e, only the scattering coefficients of the first two partial waves are required to vanish. The accuracy of these heuristic though practical approximations is hard to assess mathematically. It is also less easy to modify this method for a nonsquare array.

We shall demonstrate the use of the homogenization analysis.

Let us first use the linear dimension ℓ of the cell to define the normalized microscale coordinate $\boldsymbol{x}^\dagger$, i.e., $\boldsymbol{x} = \ell\boldsymbol{x}^\dagger$ where $a/\ell < 1$ but not $\ll 1$. Defining the small parameter

$$\epsilon = k\ell = \mathcal{O}(ka) \ll 1\,, \tag{7.1.3}$$

the normalized Helmholtz equation reads

$$\nabla^{\dagger 2}\phi^\dagger + \epsilon^2\phi^\dagger = 0\,, \quad \boldsymbol{x}^\dagger \in \Omega_f\,, \tag{7.1.4}$$

where Ω_F denotes the fluid domain outside the cylinders. Returning to physical variables but retaining the order symbol we have

$$\nabla^2\phi + \epsilon^2 k^2\phi = 0\,, \quad \boldsymbol{x} \in \Omega_f\,. \tag{7.1.5}$$

Let us define periodic cells Ω of size ℓ centered around a single cylinder and assume Eq. (7.1.3). Employing two coordinates $\boldsymbol{x}$ and $\boldsymbol{x}' = \epsilon\boldsymbol{x}$, we introduce the multiple scales expansion so that

$$(\nabla^2 + 2\epsilon\nabla\cdot\nabla' + \epsilon^2\nabla'^2)(\phi_0 + \epsilon\phi_1 + \epsilon^2\phi_2 + \cdots) + \epsilon^2k^2(\phi_0 + \cdots) = 0\,. \tag{7.1.6}$$

In the fluid region Ω_F of each unit cell, we have, at orders $\mathcal{O}(\epsilon^0)$, $\mathcal{O}(\epsilon)$, and $\mathcal{O}(\epsilon^2)$,

$$\nabla^2\phi_0 = 0\,, \tag{7.1.7}$$

$$\nabla^2\phi_1 + 2\nabla'\cdot\nabla\phi_0 = 0\,, \tag{7.1.8}$$

$$\nabla^2\phi_2 + 2\nabla'\cdot\nabla\phi_1 + \nabla'^2\phi_0 + k^2\phi_0 = 0\,, \tag{7.1.9}$$

respectively. On the boundary of each cylinder $\boldsymbol{x}_i \in B_N$, i.e., $|\boldsymbol{x} - \boldsymbol{x}_N| = a$, we assume perfect reflection so that the normal velocity vanishes,

$$\boldsymbol{n}\cdot\nabla\phi_0 = 0\,, \tag{7.1.10}$$

$$\boldsymbol{n}\cdot\nabla\phi_1 + \boldsymbol{n}\cdot\nabla'\phi_0 = 0\,, \tag{7.1.11}$$

$$\boldsymbol{n}\cdot\nabla\phi_2 + \boldsymbol{n}\cdot\nabla'\phi_1 = 0\,. \tag{7.1.12}$$

Ω-periodicity is required of all $\phi_0, \phi_1, \ldots,$.

From Eqs. (7.1.7) and (7.1.10), ϕ_0 is independent of the short scale,

$$\phi_0 = \phi_0(\boldsymbol{x}')\,. \tag{7.1.13}$$

It follows from Eq. (7.1.8) that,

$$\nabla^2\phi_1 = 0\,. \tag{7.1.14}$$

In view of Eq. (7.1.11), we let

$$\phi_1 = -S_m \frac{\partial \phi_0}{\partial x'_m} . \tag{7.1.15}$$

From Eq. (7.1.14), we have the governing equation for the vector $\{S_m\}$,

$$\frac{\partial^2 S_m}{\partial x_i \partial x_i} = 0 , \quad x_i \in \Omega_F . \tag{7.1.16}$$

On the cylinder boundary, Eq. (7.1.11) implies that

$$-n_i \frac{\partial S_m}{\partial x_i} \frac{\partial \phi_o}{\partial x'_m} = -n_i \frac{\partial \phi_0}{\partial x'_m} \delta_{mi} , \tag{7.1.17}$$

yielding

$$n_i \frac{\partial S_m}{\partial x_i} = n_i \delta_{mi} = n_m , \quad x_i \in B . \tag{7.1.18}$$

If the cylinder is circular, we have

$$\frac{\partial S_m}{\partial r} = n_m , \tag{7.1.19}$$

or

$$\frac{\partial S_x}{\partial r} = n_x = \cos\varphi , \quad \frac{\partial S_y}{\partial r} = n_y = \sin\varphi . \tag{7.1.20}$$

Together with the requirement that S_m is Ω-periodic, Eqs. (7.1.16) and (7.1.20) define the cell problem for S_m. For uniqueness of S_m we further impose

$$\langle S_m \rangle = \frac{1}{\Omega} \iint_{\Omega_F} S_m \, dA = 0 . \tag{7.1.21}$$

We now apply Green's formula to ϕ_0 and ϕ_2 over a microcell

$$\iint_{\Omega_f} (\phi_0 \nabla^2 \phi_2 - \phi_2 \nabla^2 \phi_0) dA = \int_{\partial\Omega} \left(\phi_0 \frac{\partial \phi_2}{\partial n} - \phi_2 \frac{\partial \phi_0}{\partial n} \right) ds , \tag{7.1.22}$$

where n_i is the outward unit normal from the fluid. Since Eq. (7.1.12) can be written as

$$n_i \frac{\partial \phi_2}{\partial x_i} = -n_i \frac{\partial \phi_1}{\partial x'_i} = n_i S_m \frac{\partial^2 \phi_0}{\partial x'_m \partial x'_i} . \tag{7.1.23}$$

Making use of the above result on the right-hand side and Eqs. (7.1.7) and (7.1.9) on the left-hand side of Eq. (7.1.22), we have

$$\iint_{\Omega_f} \phi_0 \left[2 \frac{\partial}{\partial x'_i} \left(\frac{\partial S_k}{\partial x_i} \frac{\partial \phi_0}{\partial x'_k} \right) - \frac{\partial^2 \phi_0}{\partial x'_i \partial x'_i} - k^2 \phi_0 \right] dA$$
$$= \left(\int_{\partial\Omega} \phi_0 n_i S_k \, ds \right) \frac{\partial^2 \phi_0}{\partial x'_i \partial x'_k} . \tag{7.1.24}$$

Note that

$$\iint_{\Omega_f} 2\frac{\partial}{\partial x_i'}\left(\frac{\partial S_m}{\partial x_i}\frac{\partial \phi_0}{\partial x_m'}\right) dA = 2\iint_{\Omega_f} \frac{\partial S_m}{\partial x_i}\frac{\partial^2 \phi_0}{\partial x_i' \partial x_m'} dA$$

$$= \iint_{\Omega_f} \left[\frac{\partial S_i}{\partial x_m} + \frac{\partial S_m}{\partial x_i}\right] \frac{\partial^2 \phi_0}{\partial x_i' \partial x_m'} dA. \quad (7.1.25)$$

Using the fact that ϕ_0 is independent of $\boldsymbol{x}$, the effective medium equation is found

$$\boxed{\beta_{im}\frac{\partial^2 \phi_0}{\partial x_i' \partial x_m'} + \theta_f k^2 \phi_0 = 0}. \quad (7.1.26)$$

The tensor β_{im} is defined by

$$\beta_{im} = \theta_f \delta_{im} + \frac{1}{\Omega}\left\{\int_{B_s} n_i S_m ds - \iint_{\Omega_f} \left[\frac{\partial S_i}{\partial x_m} + \frac{\partial S_m}{\partial x_i}\right]\right\} dA, \quad (7.1.27)$$

where B_s is the solid boundary of the cylinder. The coefficient

$$\theta_f = \frac{1}{\Omega}\iint_{\Omega_f} dA = \frac{\Omega_f}{\Omega} \quad (7.1.28)$$

is the area fraction of fluid. Both β_{im} and θ_f depend on the cell geometry. Equation (7.1.26) describes wave propagation over the scale of $\mathcal{O}(1/k)$ and is accurate to leading order for small ka. For a rectangular array with cylinder radius a and center-to-center separation d_1, d_2 with $2a < d_1 < d_2$, $\theta_f = 1 - \theta_s = 1 - \pi a^2/d_1 d_2$ where $\theta_s = \pi a^2/d_1 d_2$ is the solid fraction.

The line integral in Eq. (7.1.27) can be further manipulated by using the defining equations of the cell problem. Multiplying Eq. (7.1.16) by S_j and integrating over Ω_j

$$0 = \iint_{\Omega_f} S_j \frac{\partial^2 S_m}{\partial x_i \partial x_i} dA = \iint_{\Omega_f} \frac{\partial}{\partial x_i}\left(S_j \frac{\partial S_m}{\partial x_i}\right) dA - \iint_{\Omega_f} \frac{\partial S_j}{\partial x_i}\frac{\partial S_m}{\partial x_i} dA$$

$$= \int_{\partial\Omega} S_j \frac{\partial S_m}{\partial x_i} n_i ds - \iint_{\Omega_f} \frac{\partial S_j}{\partial x_i}\frac{\partial S_m}{\partial x_i} dA$$

$$= \int_B S_j n_m ds - \iint_{\Omega_f} \frac{\partial S_j}{\partial x_i}\frac{\partial S_m}{\partial x_i} dA, \quad (7.1.29)$$

after using Ω-periodicity and Eq. (7.1.18). After changing the indices,

$$\int_B S_m n_i ds = \iint_{\Omega_f} \frac{\partial S_m}{\partial x_j}\frac{\partial S_i}{\partial x_j} dA, \tag{7.1.30}$$

we obtain

$$\beta_{im} = \theta_f \delta_{im} + \frac{1}{\Omega}\left\{\iint_{\Omega_f} \frac{\partial S_m}{\partial x_j}\frac{\partial S_i}{\partial x_j} dA - \iint_{\Omega_f}\left[\frac{\partial S_i}{\partial x_m} + \frac{\partial S_m}{\partial x_i}\right] dA\right\}. \tag{7.1.31}$$

A further simplification can be made by letting

$$b_k = S_k - x_k, \tag{7.1.32}$$

then

$$\beta_{im} = \frac{1}{\Omega}\iint_{\Omega_f} \frac{\partial b_m}{\partial x_j}\frac{\partial b_i}{\partial x_j} dA. \tag{7.1.33}$$

Clearly $\beta_{im} = \beta_{mi}$ is symmetric, and is also positive definite as shown in Chapter 4.

For a square array $\beta_{im} = \beta\delta_{im}$ the effective medium is isotropic, and the governing equation reduces to

$$\nabla^2\phi_0 + \frac{\theta_f k^2}{\beta}\phi_0 = 0. \tag{7.1.34}$$

The effective wavenumber for a fixed ω can be defined as

$$k_e = k\sqrt{\frac{\theta_f}{\beta}}. \tag{7.1.35}$$

Since the phase velocity is $C = \omega/k$ in open space and $C_e = \omega/k_e$ in the cylinder region, the refractive index is

$$R = \frac{C}{C_e} = \frac{k_e}{k} = \sqrt{\frac{\theta_f}{\beta}}. \tag{7.1.36}$$

In the limit of no cylinders, $S_m = \theta_f = 0$ and $\beta = 1$. Equation (7.1.1) is recovered. In general the effective medium is anisotropic and a refractive index tensor can be defined as

$$R_{im} = \sqrt{\frac{\theta_f}{\beta_{im}}}. \tag{7.1.37}$$

The cell problem for S_k is of the Neumann type and can be solved by the discrete method of finite elements. The boundary-value problem is equivalent to the stationarity of the following functional.

$$\mathcal{F} = \frac{1}{2}\iint_{\Omega} \frac{\partial S_m}{\partial x_i}\frac{\partial S_m}{\partial x_i} dA - \int_{B_s} n_m S_m ds\,, \tag{7.1.38}$$

where Ω_f denotes the fluid volume in the cell. The first variation is

$$\begin{aligned}\delta\mathcal{F} &= \iint_{\Omega_f} \frac{\partial S_m}{\partial x_i}\frac{\partial \delta S_m}{\partial x_i} dA - \int_{B_s} n_m \delta S_m ds = -\iint_{\Omega_f} \delta S_m \frac{\partial^2 S_m}{\partial x_i \partial x_i} dA \\ &+ \int_{B_s} \delta S_m \left(n_i \frac{\partial S_m}{\partial x_i} - n_m \right) ds + \int_{B_c} \delta S_m n_i \frac{\partial S_m}{\partial x_i} ds\,, \end{aligned} \tag{7.1.39}$$

where B_c is the fluid part of the cell boundary. Clearly the first two integrals vanish for any δS_m. The last integral vanishes because of periodicity. Hence $\delta\mathcal{F} = 0$ for small but arbitrary δS_m. The necessary condition for equivalence is proven.

By discretizing the cell into finite elements, an algebraic equation can be obtained and solved for the nodal unknowns. Using a free software[1] which contains codes for grid generation and computations, sample results over the full range of solid fraction $\theta_s = \pi a^2/(d_1 d_2)$ for rectangular arrays with different aspect ratios $r = d_2/d_1 > 1$ are shown in Fig. 7.1. The maximum solid fraction is $\theta_s = \pi d_1/4d_2$ corresponding to $2a = d_1$. Because S_1 is odd in x_1 and even in x_2, while S_2 is even in x_1 and odd in x_2, it can be shown that β_{im} is diagonal. The refractive index tensor has only two components $R_1 \equiv R_{11}, R_2 = R_{22}$. Correspondingly, the effective wavenumber vector components for a propagating wave are $k_1 = k/R_1, k_2 = k/R_2$.

Extension to three-dimensional array is straightforward. In particular one needs only change all gradient operators from two to three dimensions. All line integrals are replaced by surface integrals and all surface integrals by volume integrals. If each cell is a rectangular box with a sphere at the center, the canonical cell problem for $S_m, m = 1,2,3$ must satisfy the following boundary condition on the sphere $r = 1$:

$$\frac{\partial S_x}{\partial r} = \cos\varphi \sin\psi\,, \quad \frac{\partial S_y}{\partial r} = \sin\varphi\sin\psi \quad \frac{\partial S_z}{\partial r} = \cos\psi \tag{7.1.40}$$

in the local spherical polar coordinates related to the Cartesian coordinates by $x = r\cos\varphi\sin\psi, y = r\sin\varphi\sin\psi$, and $z = r\cos\psi$.

[1]FreeFEM++ from http://www.freefem.org/ff++/.

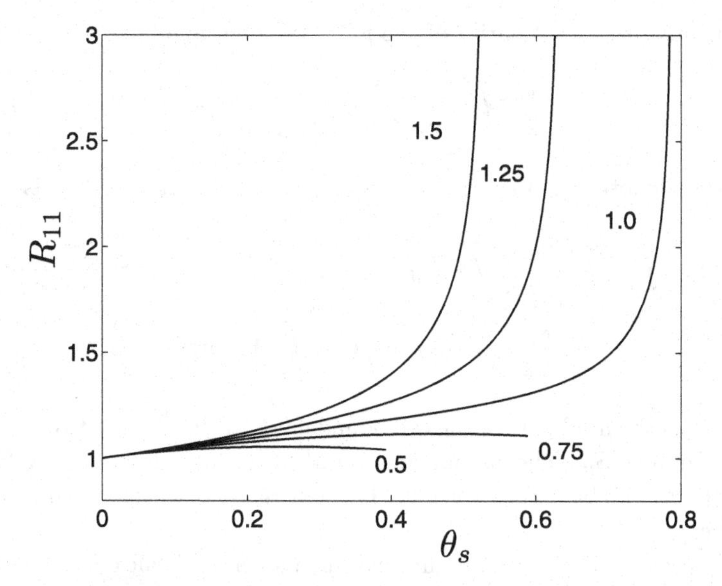

Fig. 7.1. Refractive index $R_{11} \equiv R_1$ for a rectangular array with different ratios $r = d_1/d_2$ and for all solid fractions θ_s $(= 1 - \theta_f)$ up to the maximum when neighboring circles are in contact. The same curves can also represent $R_{22} \equiv R_2$ if r is interpreted as $r' = d_2/d_1$ instead. For a square array $r = 1$, $R_{11} = R_{22} = R$. Computations are made by X. Garnaud.

7.2. Bragg Scattering of Short Waves by a Cylinder Array

In Sec. 1.2 one-dimensional Bragg scattering of monochromatic waves by weak periodic inhomogeneities in a medium was treated by the homogenization method. Here we extend the technique to sound scattered by a linear array of identical cylinders of radius a spaced at equal distance D apart along the centerline of a long wave guide (channel) of width D, see Fig. 7.2.

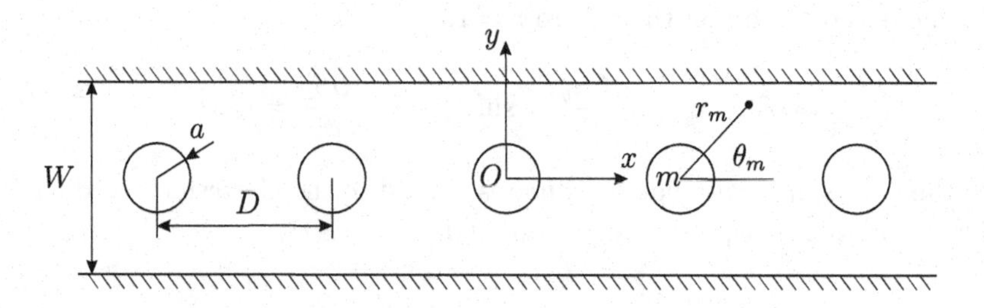

Fig. 7.2. A line of periodic circular cylinders in a channel.

The number of cylinders is large but finite. The problem is equivalent to normal incidence on one side of a long array of finite width, where the array is formed by square lattices. Strictly speaking the boundary-value problem is two-dimensional, hence is more involved than the example in Sec. 1.2.

An analogous problem for the scattering of infinitesimal water waves by a periodic array of vertical piles in a sea of constant mean depth was studied by Li and Mei (2004, 2007). Their mathematical problem is three dimensional but can be reduced to two dimensions. We shall follow their derivation closely and consider the radius of each cylinder to be very small compared to the incident wavelength, i.e., $ka = \epsilon \ll 1$. It is known from the theory of scattering for one small cylinder that the scattered wave is $\mathcal{O}(ka)^2$ times smaller than the incident wave. It is also known that scattering by periodic scatterers can be strong if $kD = m\pi$ where $m = 1, 2, 3, \ldots$. After multiple scattering by $N = \mathcal{O}(1/\epsilon)$ cylinders Bragg resonance can be so strong that the scattered wave becomes comparable to the incident wave. Let us construct an approximate theory valid in the neighborhood of Bragg resonance. Since the assumption of small ka restricts the theory to moderate m only, we shall assume $m = 1$ from here on. Modifications for $m = 2, 3$ are straightforward.

7.2.1. *Envelope Equations*

We begin with the governing equation for the sound velocity potential $\Phi(x, y, t)$

$$\Phi_{xx} + \Phi_{yy} = \frac{1}{C^2}\Phi_{tt}\,. \tag{7.2.1}$$

The normal velocity vanishes on the channel banks

$$\frac{\partial \Phi}{\partial y} = 0\,, \quad y = \pm\frac{D}{2}\,. \tag{7.2.2}$$

Let the cylinder centers (lattice points) be at $x_n = nD$, $y = 0$ with $0, 1, 2, \ldots, N$ and r, θ are the local polar coordinates centered at a lattice point $(x_n, 0)$, then,

$$\frac{\partial \Phi}{\partial r} = 0\,, \quad r = a\,. \tag{7.2.3}$$

The incident wave is given by plane wave with a slight shift from Bragg resonance,

$$\Phi_0 = A(x, t)e^{i(kx - \omega t)}\,. \tag{7.2.4}$$

The amplitude A can depend on x, t slowly to allow detuning or slow modulation. Let us define the small parameter

$$\epsilon = ka \ll 1, \tag{7.2.5}$$

and introduce fast and slow coordinates x, t; $x' = \epsilon^2 x$, $t' = \epsilon^2 t$, and the expansion

$$\Phi(x, y, t) = e^{-i\omega t}[\phi_1(x, y, x', t') + \epsilon^2 \phi_2(x, y, x', t') + O(\epsilon^4)]. \tag{7.2.6}$$

By separating orders a sequence of boundary-value problems governing the microscale variations in a unit cell will be obtained. Each cell is but one among many periods. We then invoke the Bloch theorem (Ashcroft and Mermin, 1976) in solid-state physics that the potential $\phi_n(\boldsymbol{x})$ depends on the microscale coordinate $\boldsymbol{x}$ as

$$\phi_n(\boldsymbol{x}) = e^{\pm ikx} f_n(\boldsymbol{x}), \quad \text{where } f_n(x, y) = f_n(x + D, y), \tag{7.2.7}$$

i.e., $f_n(\boldsymbol{x})$ is D-periodic in x. Near the Bragg state, $kD = \pi$, Bloch condition implies:

$$\phi_n\left(-\frac{D}{2}, y\right) = \phi_n\left(\frac{D}{2}, y\right), \quad \frac{\partial \phi_n}{\partial x}\left(-\frac{D}{2}, y\right) = -\frac{\partial \phi_n}{\partial x}\left(\frac{D}{2}, y\right). \tag{7.2.8}$$

At $\mathcal{O}(1)$, the governing equations for the fast variations of ϕ_1 are:

$$\nabla^2 \phi_1 + k^2 \phi_1 = 0, \tag{7.2.9}$$

$$\frac{\partial \phi_1}{\partial r} = 0, \quad r = a, \tag{7.2.10}$$

$$\frac{\partial \phi_1}{\partial y} = 0, \quad y = \pm \frac{D}{2}. \tag{7.2.11}$$

Because of the small size, the cylinder does not affect the wave field at the leading order, hence its boundary condition is ineffective. The solution consists of simple incident and reflected waves without scattering

$$\phi_1 = A_+(x', t') e^{ikx} + A_-(x', t') e^{-ikx}, \tag{7.2.12}$$

which automatically satisfies the Bloch condition. The parts

$$\psi^\pm \equiv e^{\pm ikx} \tag{7.2.13}$$

are two homogeneous solutions to the two-dimensional Helmholtz equation in the cell.

At the next order, ϕ_2 is governed by the inhomogeneous cell problem: $O(\epsilon^2)$:

$$\nabla^2 \phi_2 + k^2 \phi_2 = -2 \frac{\partial^2 \phi_1}{\partial x \partial x'} + \frac{2}{C^2} \frac{\partial^2 \phi_1}{\partial t \partial t'}, \tag{7.2.14}$$

$$\frac{\partial \phi_2}{\partial r} = -\frac{1}{\epsilon^2}\frac{\partial \phi_1}{\partial r}, \quad r = a, \tag{7.2.15}$$

$$\frac{\partial \phi_2}{\partial y} = 0, \quad y = \pm\frac{D}{2}. \tag{7.2.16}$$

As usual the slow variations of $A^{\pm}(x', t')$ are to be found by applying Green's formula to ϕ_1 and ϕ_2, or, equivalently, $\psi^{\pm}$ and ϕ_2 in the cell,

$$\iint_{\Omega} (\phi_2 \nabla^2 \psi^{\pm} - \psi^{\pm}\nabla^2\phi_2)\, dA = \int_{\partial\Omega}\left(\phi_2 \frac{\partial \psi^{\pm}}{\partial n} - \psi^{\pm}\frac{\partial \phi_2}{\partial n}\right) ds, \tag{7.2.17}$$

where $\partial\Omega$ is the cell boundary consisting of the cylinder surface, the sides $x = x_n \pm D/2$ and $y = \pm D/2$, and $\boldsymbol{n}$ is the unit normal pointing out of the fluid. For the area integrals on the left-hand side we first use Eq. (7.2.12) to rewrite Eq. (7.2.14) as

$$\begin{aligned}\nabla^2\phi_2 + k^2\phi_2 = &-2ik\left(\frac{\partial A_+}{\partial x'}e^{ikx} - \frac{\partial A_-}{\partial x'}e^{-ikx}\right)\\ &-\frac{2i\omega}{C^2}\left(\frac{\partial A_+}{\partial t'}e^{ikx} + \frac{\partial A_-}{\partial t'}e^{-ikx}\right).\end{aligned} \tag{7.2.18}$$

The surface integral in Eq. (7.2.17) becomes

$$\begin{aligned}&-\iint_{-D/2}^{D/2} e^{\pm ikx}\left[2ik\left(-\frac{\partial A_+}{\partial x'}e^{ikx} + \frac{\partial A_-}{\partial x'}e^{-ikx}\right)\right.\\ &\qquad\left.-\frac{2i\omega}{C^2}\left(-\frac{\partial A_+}{\partial t'}e^{ikx} - \frac{\partial A_-}{\partial t'}e^{-ikx}\right)\right] dy\, dx\\ &= \iint_{-D/2}^{D/2}\left\{2ik\left(\frac{\partial A_+}{\partial x'}\begin{bmatrix} e^{2ikx}\\ 1\end{bmatrix} - \frac{\partial A_-}{\partial x'}\begin{bmatrix}1\\ e^{-2ikx}\end{bmatrix}\right)\right.\\ &\qquad\left.+\frac{2i\omega}{C^2}\left(\frac{\partial A_+}{\partial t'}\begin{bmatrix} e^{2ikx}\\ 1\end{bmatrix} + \frac{\partial A_-}{\partial t'}\begin{bmatrix}1\\ e^{-2ikx}\end{bmatrix}\right)\right\} dy\, dx\\ &= \frac{2i\pi^2}{k}\left\{\frac{\partial A_+}{\partial x'}\begin{bmatrix}0\\1\end{bmatrix} - \frac{\partial A_-}{\partial x'}\begin{bmatrix}1\\0\end{bmatrix}\right.\\ &\qquad\left.+\frac{1}{C}\left(\frac{\partial A_+}{\partial t'}\begin{bmatrix}0\\1\end{bmatrix} + \frac{\partial A_-}{\partial t'}\begin{bmatrix}1\\0\end{bmatrix}\right)\right\}.\end{aligned} \tag{7.2.19}$$

Use has been made of $kD = \pi$.

On the right-hand side of Eq. (7.2.17) only the line integral along the cylinder surface matters. For its evaluation we need $\partial\phi_2/\partial r|_{r=a}$

and $\phi_2(a,\theta)$. From Eq. (7.2.15), $\partial\phi_2/\partial r|_{r=a}$ is found by expanding $e^{\pm ikx} = e^{\pm ikx_n \pm ikr\cos\theta}$ for small kr and evaluating at $r = a$,

$$\left.\frac{\partial\phi_2}{\partial r}\right|_{r=a} = -\frac{1}{\epsilon^2}\frac{\partial\phi_1}{\partial r} = \frac{k^2 a}{2\epsilon^2}(1+\cos 2\theta)[A_+ e^{ikx_n} + A_- e^{-ikx_n}]$$
$$+\frac{ik}{\epsilon^2}\cos\theta[A_+ e^{ikx_n} - A_- e^{-ikx_n}] + \mathcal{O}(\epsilon^2)\,, \quad r = a\,. \tag{7.2.20}$$

This result provides a boundary condition for finding $\phi_2(r,\theta)$ for small $r = \mathcal{O}(a)$. Since near the cylinder all spatial derivatives are of order $\mathcal{O}(1/a)$, Eq. (7.2.18) is dominated by Laplace's equation

$$\nabla^2\phi_2 = \mathcal{O}(\epsilon^2)\,. \tag{7.2.21}$$

The solution is obtained by separation of variables,

$$\phi_2(r,\theta) \cong \frac{1}{2}\left(A_+ e^{ikx_n} + A_- e^{-ikx_n}\right)\ln\frac{r}{a}$$
$$+\frac{i}{kr}(A_+ e^{ikx_n} + A_- e^{-ikx_n})\cos\theta$$
$$-\frac{a^2}{4r^2}(A_+ e^{ikx_n} + A_- e^{-ikx_n})\cos 2\theta + \cdots\,, \tag{7.2.22}$$

which gives the boundary value

$$\phi_2(a,\theta) \cong \frac{i}{ka}(A_+ e^{ikx_n} - A_- e^{-ikx_n})\cos\theta$$
$$-\frac{1}{4}(A_+ e^{ikx_n} + A_- e^{-ikx_n})\cos 2\theta + \cdots\,. \tag{7.2.23}$$

Approximating $\psi^{\pm}$ and $\partial\psi^{\pm}/\partial r$ for small kr by

$$\psi^{\pm}(r,\theta) \cong e^{\pm ikx_n}[1 \pm ikr\cos\theta + \mathcal{O}(\epsilon^2)]\,, \tag{7.2.24}$$

$$\frac{\partial\psi\pm}{\partial r} \cong ke^{\pm ikx_n}\left[\pm i\cos\theta - \frac{kr}{2}(1+\cos 2\theta) + \mathcal{O}(\epsilon^2)\right], \tag{7.2.25}$$

we find after invoking $2kx_n = 2knD = 2n\pi$,

$$a\int_0^{2\pi}\left(\phi_2\frac{\partial\psi^{\pm}}{\partial r}\right)_{r=a} d\theta \cong \mp\pi(A_+ - A_-)\,, \tag{7.2.26}$$

and

$$a\int_0^{2\pi}\left(\psi^{\pm}\frac{\partial\phi_2}{\partial r}\right)_{r=a} d\theta \cong \pi(A_{\pm} - 3A_{\pm})\,. \tag{7.2.27}$$

Substituting Eqs. (7.2.19), (7.2.26) and (7.2.27) into Eq. (7.2.17) we get a pair of equations coupling the envelopes of the incident and reflected waves

$$\boxed{\frac{\partial A_+}{\partial t'} + C\frac{\partial A_+}{\partial x'} = i\Omega_0(A_+ - 3A_-)}\,, \tag{7.2.28}$$

$$\boxed{\frac{\partial A_-}{\partial t'} - C\frac{\partial A_-}{\partial x'} = i\Omega_0(A_- - 3A_+)}\,, \tag{7.2.29}$$

with the coupling constant

$$\Omega_0 = \frac{\omega}{2\pi}\,. \tag{7.2.30}$$

Use has been made of $\omega = kC$. In physical coordinates they read:

$$\frac{\partial A_+}{\partial t} + C\frac{\partial A_+}{\partial x} = ik^2a^2\Omega_0(A_+ - 3A_-)\,, \tag{7.2.31}$$

$$\frac{\partial A_-}{\partial t} - C\frac{\partial A_-}{\partial x} = ik^2a^2\Omega_0(A_- - 3A_+)\,. \tag{7.2.32}$$

By cross-differentiation, the two equations can be combined into one

$$\left(\frac{\partial^2}{\partial t'^2} - C^2\frac{\partial^2}{\partial x'^2} + 2\Omega_0^2 - i\Omega_0\frac{\partial}{\partial t'}\right)A_\pm = 0\,. \tag{7.2.33}$$

Without the last term, the equation is the Klein–Gordon equation in physics.

7.2.2. Dispersion Relation for a Detuned Wave Train

Consider a sinusoidally modulated envelopes propagating in an infinitely long wave guide

$$A = A_0 e^{i(Kx' - \Omega t')} = A_0 e^{i(\epsilon^2 Kx - \epsilon^2 \Omega t)}\,. \tag{7.2.34}$$

The corresponding potentials are

$$Ae^{i(kx-\omega t)} = A_0 e^{i((k+\epsilon^2 K)x - (\omega + \epsilon^2\Omega)t)}\,, \tag{7.2.35}$$

which represent incident and reflected wave trains with wavenumbers and frequencies slightly detuned by $\epsilon^2 K$ and $\epsilon^2\Omega$, respectively. Substituting Eq. (7.2.34) into Eq. (7.2.33), we find the dispersion relation between K and Ω,

$$K = \frac{\Omega_0}{C}\sqrt{\left(\frac{\Omega}{\Omega_0} + 2\right)\left(\frac{\Omega}{\Omega_0} - 1\right)}\,. \tag{7.2.36}$$

For real Ω, K is real if $\Omega/\Omega_0 > 1$ or < -2, the envelopes are propagating waves themselves. However, within the band $-2 < \Omega/\Omega_0 < 1$,

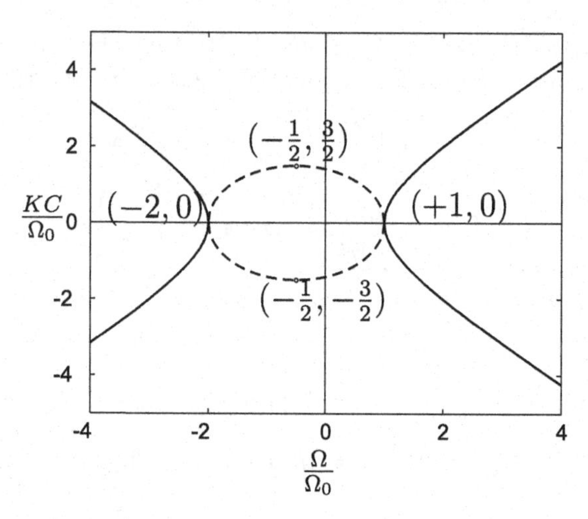

Fig. 7.3. Dispersion relation of progressive waves through a line of periodic cylinders. Solid curve: $\mathcal{R}e\,(KC/\Omega_0)$; dashed curve: $\mathcal{I}m\,(KC/\Omega_0)$. From Li and Mei (2007), *J. Fluid Mech.*

K is imaginary; signifying exponential decay as $x \to \pm\infty$. This band is called the bandgap within which propagation is forbidden. For large $|\Omega|/\Omega_0$, $K \to \Omega/C$, hence $k + \epsilon^2 K = (\omega + \epsilon^2\Omega)/C$ far outside the bandgap. The numerical values of $\mathcal{R}e\,(KC/\Omega_0)$ and $\mathcal{I}m\,(KC/\Omega_0)$ are plotted in Fig. 7.3.

7.2.3. *Scattering by a Finite Strip of Periodic Cylinders*

For assessing the accuracy of the asymptotic theory and to illustrate the effects of bandgap, we consider a finite but large number of periodic cylinders in the strip $0 \le x' \le L$, bounded by open air in $x' < 0$ and $x' > L$. A slightly detuned incident wave arrives from $x = -\infty$,

$$A_+(x', t') = A_0 e^{i(Kx' - \Omega t')}\,. \tag{7.2.37}$$

In the cylinder region, we assume the envelopes to be of the form

$$A_+(x',') = A_0 T(x') e^{-i\Omega t'}\,, \quad A_-(x', t') = A_0 R(x') e^{-i\Omega t'}\,, \tag{7.2.38}$$

where $T(x_1)$ and $R(x_1)$ are the local transmission and reflection coefficients, respectively. From Eq. (7.2.33) we find

$$\frac{d^2}{dx'^2}\begin{pmatrix} T \\ R \end{pmatrix} + \frac{\Omega_0^2}{C^2}\left(\frac{\Omega}{\Omega_0} + 2\right)\left(\frac{\Omega}{\Omega_0} - 1\right)\begin{pmatrix} T \\ R \end{pmatrix} = 0\,. \tag{7.2.39}$$

It is evident that both $T(x')$ and $R(x')$ are oscillatory in x' if Ω/Ω_0 lies outside the bandgap and monotonic within. We impose the boundary condition $T(0) = 1$ at the entrance and $R(L) = 0$ at the exit. From the envelope equations we also get

$$R = \frac{2}{3}\left(\frac{\Omega}{\Omega_0} + \frac{1}{2}\right)T + \frac{2iC}{3\Omega_0}\frac{dT}{dx'}, \tag{7.2.40}$$

which gives the boundary condition for T at $x_1 = L$.

The solution of the boundary-value problem is straightforward. We only give the transmission coefficient at the exit and the reflection coefficient at the entry.

Outside the bandgap, we denote by P the real quantity

$$P = \sqrt{\left(\frac{\Omega}{\Omega_0} + 2\right)\left(\frac{\Omega}{\Omega_0} - 1\right)}. \tag{7.2.41}$$

The transmission coefficient at the exit is

$$T(L) = \frac{2iP}{(1 + 2(\Omega/\Omega_0))\sin(P\Omega_0 L/C) + 2iP\cos(P\Omega_0 L/C)}, \tag{7.2.42}$$

and the reflection coefficient at the entrance is

$$R(0) = \frac{3\sin(P\Omega_0 L/C)}{(1 + 2(\Omega/\Omega_0))\sin(P\Omega_0 L/C) + 2iP\cos(P\Omega_0 L/C)}. \tag{7.2.43}$$

In the bandgap $-2 < \Omega/\Omega_0 < 1$, we denote

$$Q = -iP = \sqrt{\left[\frac{\Omega}{\Omega_0} + 2\right]\left[1 - \frac{\Omega}{\Omega_0}\right]} \tag{7.2.44}$$

is real.

The transmission coefficient at the exit is

$$T(L) = \frac{2iQ}{(1 + 2(\Omega/\Omega_0))\sinh(Q\Omega_0 L/C) + 2iQ\cosh(Q\Omega_0 L/C)}, \tag{7.2.45}$$

and the reflection coefficient at the entrance is

$$R(0) = \frac{3\sinh(Q\Omega_0 L/C)}{(1 + 2(\Omega/\Omega_0))\sinh(Q\Omega_0 L/C) + 2iQ\cosh(Q\Omega_0 L/C)}. \tag{7.2.46}$$

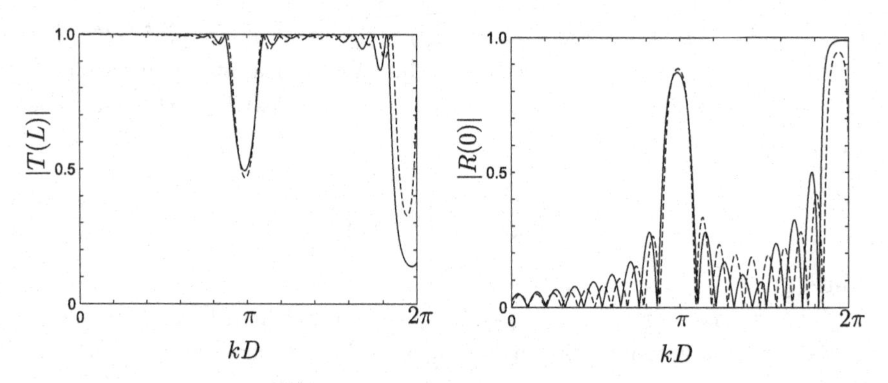

Fig. 7.4. Transmission and reflection coefficients by a row of 21 circular cylinders of radius $a/D = 0.10$. Solid curve: asymptotic solution; dashed curve: finite element solution. From Li and Mei (2007), *J. Fluid Mech.*

Effects of detuning are plotted in Fig. 7.4 which shows strong reflection and weak transmission inside the bandgap.

In a similar theory for water waves through vertical piles, Li and Mei (2007a) have checked the accuracy of the asymptotic approximation by direct numerical computation using finite element method for 21 cylinders. The comparison is shown in Fig. 7.4 for the range of $0 < kD < 2\pi$ which includes the entire neighborhood of the first Bragg resonance peak $kD = \pi$ and part of the next peak at $kD = 2\pi$. The agreement is very good at and around the first peak. Near the second peak the agreement is understandably less good since ka is no longer very small. Modification of the asymptotic theory for the second peak is straightforward.

7.3. Sound Propagation in a Bubbly Liquid

Bubbles are present near the crests of breaking waves on the ocean surface and are around ships especially behind ship propellers. Acoustic remote sensing of the ocean surface and sonar detection of submarines require the understanding of wave scattering by bubble clouds. The effective equation for sound must account for the oscillation of the much smaller bubbles which may radiate sound. While the result can be derived by phenomenological reasoning (van Wijngaarden (1968, 1972)), the method of homogenization can give more physical insight into the interplay between different scales. The derivation to be described here was first applied to a periodic array of bubbles by Calflisch *et al.* (1985b).

7.3.1. *Scale and Order Estimates*

We consider bubbles sparsely populated in a large body of water, and denote by $\boldsymbol{u}$ the fluid velocity, p the hydrodynamic pressure, ρ the density, and C_o the speed of sound, all of the surrounding water. The radius of a typical bubble R_o is assumed to be much smaller than inter-bubble distance ℓ, which is in turn much smaller than the typical sound wavelength $\sim k^{-1}$ in the far field. Two ratios of the three length scales can be formed

$$\delta = kR_o\,, \qquad \epsilon = k\ell\,; \tag{7.3.1}$$

so that

$$\delta \ll \epsilon \ll 1\,. \tag{7.3.2}$$

As will be checked later, the oscillation amplitude away from a bubble is expected to be small enough to justify the linearized conservation laws of mass

$$\frac{\partial p}{\partial t} + \rho C_o^2 \nabla \cdot \boldsymbol{u} = 0\,, \tag{7.3.3}$$

and momentum

$$\rho \frac{\partial \boldsymbol{u}}{\partial t} = -\nabla p\,. \tag{7.3.4}$$

We shall define three regions centered around a bubble: the *near field* $r = \mathcal{O}(a)$, the *intermediate field* $r = \mathcal{O}(\ell)$, and the *far field* $r = \mathcal{O}(2\pi/k)$. Let us first estimate heuristically the order of magnitude of the physical quantities in various regions.

First the far field. Let the characteristic amplitudes of the fluid displacement in the sound wave be A_o. The defining length scale and the scales of other physical quantities are

$$\boldsymbol{x} \sim \frac{1}{k}\,, \quad t \sim \frac{1}{\omega}\,, \quad \boldsymbol{u} \sim \omega A_o\,, \quad p \sim \frac{\rho\omega^2 A_o}{k}\,, \tag{7.3.5}$$

where p represents the change of pressure and not its absolute magnitude.

In the intermediate field, the spatial scale is changed to $\boldsymbol{x} \sim \ell$, while the time scale remains the same as in other regions. The scales of velocity and pressure are the same as the far field sound for continuity, hence, we have

$$\boldsymbol{x} \sim \ell\,, \quad t \sim \frac{1}{\omega}\,, \quad \boldsymbol{u} \sim \omega A_o\,, \quad p \sim \frac{\rho\omega^2 A_o}{k}\,. \tag{7.3.6}$$

In the *near field* of the bubble, the dynamics is characterized by the amplitude of bubble oscillations a_o. The appropriate scales are:

$$\boldsymbol{x} \sim R_o\,, \quad t \sim \frac{1}{\omega}\,, \quad \boldsymbol{u} \sim \omega a_o\,, \quad p \sim \rho R_o \omega^2 a_o\,. \tag{7.3.7}$$

7.3.2. Near Field of a Spherical Bubble

Let us first show that, at the leading order, compressibility is unimportant near a bubble. Consider mass conservation. The total rate of dilatation in the unit cell of volume ℓ^3 is of the order $\ell^3 \nabla \cdot \boldsymbol{u} \sim k\omega A_o \ell^3$. On the other hand the volume change of the bubble is $4\pi R_o^2 \partial a/\partial t \sim \omega a_o R_o^2$, since $S_b = 4\pi R_o^2$ is the surface area of the bubble. The two must be comparable, hence we have the following order relation:

$$\frac{A_o}{a_o} \sim \left(\frac{R_o}{\ell}\right)^2 \frac{1}{k\ell} \sim \frac{\delta^2}{\epsilon^3} \quad \text{or} \quad a_o \sim \frac{k\ell A_o}{(R_o/\ell)^2}\,. \tag{7.3.8}$$

On the other hand, the pressure variation in all regions must be comparable. Equating the orders of magnitudes of the pressures in the near and far fields, we get

$$\frac{A_o}{a_o} \sim kR_o \sim \delta\,. \tag{7.3.9}$$

The last two order relations together imply that

$$\delta \sim \epsilon^3 \tag{7.3.10}$$

(Calflisch *et al.*, 1985a). This is the same order relation for viscous flow through a moderately dilute suspension of small particles (Levy and Sanchez-Palencia, 1983).

With the scales given in Eq. (7.3.8) it is easy to see that

$$\frac{1}{\rho C_o^2} \frac{\partial p/\partial t}{\nabla \cdot \boldsymbol{u}} \sim (kR_o)^2\,. \tag{7.3.11}$$

With a very small error of $(kR_o)^2 = \delta^2 = \mathcal{O}(\epsilon^6)$, the mass conservation law reduces to

$$\nabla \cdot \boldsymbol{u} \cong 0\,, \tag{7.3.12}$$

i.e., water is essentially incompressible. Before restricting the bubble amplitude the momentum equation can still be nonlinear,

$$\rho\left(\frac{\partial \boldsymbol{u}}{\partial t} + \boldsymbol{u} \cdot \nabla \boldsymbol{u}\right) = -\nabla p\,. \tag{7.3.13}$$

Introducing the velocity potential ϕ by

$$\boldsymbol{u} = \nabla \phi\,. \tag{7.3.14}$$

It follows from Eq. (7.3.12) that

$$\nabla^2 \phi = 0 \,. \tag{7.3.15}$$

Assuming that the bubble surface oscillates only radially, the kinematic condition on the instantaneous surface is

$$\boldsymbol{u} \cdot \boldsymbol{n} = \frac{\partial R}{\partial t} \quad \text{on } r = R(t) \,. \tag{7.3.16}$$

To satisfy the preceding two conditions, the solution must be

$$\phi = -\frac{\partial R}{\partial t}\frac{R^2}{r} \,, \qquad \boldsymbol{u} = \nabla \phi = \frac{\partial R}{\partial t}\frac{R^2}{r^2}\boldsymbol{e}_r \,, \tag{7.3.17}$$

with r being the local radial coordinate from the bubble center. The momentum equation in water can be integrated to give the Bernoulli equation,

$$p + \rho \frac{\partial \phi}{\partial t} + \frac{\rho}{2}(\nabla \phi)^2 = p_o' \,, \tag{7.3.18}$$

where p_o' is the sound pressure far away from the bubble, i.e., in the intermediate field.

Accounting for surface tension, the water pressure p and the bubble pressure p_B must be related by

$$p = p_B + \frac{2T}{R} \tag{7.3.19}$$

at any time and

$$p_{Bo} = p_0' + \frac{2T}{R_o} \tag{7.3.20}$$

at static equilibrium, where T denotes the coefficient of surface tension. Let us assume that air in the bubble obeys the perfect gas law, i.e.,

$$\frac{p_B}{p_{Bo}} = \left(\frac{R_o}{R}\right)^{3\gamma} \tag{7.3.21}$$

where γ is the ratio of specific heats of the gas. From these equations we get in general the well-known Rayleigh–Plesset equation,

$$\boxed{p_{Bo}\left[\left(\frac{R_o}{R}\right)^{3\gamma} - 1\right] - 2T\left(\frac{1}{R} - \frac{1}{R_o}\right) - p_0' = \rho\left[R\frac{\partial^2 R}{\partial t^2} + \frac{3}{2}\left(\frac{\partial R}{\partial t}\right)^2\right]} \tag{7.3.22}$$

on the bubble surface.

If the bubble amplitude is small compared to its mean radius we write $R = R_o + a$ with $a \ll R_o$. Equation (7.3.22) may be linearized to give,

$$-\frac{p_0'}{\rho R_o} = \frac{\partial^2 a}{\partial t^2} + \omega_o^2 a \tag{7.3.23}$$

on the mean bubble surface, where

$$\omega_o^2 \equiv \frac{3\gamma p_{B_o}}{\rho R_o} - \frac{2T}{\rho R^2}, \tag{7.3.24}$$

so that ω_o is the natural frequency of a spherical bubble.

Although in subsequent discussions, we shall limit ourselves to infinitesimal oscillations of the bubble surface, it is interesting to note that nonlinearity is more important in the near field of the bubble than in the far field. This is because the bubble nonlinearity is measured by a/R whereas the sound nonlinearity is measured by kA. From Eqs. (7.3.9) and (7.3.10) it can be seen that their ratio is very small indeed

$$\frac{kA}{a/R} \sim \epsilon^2 (kR)^2 \sim \epsilon^4, \tag{7.3.25}$$

which justifies the linearization in the intermediate and far fields.

7.3.3. The Intermediate Field

Let us distinguish the unknowns in the intermediate field by primes. With the scales given by Eq. (7.3.7), the relative magnitudes of terms in the governing equations (7.3.3) and (7.3.4) can be estimated in dimensionless variables, and rewritten in physical form as

$$\frac{\epsilon}{\rho C_o^2}\frac{\partial p'}{\partial t} + \nabla' \cdot \boldsymbol{u}' = 0, \tag{7.3.26}$$

$$\epsilon\rho\frac{\partial \boldsymbol{u}'}{\partial t} = -\nabla' p'. \tag{7.3.27}$$

Since $R_o/\ell \sim \epsilon^2$ and $k\ell \sim \epsilon$, let us employ two space coordinates in the intermediate field

$$\boldsymbol{x}' = \epsilon^2 \boldsymbol{x} \quad \text{and} \quad \boldsymbol{x}'' = \epsilon \boldsymbol{x}' = \epsilon^3 \boldsymbol{x}, \tag{7.3.28}$$

and assume the expansions:

$$\boldsymbol{u}' = \boldsymbol{u}_o' + \epsilon \boldsymbol{u}_1' + \cdots, \quad p' = p_0' + \epsilon p_1' + \cdots,$$

with $\boldsymbol{u}_i' = \boldsymbol{u}_i'(\boldsymbol{x}', \boldsymbol{x}'', t)$, $p_i' = p_i'(\boldsymbol{x}', \boldsymbol{x}'', t)$. For the first two orders, the perturbation equations are

$$\nabla' \cdot \boldsymbol{u}_0' = 0\,, \tag{7.3.29}$$

$$\frac{1}{\rho C_o^2}\frac{\partial p_0'}{\partial t} + \nabla' \cdot \boldsymbol{u}_1' + \nabla'' \cdot \boldsymbol{u}_0' = 0 \tag{7.3.30}$$

from mass conservation (Eq. (7.3.3)) and

$$0 = -\nabla' p_o'\,, \tag{7.3.31}$$

$$\rho\frac{\partial \boldsymbol{u}_0'}{\partial t} = -\nabla' p_1' - \nabla'' p_0' \tag{7.3.32}$$

from momentum conservation (Eq. (7.3.4)). Equation (7.3.31) implies that p_0' is independent of $\boldsymbol{x}'$ and depends only on $\boldsymbol{x}''$ and t. Equations (7.3.29) and (7.3.32) govern $\boldsymbol{u}_0'$ and p_1' in the Ω' cell of volume $\Omega' = \mathcal{O}(\ell^3)$. We further impose the condition that $\boldsymbol{u}_0$ and p_1' are periodic from cell to cell (Ω'-periodic).

At the outer edge of the near field, the rate of radially outward volume flux is

$$q = \frac{\partial}{\partial t}\left(\frac{4}{3}\pi R^3\right). \tag{7.3.33}$$

Let $\partial\Omega_b'$ be a spherical shell at this outer edge and concentric to the bubble, and $\boldsymbol{n}$ denote a unit normal vector pointing radially outward from $\partial\Omega_b$. Approaching the bubble from the intermediate field, an observer sees the following radial flux

$$q = -\lim_{r'\to 0}\iint_{\partial\Omega_b'} \boldsymbol{u}' \cdot \boldsymbol{n}\, dS' = -\lim_{r'\to 0}\iint_{\partial\Omega_b'} (\boldsymbol{u}_0' + \epsilon\boldsymbol{u}_1') \cdot \boldsymbol{n}\, dS'\,. \tag{7.3.34}$$

Now the contribution from $\boldsymbol{u}_0'$ to the surface integral above vanishes because of Eq. (7.3.29) (near incompressibility). Equating Eqs. (7.3.33) and (7.3.34) we get a boundary condition for $\boldsymbol{u}_1'$:

$$\lim_{r'\to 0}\iint_{\partial\Omega_b'} \boldsymbol{u}_1' \cdot \boldsymbol{n}\, dS' = -4\pi R^2\frac{\partial R}{\partial t}\,. \tag{7.3.35}$$

Thus, while the bubble deforms, its volume changes by only $\mathcal{O}(\epsilon)$. An equivalent statement is

$$\lim_{r'\to 0} \boldsymbol{u}_1' = -\frac{dR}{dt}\,. \tag{7.3.36}$$

7.3.4. The Macroscale Equation

Let the Ω'-cell average of any function f be defined by

$$\langle f\rangle' = \frac{1}{\Omega'}\iiint_{\Omega'_w} f\,d\Omega'\,, \tag{7.3.37}$$

where Ω'_w is the volume occupied by water in the Ω'-cell, excluding the bubble or the small sphere inside $\partial\Omega'_b$. The cell average of Eq. (7.3.30) gives

$$\begin{aligned}(1-\theta_b)&\frac{1}{\rho C_o^2}\frac{\partial p'_0}{\partial t} + \nabla''\cdot\langle \boldsymbol{u}'_0\rangle' \\ &= -\iiint_{\Omega'_w}\nabla'\cdot\boldsymbol{u}'_1\,d\Omega' = -n\Omega'\lim_{r'\to 0}\frac{1}{\Omega'}\iint_{\partial\Omega'_w}\boldsymbol{u}'_1\cdot\boldsymbol{n}\,dS' \\ &= 4\pi nR^2\frac{\partial R}{\partial t}\end{aligned} \tag{7.3.38}$$

in view of Eq. (7.3.35), where

$$\theta_b \equiv \frac{(4/3)\pi R_o^3}{\ell^3} \tag{7.3.39}$$

denotes the volume fraction of bubbles per unit cell. Since

$$\theta_b = \mathcal{O}\left(\frac{R_o^3}{\ell^3}\right) = \mathcal{O}(k^2R_o^2) \ll 1\,, \tag{7.3.40}$$

θ_b may be neglected relative to unity in Eq. (7.3.38), yielding

$$\boxed{\frac{1}{\rho C_0^2}\frac{\partial p'_0}{\partial t} + \nabla''\cdot\langle \boldsymbol{u}'_0\rangle' = 4\pi nR^2\frac{\partial R}{\partial t}}\,. \tag{7.3.41}$$

Because of the assumed periodicity, the Ω'-average of Eq. (7.3.32) gives

$$\boxed{\rho\frac{\partial\langle \boldsymbol{u}'_0\rangle'}{\partial t} = -\nabla''p'_0}\,. \tag{7.3.42}$$

Together with the Rayleigh–Plesset equation (7.3.22), Eqs. (7.3.41) and (7.3.42) govern the coupled dynamics of the sound field and bubble oscillations. The last two equations are certainly obvious from the physical viewpoint, and were first deduced by Calflisch *et al.* (1985b) by the homogenization theory.

If the bubble oscillations are small we can linearize Eq. (7.3.38) and then eliminate $\langle \boldsymbol{u}'_0\rangle$ from Eqs. (7.3.41) and (7.3.42). The result is

$$\frac{1}{C_0^2}\frac{\partial^2 p'_0}{\partial t^2} - \nabla'^2 p'_0 = \rho n 4\pi R_o^2\frac{\partial^2 a}{\partial t^2}\,. \tag{7.3.43}$$

Finally, a may be eliminated from Eqs. (7.3.23) and (7.3.43) to yield a single equation for p_o',

$$\boxed{\left(\frac{\partial^2}{\partial t^2}+\omega_o^2\right)\left(\nabla''^2 p_0'-\frac{1}{C_o^2}\frac{\partial^2 p_0'}{\partial t^2}\right)=4\pi n R_o\frac{\partial^2 p_0'}{\partial t^2}=\frac{3\theta_b}{R_o^2}\frac{\partial^2 p_0'}{\partial t^2}}\,. \tag{7.3.44}$$

This equation governs sound propagation of infinitesimal amplitude in water populated by a periodic array of bubbles.

For simple harmonic motion with frequency ω, Eq. (7.3.44) reduces to the familiar Helmholtz equation:

$$\nabla''^2 p_0' + \frac{\omega^2}{C^2}p_0' = 0\,, \tag{7.3.45}$$

where the effective sound speed C is given by

$$\frac{1}{C^2}=\frac{1}{C_o^2}+\frac{3\theta_b/R_o^2}{\omega_o^2-\omega^2}, \quad \text{or} \quad \frac{C}{C_o}=\left(\frac{1-(\omega^2/\omega_o^2)}{1+\sigma-(\omega^2/\omega_o^2)}\right)^{1/2}, \tag{7.3.46}$$

with

$$\sigma=\frac{3\theta_b}{k_o^2R_o^2}=\frac{4\pi R_o}{k_o^2\ell^3}=\mathcal{O}\left(\frac{4\pi\delta}{\epsilon^3}\right), \tag{7.3.47}$$

where $k_o \equiv \omega_o/C_o$. Let $k=\omega/C$ be the wavenumber of a plane wave in the bubbly liquid. Equation (7.3.46) also implies

$$\frac{k}{k_o}=\frac{\omega}{\omega_o}\left(\frac{1+\sigma-(\omega^2/\omega_o^2)}{1-(\omega^2/\omega_o^2)}\right)^{1/2}.$$

Clearly, C and k are imaginary if the frequency falls in the bandgap

$$1<\frac{\omega}{\omega_o}<\sqrt{1+\sigma}\,.$$

Hence a plane sound wave can only propagate outside this bandgap, i.e., $\omega/\omega_o < 1$ or $\omega/\omega_o > \sqrt{1+\sigma}$. Since the factor σ is of order unity, the effective sound speed in a bubble cloud can be significantly reduced from C_o, even for a small volume fraction θ_b. This is a physically remarkable result. When the volume fraction is extremely small so that the ratio σ is also small but ω/ω_o is not close to unity, Eq. (7.3.46) can be approximated by

$$\frac{C}{C_o}\approx 1+\frac{3\theta_b}{2k_o^2R_o^2}\frac{\omega_o^2}{\omega^2}\,. \tag{7.3.48}$$

If the volume fraction is small and the frequency is also low, then,

$$\frac{C}{C_o} \approx 1 - \frac{3\theta_b}{k_o^2 R_o^2} \tag{7.3.49}$$

(see, e.g., van Wijngaarden (1972)).

On the other hand for very low frequencies but $\sigma \gg 11$, we get

$$\frac{C}{C_o} = \sigma^{-1/2} = \left(\frac{3\theta_b}{k_o^2 R_o^2}\right)^{-1/2} \tag{7.3.50}$$

which is the limit of Eq. (7.3.46) by ignoring the compressibility of pure water. Thus, as long as $\theta_b \ll 1$, Eq. (7.3.46) encompasses a wide range of σ although it is deduced on the basis of $\sigma = \mathcal{O}(1)$. Equation (7.3.45) can be used to examine sound scattering by a bubble cloud of certain size and geometry.

7.4. One-Dimensional Sound Through a Weakly Random Medium

There are numerous situations where one needs to know how waves propagate through a medium with random impurities. Examples include light through sky with dust particles, sound through water with bubbles, sea waves over a irregular topography, or elastic waves through a solid with cracks, fibers, cavities, hard or soft grains, etc. For one-dimensional propagation in a large region of random inhomogeneities, it is known that multiple scattering yields a small complex shift of the propagation constant with the real part corresponding to a change of wavenumber while the imaginary part to a gradual attenuation in space (Keller, 1964; Karal and Keller, 1964). In particular, the spatial attenuation due to random scattering is effective for a broad range of incident wave frequencies, unlike the effects of periodic inhomogeneities which cause strong scattering only for certain frequency bands (Bragg scattering, see, e.g., Chapter 1). In the context of solid-state physics, there is a related theory called *Anderson localization* (Anderson, 1958) which predicts that a metal conductor can behave like an insulator if the microstructure is sufficiently disordered. A survey of localization in many types of classical waves based on linearized theories can be found in Sheng (1995) where diagrammatic techniques are used extensively.

In this and subsequent sections we shall show that the method of homogenization can be readily modified for treating long-range effects of weak random inhomogeneities. To introduce the key ideas let us discuss first the simplest example of one-dimensional sound and begin with the Helmholtz equation for sinusoidal waves,

$$\frac{d^2U}{dx^2} + k^2(1+\epsilon M(x))^2 U = 0\,, \quad -\infty < x < \infty\,, \tag{7.4.1}$$

where $M(x)$ is a random function of x with zero stochastic mean. An incident wave train

$$U_{inc} = A_0 e^{ikx} \tag{7.4.2}$$

arrives from the left-infinity where there is no disorder. What will happen, on the average, to waves after they enter the region of disorder?

Consider an ensemble of random media. For each realization, the index of refraction fluctuates about the mean k by the amount of order $\mathcal{O}(\epsilon)$. Since $\langle M \rangle = 0$, we can expect that the wave phase is affected only by the mean-square of ϵM, which is of the order $\mathcal{O}(\epsilon^2)$. With this estimate, it follows that slow variations will become significant after the long distance of $\mathcal{O}(1/\epsilon^2 k)$. Hence we shall assume that the disorder has fast fluctuations over $x = \mathcal{O}(1/k)$ and is extended over the distance $x = \mathcal{O}(1/\epsilon^2 k)$. Accordingly, it is natural to introduce the slow coordinate $x'' = \epsilon^2 x$ so that

$$M = M(x, x'')\,. \tag{7.4.3}$$

For simplicity we shall further assume that M is stationary with respect to the short scale

$$\langle M(x, x'')M(x_o, x'')\rangle = C_M(|x - x_o|, x'')\,, \tag{7.4.4}$$

where $\langle f \rangle$ denotes the ensemble average of f so that C_M is the covariance of M.

To solve the stochastic differential equation (7.4.1), let us try the following expansion,

$$U = U_0(x, x'') + \epsilon U_1(x, x'') + \epsilon^2 U_2(x, x'') + \cdots\,. \tag{7.4.5}$$

Substituting Eq. (7.4.5) into Eq. (7.4.1), the following perturbation equations are found,

$$\frac{\partial^2 U_0}{\partial x^2} + k^2 U_0 = 0\,, \tag{7.4.6}$$

$$\frac{\partial^2 U_1}{\partial x^2} + k^2 U_1 = -2k^2 M U_0\,, \tag{7.4.7}$$

$$\frac{\partial^2 U_2}{\partial x^2} + k^2 U_2 = -2\frac{\partial U_0}{\partial x \partial x''} - k^2(2MU_1 + M^2 U_0)\,. \tag{7.4.8}$$

The solution at the leading order is

$$U_0 = A(x'')e^{ikx} \quad \text{where } A(0) = A_o\,. \tag{7.4.9}$$

At the next order the inhomogeneous equation is solved by Green's function $G(x, x_o)$ defined by

$$\frac{\partial^2 G}{\partial x^2} + k^2 G = \delta(x - x_o)\,, \tag{7.4.10}$$

where G is outgoing at infinities. The solution is easily found to be

$$G = -\frac{i}{2k}e^{ik|x-x_o|}\,. \tag{7.4.11}$$

The solution for U_1 is

$$\begin{aligned} U_1 &= -\int_{-\infty}^{\infty} G(x, x_o)[2k^2 M(x_o, x'')U_0(x_o, x'')]dx_o \\ &= ikA(x'')\int_{-\infty}^{\infty} M(x_o, x'')e^{ikx_o}e^{ik|x-x_o|}dx_o\,, \end{aligned} \tag{7.4.12}$$

which is random with zero mean.

For the $\mathcal{O}(\epsilon^2)$ problem governed by Eq. (7.4.8), we note that the inhomogeneous terms can be rewritten as

$$2\frac{\partial^2 U_0}{\partial x \partial x''} = e^{ikx}\, 2ik\frac{\partial A}{\partial x''}\,,$$

$$2k^2 M U_1 = e^{ikx}\, 2ikA(x'')\int_{-\infty}^{\infty} M(x, x'')M(x_o, x'')e^{ik|x-x_o|}e^{-ik(x-x_o)}dx_o\,,$$

$$k^2 M^2 U_0 = e^{ikx}\, k^2 M(x, x'')M(x, x'')A(x'')\,.$$

We now take the ensemble average of Eq. (7.4.8), and get

$$\begin{aligned} &\frac{\partial^2 \langle U_2\rangle}{\partial x^2} + k^2\langle U_2\rangle \\ &\quad = -e^{ikx}\left\{2ik\frac{\partial A}{\partial x''} + k^2 A(x'')\langle M^2(x, x'')\rangle\right. \\ &\qquad \left. + 2ik^3 A(x'')\int_{-\infty}^{\infty}\langle M(x, x'')M(x_o, x'')\rangle e^{ik|x-x_o|}e^{-ik(x-x_o)}dx_o\right\}. \end{aligned}$$

In view of Eq. (7.4.4), the integrand above contains the correlation $C_M(|\xi|, x'')$ where $\xi = x - x_o$ while $\langle M^2(x, x'')\rangle = C_M(0, x'')$. Since the right-hand side is proportional to the homogeneous solution e^{ikx}, we must set its coefficient to zero to avoid unbounded resonance, i.e.,

$$\boxed{\frac{\partial A}{\partial x''} + \beta A = 0}\,, \tag{7.4.13}$$

where β is the complex coefficient

$$\beta(x'') = \beta_r + i\beta_i = k^2 \int_{-\infty}^{\infty} C_M(|\xi|, x'') e^{ik|\xi|} e^{-ik\xi} d\xi - \frac{ik}{2} C_M(0, x'')\,. \tag{7.4.14}$$

Clearly the integral above is just a known function of x'' once the correlation function C_M is prescribed. Equation (7.4.13) governs the slow evolution of A in space. For given C_M β can be readily evaluated and the amplitude $A(x'')$ solved.

To see the physical consequences, let the region of disorder be semi-infinite so that $\beta = 0$, $x'' < 0$ and $\beta =$ constant, $x'' > 0$, then the solution is simply

$$A = A(0)e^{-\beta x''} = A(0)e^{-\beta_r x''} e^{-i\beta_i x''}\,. \tag{7.4.15}$$

Thus the wave amplitude decays exponentially over the distance $\mathcal{O}(L_{loc})$ where

$$L_{loc} = \frac{1}{\beta_r \epsilon^2} \tag{7.4.16}$$

characterize the distance beyond which the wave train cannot penetrate. We shall refer to L_{loc} as the localization distance. For simple correlation functions, the integral for β can be explicitly evaluated. For example let

$$C_M(|x - x_o, x''|) = \sigma^2(x'') e^{-\alpha|x - x_o|}\,, \tag{7.4.17}$$

where $\epsilon\sigma$ is the root-mean-square (RMS) height of disorder, and $1/\alpha$ is the correlation length. It is straightforward to show that

$$\int_{-\infty}^{\infty} e^{ik|\xi|} e^{-\alpha|\xi|} e^{-ik\xi} d\xi = \frac{2(\alpha^2 + 2k^2)}{\alpha(\alpha^2 + 4k^2)} + \frac{2ik}{\alpha^2 + 4k^2}\,, \tag{7.4.18}$$

(Soong, 1973) so that

$$\beta = \beta_r + i\beta_i = 2k^2\sigma^2 \frac{\alpha^2 + 2k^2}{\alpha(\alpha^2 + 4k^2)} - \frac{ik\sigma^2}{2} \frac{\alpha^2}{\alpha^2 + 4k^2}\,. \tag{7.4.19}$$

Returning to the natural coordinate the leading-order wave is then

$$U_0 = A_0 \exp\left\{ ik \left[1 + \frac{\epsilon^2\sigma^2}{2} \frac{\alpha^2}{\alpha^2 + 4k^2} \right] x \right\} \times \exp\left\{ -\frac{2\epsilon^2 k^2 \sigma^2}{\alpha} \left(\frac{\alpha^2 + 2k^2}{\alpha^2 + 4k^2} \right) x \right\}, \quad x > 0\,. \tag{7.4.20}$$

Corresponding to the real part β_r, the dimensionless localization distance is

$$kL_{loc} = \frac{1}{2\epsilon^2\sigma^2}\frac{1+4k^2/\alpha^2}{(k/\alpha)(1+2k^2/\alpha^2)}, \tag{7.4.21}$$

which decreases with increasing RMS height σ. If the waves are much longer than the correlation length, $k/\alpha \ll 1$, kL_{loc} increases without bound and localization is weak. If the waves are much shorter than the correlation length $k/\alpha \gg 1$, kL_{loc} decreases, waves cannot penetrate deeply into the disordered region. On the other hand, the imaginary part β_i leads to a change of phase. For a stronger disorder (larger σ), the wavenumber increases, hence the waves become shorter and the phase velocity $(C = \omega/(k+\epsilon^2\beta_i))$ becomes smaller.

Thus for weak disorder the method of homogenization still applies, as long as the cell-average is replaced by the stochastic average. Other aspects of analytical techniques for localization in linearized wave propagation can be found in Sheng (1995), Nachbin (1997), and Nachbin and Papanicolaou (1992). For general theories of linear waves in random media, see Chernov (1960), Frisch (1968), and Ishimaru (1997). For extensions of homogenization theory to random media in diffusion and other problems references are made to Papanicolaou and Varadhan (1995) and Papanicolaou (1995).

We now turn to examples involving weak nonlinearity in addition.

7.5. Weakly Nonlinear Dispersive Waves in a Random Medium

7.5.1. *Envelope Equation*

We consider the lateral displacement of a nonlinear string, which is buried in a weakly random elastic medium,[2]

$$\rho\frac{\partial^2 V}{\partial t^2} - T\frac{\partial^2 V}{\partial x^2} + \mathsf{K}(1+\epsilon M(x))V + \epsilon^2\delta V^3 = 0\,, \tag{7.5.1}$$

V denotes the lateral displacement of the string, ρ the mass per unit length, T tension in the string, K the mean spring constant of the surrounding medium, $\epsilon\mathsf{K}M(x)$ the random fluctuations of the spring force, and $\epsilon^2\delta V^2$ the effect of string nonlinearity. Positive (or negative) δ corresponds to hard (or soft) springs. We assume that M has zero mean and the correlation length scale of $\mathcal{O}(1/k)$.

[2]The material here is extracted from Mei and Pihl (2002).

While the random fluctuation requires the slow coordinate $x'' = \epsilon^2 x$, nonlinearity and dispersion for narrow-banded waves need the intermediate coordinate $x' = \epsilon x$ and times $t' = \epsilon t, t'' = \epsilon^2 t$. We therefore introduce the three-variable expansions,

$$V = V_0 + \epsilon V_1 + \epsilon^2 V_2 + \cdots , \quad \text{with } V_n = V_n(x, x'x'', t, t', t''), \quad n = 0, 1, 2, \ldots . \tag{7.5.2}$$

At orders $\mathcal{O}(\epsilon^0)$, $\mathcal{O}(\epsilon)$, and $\mathcal{O}(\epsilon^2)$, the following perturbation equations are obtained

$$\rho \frac{\partial^2 V_0}{\partial t^2} - T \frac{\partial^2 V_0}{\partial x^2} + \mathsf{K} V_0 = 0 , \tag{7.5.3}$$

$$\rho \frac{\partial^2 V_1}{\partial t^2} - T \frac{\partial^2 V_1}{\partial x^2} + \mathsf{K} V_1 + \mathsf{K} M V_0 + 2\rho \frac{\partial^2 V_0}{\partial t \partial t'} - 2T \frac{\partial V_0}{\partial x \partial x'} = 0 , \tag{7.5.4}$$

$$\rho \frac{\partial V_2}{\partial t^2} - T \frac{\partial^2 V_2}{\partial x^2} + \mathsf{K} V_2 + \rho \left(\frac{\partial^2 V_0}{\partial t'^2} + 2 \frac{\partial^2 V_0}{\partial t \partial t''} + 2 \frac{\partial^2 V_1}{\partial t \partial t'} \right)$$
$$+ \mathsf{K} M V_1 - T \left(\frac{\partial^2 V_0}{\partial x'^2} + 2 \frac{\partial^2 V_0}{\partial x \partial x''} + 2 \frac{\partial^2 V_1}{\partial x \partial x'} \right) + \delta V_0^3 = 0 . \tag{7.5.5}$$

Let us take the leading-order solution to be a rightward sinusoidal wave train

$$V_0 = A\,(x', x''; t', t'')\, e^{i(kx - \omega t)} . \tag{7.5.6}$$

It follows from Eq. (7.5.3) that k and ω are related by

$$\omega = \left(\frac{Tk^2 + \mathsf{K}}{\rho} \right)^{1/2} . \tag{7.5.7}$$

Since the relation is nonlinear, the wave is dispersive. The preceding relation can also be written as

$$k = \left(\frac{\rho\omega^2 - \mathsf{K}}{T} \right)^{1/2} . \tag{7.5.8}$$

Now k is real (propagating wave) only if $\omega > \sqrt{\mathsf{K}/\rho}$, and imaginary (evanescent wave) if $\omega < \sqrt{K/\rho}$. We shall restrict our attention to propagating waves only.

At the order $\mathcal{O}(\epsilon)$ Eq. (7.5.4) can be written as

$$\rho \frac{\partial^2 V_1}{\partial t^2} - T \frac{\partial^2 V_1}{\partial x^2} + \mathsf{K} V_1 + \mathsf{K} M(x, x'') A e^{ikx - i\omega t}$$
$$+ e^{ikx - i\omega t} \left\{ -2i\omega\rho \frac{\partial A}{\partial t'} - 2ikT \frac{\partial A}{\partial x'} \right\} = 0 . \tag{7.5.9}$$

Taking the stochastic average we have

$$\rho\frac{\partial^2\langle V_1\rangle}{\partial t^2} - T\frac{\partial^2\langle V_1\rangle}{\partial x^2} + \mathsf{K}\langle V_1\rangle + e^{ikx-i\omega t}\left\{-2i\omega\rho\frac{\partial A}{\partial t'} - 2ikT\frac{\partial A}{\partial x'}\right\} = 0\,. \tag{7.5.10}$$

To avoid unbounded resonance we must not allow any forcing proportional to the homogeneous solution at all x, hence

$$\frac{\partial A}{\partial t'} + \frac{T\mathsf{K}}{\rho\omega}\frac{\partial A}{\partial x'} = \frac{\partial A}{\partial t'} + C_g\frac{\partial A}{\partial x'} = 0\,, \tag{7.5.11}$$

where

$$C_g = \frac{TK}{\rho\omega} = \frac{\partial\omega}{\partial k} \tag{7.5.12}$$

is the group velocity. The general solution of Eq. (7.5.11) is of the form $A = A(x' - C_g t')$. Hence, in the region of intermediate scale $(x', t') = \mathcal{O}(1)$ the wave envelope propagates at the group velocity without changing form. Now Eq. (7.5.9) reduces to

$$\rho\frac{\partial^2 V_1}{\partial t^2} - T\frac{\partial^2 V_1}{\partial x^2} + \mathsf{K}V_1 = -\mathsf{K}M(x, x'')Ae^{ikx-i\omega t}\,, \tag{7.5.13}$$

where the forcing term on the right-hand side is a random function of x. Let

$$V_1 = \widetilde{V}_1 e^{-i\omega t}\,, \tag{7.5.14}$$

then

$$\frac{\partial^2\widetilde{V}_1}{\partial x^2} + k^2\widetilde{V}_1 = \frac{\mathsf{K}}{T}M(x, x'')A(x'')e^{ikx}\,, \tag{7.5.15}$$

where use is made of Eq. (7.5.8). Equation (7.5.15) can be solved by using the Green function of Eq. (7.4.10), with the result

$$\widetilde{V}_1 = \frac{-i\mathsf{K}A}{2kT}\int_{-\infty}^{\infty} e^{ik|x-x_o|}M(x)e^{ikx_o}dx_o\,, \tag{7.5.16}$$

so that

$$V_1 = \frac{-i\mathsf{K}A}{2kT}e^{ikx-i\omega t}\int_{-\infty}^{\infty} e^{ik|x-x_o|}e^{-ik(x-x_o)}M(x)dx_o\,, \tag{7.5.17}$$

which behaves as outgoing waves as $x - x_o \to \pm\infty$.

Since the ensemble average of $M(x)$ vanishes, i.e., $\langle M(x)\rangle = 0$, we have,

$$\langle V_1\rangle = \langle\widetilde{V}_1\rangle e^{-i\omega t} = 0\,. \tag{7.5.18}$$

The ensemble average of Eq. (7.5.5) becomes

$$\rho \frac{\partial^2 \langle V_2 \rangle}{\partial t^2} + -T \frac{\partial^2 \langle V_2 \rangle}{\partial x^2} + \mathsf{K}\langle V_2 \rangle + \rho \frac{\partial^2 V_0}{\partial t'^2} - T \frac{\partial V_0}{\partial x'^2}$$
$$+ \left[\rho \left(-2i\omega \frac{\partial A}{\partial x''} \right) - T \left(2ik \frac{\partial V_0}{\partial x''} \right) \right] e^{ikx - i\omega t}$$
$$+ \mathsf{K}A \left\{ \frac{-i\mathsf{K}}{2kT} \int e^{ik|x - x_o|} \langle M(x) M(x_o) \rangle e^{-ik(x - x_o)} dx_o \right\} e^{ikx - i\omega t}$$
$$+ 3\delta |A|^2 A e^{ikx - i\omega t} + NRT = 0 \,, \tag{7.5.19}$$

where NRT stands for all nonresonating terms and $\langle M(x)M(x')\rangle$ is the correlation function of the irregularities. Using Eqs. (7.5.6) and (7.5.11), one can show that

$$\rho \frac{\partial^2 V_0}{\partial t_1^2} - T \frac{\partial V_0}{\partial x'^2} = -\rho\omega \frac{d^2\omega}{dk^2} \frac{\partial^2 A}{\partial x'^2} e^{ikx - i\omega t} \,, \tag{7.5.20}$$

where

$$\frac{d^2\omega}{dk^2} = \frac{dC_g}{dk} = \frac{T\mathsf{K}}{\rho^2\omega^3} > 0 \,. \tag{7.5.21}$$

Equating to zero the sum of all resonance-forcing terms which are proportional to $e^{ikx - i\omega t}$, we get the evolution equation for A,

$$-2i\omega\rho \left(\frac{\partial A}{\partial t''} + C_g \frac{\partial A}{\partial x''} \right) - \omega\rho \frac{d^2\omega}{dk^2} \frac{\partial^2 A}{\partial x'^2} + 3\delta |A|^2 A$$
$$- i \frac{\mathsf{K}^2 A}{2kT} \int_{-\infty}^{\infty} \langle M(x) M(x_o) \rangle e^{ik|x - x_o|} e^{-k(x - x_o)} dx_o = 0 \,, \tag{7.5.22}$$

which can be rewritten as

$$2i \left(\frac{\partial A}{\partial t''} + C_g \frac{\partial}{\partial x''} \right) + \frac{d^2\omega}{dk^2} \frac{\partial^2 A}{\partial x'^2} - 3\hat{\delta} |A|^2 A + 2\beta A = 0 \,, \tag{7.5.23}$$

where

$$\hat{\delta} = \frac{\delta}{\rho\omega} \,, \tag{7.5.24}$$

and

$$\beta = \beta_r + i\beta_i = \frac{i\mathsf{K}^2}{4\omega\rho kT} \int_{-\infty}^{\infty} \langle M(x) M(x_o) \rangle e^{ik|x - x_o|} e^{-k(x - x_o)} dx_o \tag{7.5.25}$$

is a complex function of x'' only due to the stationarity assumption. Equation (7.5.23) governs the evolution of A over the much larger domain of $x'', t'' = \mathcal{O}(1)$. Finally Eqs. (7.5.11) and (7.5.22) can be superposed. Removing the artifice of multiple scale coordinates by noting

$$\frac{\partial}{\partial t'} + \epsilon \frac{\partial}{\partial t''} \to \frac{\partial}{\partial t'}, \quad \frac{\partial}{\partial x'} + \epsilon \frac{\partial}{\partial x''} \to \frac{\partial}{\partial x'}, \tag{7.5.26}$$

the final equation

$$\boxed{2i \left(\frac{\partial A}{\partial t'} + C_g \frac{\partial}{\partial x'} \right) + \epsilon \left\{ \frac{d^2\omega}{dk^2} \frac{\partial^2 A}{\partial x'^2} - 3\hat{\delta}|A|^2 A + 2\beta A \right\} = 0} \tag{7.5.27}$$

is uniformly valid in $(kx', \omega t') = \mathcal{O}(1/\epsilon)$, i.e., $(kx, \omega t) = \mathcal{O}(1/\epsilon^2)$.

By changing to a set of moving coordinates

$$\xi = x' - C_g t', \quad \tau = \epsilon t'. \tag{7.5.28}$$

Equation (7.5.27) becomes the nonlinear Schrödinger equation:

$$\boxed{2i \frac{\partial A}{\partial \tau} + \frac{d^2\omega}{dk^2} \frac{\partial^2 A}{\partial \xi^2} - 3\hat{\delta}|A|^2 A + 2i\beta A = 0}, \tag{7.5.29}$$

where the last term represents the effect of disorder. Note from Eq. (7.5.7) that $d^2\omega/dk^2 < 0$.

To see an immediate physical effect of random scattering let us consider a wave packet where the envelope is nonzero only in a finite region. Multiplying Eq. (7.5.29) by A^*, subtracting from the resulting equation its complex conjugate, and integrating the difference from $\xi \sim -\infty$ to $\xi \sim \infty$, we get

$$\frac{d}{d\tau} \int_{-\infty}^{\infty} |A|^2 d\xi + 2 \int_{-\infty}^{\infty} \beta_r |A|^2 d\xi = 0. \tag{7.5.30}$$

Thus, the total energy of the packet must be dissipated after a time of the order of $\mathcal{O}(1/\epsilon^2 \beta_r)$, relative to the wave period, or after propagating over the distance of $\mathcal{O}(C_g/\epsilon^2 \beta_r)$.

With the transformation

$$A = B e^{-i\beta_i \tau}. \tag{7.5.31}$$

Equation (7.5.29) can be further reduced to

$$\boxed{i \frac{\partial B}{\partial \tau} + \frac{1}{2} \frac{d^2\omega}{dk^2} \frac{\partial^2 B}{\partial \xi^2} - \frac{3\hat{\delta}}{2} |B|^2 B + i\beta_r B = 0}. \tag{7.5.32}$$

In the limit of no disorder $\beta = 0$, Eqs. (7.5.29) and (7.5.32) reduce to the classical nonlinear Schrödinger equation which governs the phenomenon of modulational instability, soliton envelopes and nonlinear evolution of many types of dispersive waves in a lossless medium (Ablowitz and Segur, 1981), and has applications in fields as diverse as ocean waves (Mei *et al.*, 2005) and fiber optics (Maloney and Newell, 2004). Mei and Hancock (2003) have shown that Eq. (7.5.29) applies in form to water waves over a randomly rough seabed with a constant mean depth. The amplitude A here corresponds to the amplitude of a slowly varying wave train, while the coefficients are functions of wavelength and sea depth. They examined how disorder affects the sideband instability of a uniform wave train, called Stokes wave, as follows.

7.5.2. Modulational Instability

We shall only consider the case of constant $\beta = \beta_r + i\beta_i$. Let the amplitude of the initially uniform wave train be A_0 and introduce the following normalization:

$$B = A_0 a\,, \quad \tau = \frac{2T}{3|\widehat{\delta}|A_0}\,, \quad \xi = \sqrt{\frac{1}{3|\widehat{\delta}|}\frac{d^2\omega}{dk^2}\frac{X}{A_0}}\,, \quad S = \frac{2\beta_r}{3|\widehat{\delta}|A_0}\,. \tag{7.5.33}$$

Equation (7.5.32) becomes

$$i\frac{\partial a}{\partial T} + \frac{\partial^2 a}{\partial X^2} - \operatorname{sgn}\widehat{\delta}|a|^2 a + iSa = 0\,, \tag{7.5.34}$$

where S represents the effects of disorder.

First let us find the spatially uniform wave train (Stokes wave) with $a = \bar{a}(T)$. Let

$$\bar{a} = |\bar{a}(T)|e^{i\Theta(T)}\,, \tag{7.5.35}$$

where Θ is the phase of $\bar{a}$ and is the phase increment of V_0. Substituting into Eq. (7.5.32) and separating the real and imaginary parts, we get

$$\frac{d|\bar{a}|}{dT} + S|\bar{a}| = 0\,, \qquad \frac{d\Theta}{dT} + \frac{3\hat{\delta}}{2}|\bar{a}|^2 = 0\,.$$

The uniform wave train is

$$\bar{a}(T) = e^{-ST}\exp\left[i\frac{\operatorname{sgn}\widehat{\delta}}{2S}(1 - e^{-2ST})\right]\,, \tag{7.5.36}$$

or

$$\bar{B}(\tau) = A_0 \exp\left[-\beta_r\tau + i\frac{3\widehat{\delta}A_0^2}{2\beta_r}(1 - e^{-2\beta_r\tau})\right] \tag{7.5.37}$$

in physical variables. In the moving coordinates, the amplitude dies out after the time $\mathcal{O}(1/\epsilon^2\beta_r)$. The wave phase is proportional to the square of the initial amplitude and also decays in time.

Let an infinitesimal disturbance be added, i.e., $a(X,T) = \bar{a}(T)(1 + a'(X,T))$ where $a' \ll 1$. Equation (7.5.32) gives after linearization,

$$i\frac{\partial a'}{\partial T} + \frac{\partial^2 a'}{\partial X^2} - \operatorname{sgn}\,\widehat{\delta}|\bar{a}|^2(a' + a'^*) = 0\,. \tag{7.5.38}$$

The real and imaginary parts are

$$-\frac{\partial I}{\partial T} + \frac{\partial^2 R}{\partial X^2} - 2\operatorname{sgn}\widehat{\delta}\,e^{-2ST}R = 0\,, \quad \frac{\partial R}{\partial T} + \frac{\partial^2 I}{\partial X^2} = 0\,, \tag{7.5.39}$$

where $R = \mathcal{R}e\,(a')$, $I = \mathcal{I}m\,(a')$. Assuming the disturbance to be two sidebands with slight shifts of wavenumbers,

$$\begin{Bmatrix} R \\ I \end{Bmatrix} = \frac{1}{2}\begin{Bmatrix} \widetilde{R}(T) \\ \widetilde{I}(T) \end{Bmatrix}(e^{iKX} + e^{-iKX})\,, \tag{7.5.40}$$

we find from Eq. (7.5.39),

$$\frac{dI}{dT} + K^2R + 2\operatorname{sgn}\widehat{\delta}e^{-ST}R = 0\,, \quad \frac{dR}{dT} - K^2I = 0\,, \tag{7.5.41}$$

which can be combined to give

$$\frac{d^2}{dT^2}\begin{Bmatrix} R \\ I \end{Bmatrix} + K^2(2\operatorname{sgn}\widehat{\delta}e^{-2ST} + K^2)\begin{Bmatrix} R \\ I \end{Bmatrix} = 0\,. \tag{7.5.42}$$

Thus, only if $\widehat{\delta} < 0$, and if the wavenumber lies within the range $0 < K < \sqrt{2}e^{-2ST}$, the sidebands are unstable. This range shrinks with time.

Mei and Hancock (2003) studied by numerical solution of Eq. (7.5.32) the nonlinear evolution of unstable Stokes waves. A sample plot is shown in Fig. 7.5. For small T, the unstable sidebands and their harmonics grow at the expense of the carrier wave. At large T the carrier wave and the sidebands attenuate to zero. Related works can be found in Mei and Pihl (2002) and Pihl *et al.* (2002).

7.6. Harmonic Generation in Random Media

The propagation of high-intensity light through matter can be strongly affected by the nonlinear properties of the medium. A striking example is the change of color when light of high intensity shines through a quartz crystal. Franken *et al.* (1961) found that when a ruby laser beam with wavelength of 6940 Å passes through a quartz crystal, ultraviolet light with

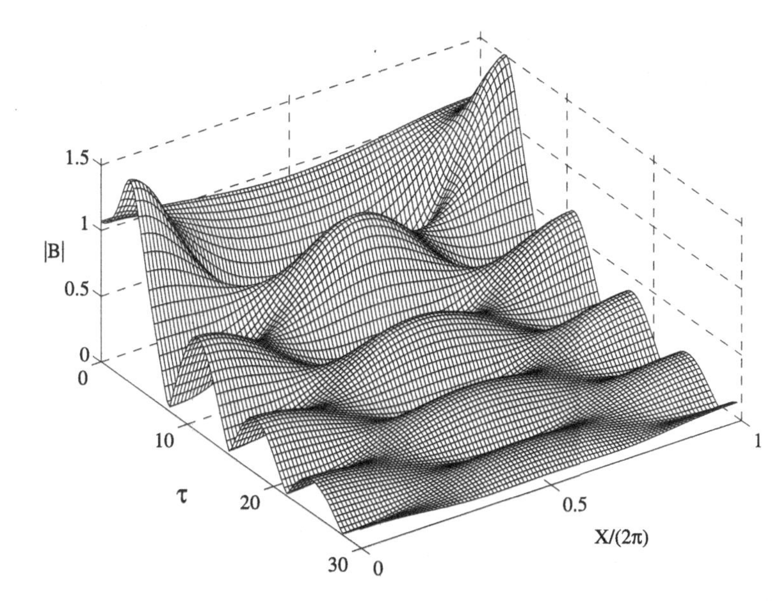

Fig. 7.5. Nonlinear evolution of an initially uniform wave train perturbed by $a' = 0.1e^{i\pi/4}\cos X$. The dimensionless disorder is $S = 0.075$. From Mei and Hancock (2003), *J. Fluid Mech.*

half the wavelength (3470 Å) was detected on the transmission side. This phenomenon is called *second-harmonic generation.* Nonlinear effects can be even more stunning when ocean waves approach the shore. Over a mildly sloping seabed, a narrow-banded swell from deep sea first experiences an amplification of height and decrease in wavelength due to refraction. As waves come closer to the shore, nonlinearity becomes more important. The free surface tends to steepen and wave energy is spread from lower to ever higher harmonics. Over a smooth beach the joint action of nonlinearity and frequency dispersion broadens the sea spectrum and may lead to wave breaking. Over a rough bottom, however, this tendency can be counteracted by multiple scattering, or bottom friction, or both.

In this section we shall only focus on the effects of random depth fluctuations on harmonic generation (Grataloup and Mei, 2003), and assume for simplicity that the mean slope of the seabed is zero.

7.6.1. *Long Waves in Shallow Water*

Let us restrict to one-dimensional propagation. It is well known for long waves in shallow water, the pressure is nearly hydrostatic, the velocity is mainly horizontal, and dispersion is weak. Let $h(x)$ denote the local depth beneath the still water level, $\eta(x,t)$ the free surface displacement above, and $u(x,t)$ the depth-averaged horizontal velocity of water. If both nonlinearity and dispersion are kept to the leading order, the laws of mass and momentum conservation are approximated by the following equations due to Boussinesq (see, e.g., Mei *et al.*, 2005),

$$\frac{\partial \eta}{\partial t} + \frac{\partial}{\partial x}[(h+\eta)u] = 0\,, \tag{7.6.1}$$

$$\frac{\partial u}{\partial t} + u\frac{\partial u}{\partial x} + g\frac{\partial \eta}{\partial x} = \frac{h}{2}\frac{\partial^2}{\partial x^2}\left(h\frac{\partial u}{\partial t}\right) - \frac{h^2}{6}\frac{\partial^3 u}{\partial x^2 \partial t}\,. \tag{7.6.2}$$

To display the relative importance of each term, let us employ the following dimensionless variables:

$$x^\dagger = k_o x\,, \quad t^\dagger = tk_o\sqrt{gH}\,, \quad \eta^\dagger = \frac{\eta}{a}\,, \quad h^\dagger = \frac{h}{H}\,, \quad u^\dagger = \frac{u}{a\sqrt{g/H}}\,, \tag{7.6.3}$$

where k_o, a, and H are, respectively, the typical wavenumber, wave amplitude, and mean depth. In normalized form Eqs. (7.6.1) and (7.6.2) are:

$$\frac{\partial \eta^\dagger}{\partial t^\dagger} + \frac{\partial}{\partial x^\dagger}[(h^\dagger + \epsilon\eta^\dagger)u^\dagger] = 0\,, \tag{7.6.4}$$

$$\frac{\partial u^\dagger}{\partial t^\dagger} + \epsilon u^\dagger \frac{\partial u^\dagger}{\partial x^\dagger} + \frac{\partial \eta^\dagger}{\partial x^\dagger} = \frac{\mu^2 h^\dagger}{2}\frac{\partial^2}{\partial {x^\dagger}^2}\left(h^\dagger \frac{\partial u^\dagger}{\partial t^\dagger}\right) - \frac{\mu^2 {h^\dagger}^2}{6}\frac{\partial^3 u^\dagger}{\partial {x^\dagger}^2 \partial t^\dagger}\,, \tag{7.6.5}$$

where the two parameters

$$\epsilon = \frac{a}{H} \ll 1\,, \quad \mu = k_o H \ll 1 \tag{7.6.6}$$

characterize respectively nonlinearity and dispersion. The ratio

$$\nu = \frac{\mu^2}{\epsilon} \tag{7.6.7}$$

measures the relative importance of dispersion versus nonlinearity. In order to avoid the need for higher-order terms, we assume that $h^\dagger$ fluctuates from the constant 1 by $\sqrt{\epsilon}\, b^\dagger(x^\dagger)$, i.e.,

$$h^\dagger(x^\dagger) = 1 - \sqrt{\epsilon}\, b^\dagger(x^\dagger)\,, \tag{7.6.8}$$

where $b^\dagger$ is a stationary random function of $x^\dagger$ with zero mean, $\langle b^\dagger(x^\dagger)\rangle = 0$.

For the sake of brevity we shall omit daggers for all dimensionless quantities from here on. Equation (7.6.4) becomes

$$\frac{\partial \eta}{\partial t} + \frac{\partial}{\partial x}[(1 - \sqrt{\epsilon} b + \epsilon\eta)u] = 0\,. \tag{7.6.9}$$

By retaining terms only up to $\mathcal{O}(\epsilon)$ and $\mathcal{O}(\mu^2)$, Eq. (7.6.5) reduces to

$$\frac{\partial u}{\partial t} + \epsilon u \frac{\partial u}{\partial x} + \frac{\partial \eta}{\partial x} = \frac{\mu^2}{3} \frac{\partial^3 u}{\partial x^2 \partial t}, \tag{7.6.10}$$

which can be used to combine with Eq. (7.6.9) to yield the stochastic differential equation,

$$\boxed{\frac{\partial^2 \eta}{\partial t^2} - \frac{\partial^2 \eta}{\partial x^2} = -\sqrt{\epsilon}\, b \frac{\partial \eta}{\partial x} + \frac{\epsilon}{2}\left(\frac{\partial^2 u^2}{\partial x^2} + \frac{\partial^2 u^2}{\partial t^2} + \frac{\partial^2 \eta^2}{\partial t^2}\right) + \frac{\mu^2}{3}\frac{\partial^4 \eta}{\partial x^4}}\,. \tag{7.6.11}$$

Our objective is to seek approximations for the propagation of an initially simple harmonic wave train entering a large region of disorder from $x \sim -\infty$.

7.6.2. *Harmonic Amplitudes*

As in previous examples the localization distance should be inversely proportional to the mean-square height of the random perturbations, i.e., $\mathcal{O}(1/(\sqrt{\epsilon})^2 k_o)$ under the assumption of Eq. (7.6.8). We therefore introduce the multiple scale coordinates x and $x' = \epsilon x$, assume $b = b(x, x')$, and expand η and u in ascending powers of $\sqrt{\epsilon}$:

$$\eta = \eta_0 + \epsilon^{1/2}\eta_1 + \epsilon\eta_2 + \cdots, \quad u = u_0 + \epsilon^{1/2}u_1 + \epsilon u_2 + \cdots, \tag{7.6.12}$$

where $\eta_j = \eta_j(x, x', t)$ and $u_j = u_j(x, x', t)$. From Eq. (7.6.11) we find the perturbation equations for the first three orders,

$$\frac{\partial^2 \eta_0}{\partial t^2} - \frac{\partial^2 \eta_0}{\partial x^2} = 0, \tag{7.6.13}$$

$$\frac{\partial^2 \eta_1}{\partial t^2} - \frac{\partial^2 \eta_1}{\partial x^2} = -\frac{\partial}{\partial x}\left(b \frac{\partial \eta_0}{\partial x}\right), \tag{7.6.14}$$

$$\frac{\partial^2 \eta_2}{\partial t^2} - \frac{\partial^2 \eta_2}{\partial x^2} = -\frac{\partial}{\partial x}\left(b \frac{\partial \eta_1}{\partial x}\right) + 2\frac{\partial^2 \eta_0}{\partial x \partial x'} + \frac{3}{2}\frac{\partial^2 \eta_0^2}{\partial t^2} + \frac{\nu}{3}\frac{\partial^4 \eta_0}{\partial x^4}. \tag{7.6.15}$$

The right-hand side of Eq. (7.6.15) has been simplified by using the leading-order approximation.

As is readily seen from Eq. (7.6.11), if one neglects nonlinearity, dispersion, and bed roughness totally, a progressive wave $(\eta, u) \propto e^{\pm i(kx - \omega t)}$ is nondispersive since $\omega = k$. With nonlinearity higher harmonics $e^{\pm im(kx-\omega t)}$, with $m = 1, 2, 3\ldots$ will be resonated. To anticipate the growth of higher harmonics we assume

$$\eta_0 = \frac{1}{2}\sum_{m=-\infty}^{\infty} A_m(x')e^{i\theta_m}, \quad u_0 = \frac{1}{2}\sum_{m=-\infty}^{\infty} B_m(x')e^{i\theta_m}. \tag{7.6.16}$$

In order to satisfy Eq. (7.6.13) we require

$$\theta_m = m(kx - \omega t), \quad \omega = k. \tag{7.6.17}$$

To ensure that η_0 is real we also require

$$A_{-m} = A_m^* \quad \text{and} \quad B_{-m} = B_m^*, \tag{7.6.18}$$

where A^* denotes the complex conjugate of A. In order that the normalized mean depth is unity, we set $A_0 = 0$. For later convenience we denote

$$k_m \equiv mk, \quad \omega_m \equiv m\omega. \tag{7.6.19}$$

From Eq. (7.6.3) the implied dimensionless wavenumbers and frequencies are defined by $k_m = k_m^*/k_o$ and $\omega_m = \omega_m^*/k_o\sqrt{gH}$.

At $\mathcal{O}(\sqrt{\epsilon})$, the forcing terms in Eq. (7.6.14) can be expanded and separated into time harmonics.

$$-\frac{\partial}{\partial x}\left(b\frac{\partial \eta_0}{\partial x}\right) = -\sum_{m=-\infty}^{\infty} F_m e^{-i\omega_m t}, \tag{7.6.20}$$

where the coefficients F_m are random functions of x,

$$F_m = \frac{1}{2} i k_m A_m(x') \frac{d}{dx}[b(x)e^{ik_m x}], \tag{7.6.21}$$

and $F_0 = 0$. The dependence of b on x' is suppressed for the time being. Let the solution of Eq. (7.6.14) be written in the form:

$$\eta_1 = \sum_{m=-\infty}^{\infty} \eta_1^{(m)} e^{-i\omega_m t}, \quad \eta_1^{(0)} = 0, \tag{7.6.22}$$

then $\eta_1^{(m)}$ where $m \neq 0$, is governed by

$$\frac{d^2\eta_1^{(m)}}{dx^2} + k_m^2 \eta_1^{(m)} = F_m(x). \tag{7.6.23}$$

For $m = 0$, $\eta_1^{(0)} = 0$. With the help of the Green function

$$G_m(|x - x_o|) = \frac{e^{ik_m|x-x_o|}}{2ik_m}, \tag{7.6.24}$$

the solution for η_1 is easily found

$$\eta_1 = \sum_{m=-\infty}^{\infty} \frac{A_m(x')}{2} e^{-i\omega_m t} \int_{-\infty}^{\infty} ik_m G_m(|x - x_o|) \frac{d}{dx_o} [b(x_o)e^{ik_m x_o}]\, dx_o\,, \tag{7.6.25}$$

which behaves as outgoing waves at infinities and has zero mean.

Finally at $\mathcal{O}(\epsilon)$, let us take the ensemble average of Eq. (7.6.15),

$$\frac{\partial^2 \langle \eta_2 \rangle}{\partial t^2} - \frac{\partial^2 \langle \eta_2 \rangle}{\partial x^2} = -\frac{\partial}{\partial x} \left\langle b \frac{\partial \eta_1}{\partial x} \right\rangle + 2 \frac{\partial^2 \eta_0}{\partial x' \partial x} + \frac{3}{2} \frac{\partial^2 \eta_0^2}{\partial t^2} + \frac{\nu}{3} \frac{\partial^4 \eta_0}{\partial x^4}\,. \tag{7.6.26}$$

Using the known solution for $\eta_1^{(m)}$, the first forcing term on the right-hand side can be written as:

$$\left\langle b \frac{\partial \eta_1^{(m)}}{\partial x} \right\rangle = -\frac{A_m}{2} \int_{-\infty}^{\infty} k_m^2 \,\mathrm{sgn}(x - x_o) G_m(|x - x_o|) \frac{d}{dx_o} \times [\langle b(x) b(x_o) \rangle e^{ik_m x_o}\, dx_o]\,.$$

Using the assumption that b is a stationary random function of x on the fast scale

$$\langle b(x, x') b(x_o, x') \rangle = \sigma^2(x') \Gamma(\xi)\,, \quad \xi = x - x_o\,, \tag{7.6.27}$$

where $\sqrt{\epsilon}\sigma(x')$ is the root-mean-square height of the roughness, and $\Gamma(\xi)$ is the autocorrelation coefficient of the bed roughness. It then follows that

$$\begin{aligned} \left\langle b \frac{\partial \eta_1^{(m)}}{\partial x} \right\rangle &= ik_m \frac{A_m}{4} e^{ik_m x} \int_{-\infty}^{\infty} dx_o \,\mathrm{sgn}(x - x_o) e^{ik_m(|x - x_o|)} \frac{d}{dx_o} \\ &\quad \times [\Gamma(x - x_o) e^{ik_m(x_o - x)}] \\ &= -ik_m A_m \frac{\sigma^2}{4} e^{ik_m x} \int_{-\infty}^{\infty} d\xi \,\mathrm{sgn}(\xi) e^{ik_m|\xi|} \frac{d}{d\xi} [\Gamma(\xi) e^{-ik_m \xi}]\,. \end{aligned} \tag{7.6.28}$$

Note that Eq. (7.6.28) is zero for $m = 0$, and terms associated with $m < 0$ is the complex conjugate of the terms associated with $m > 0$, hence

$$-\frac{\partial}{\partial x} \left[\sum_{m=-\infty}^{\infty} e^{-i\omega_m t} \left\langle b \frac{\partial \eta_1^{(m)}}{\partial x} \right\rangle \right] = \sum_{m=1}^{\infty} ik_m A_m(x_o') \beta_m e^{i\theta_m} + \text{c.c.}\,, \tag{7.6.29}$$

where the complex coefficient $\beta_m = \mathcal{R}e\beta_m + i\mathcal{I}m\beta_m$ is defined by,

$$\beta_m = \frac{\sigma^2}{4} ik_m \int_{-\infty}^{\infty} \text{sgn}(\xi) \left(\frac{d\Gamma}{d\xi} - ik_m \Gamma \right) e^{ik_m(|\xi|-\xi)} d\xi \,. \tag{7.6.30}$$

Finally the third forcing term in Eq. (7.6.26) can be shown to be

$$\frac{3}{2}\frac{\partial^2 \eta_0^2}{\partial t^2} = \sum_{m=1}^{\infty} -\frac{3}{8}\omega_m^2 e^{i\theta_m} \left[\sum_{l=1}^{\infty} 2A_l^* A_{m+l} + \sum_{l=1}^{[m/2]} \alpha_l A_l A_{m-l} \right] + \text{c.c.}\,, \tag{7.6.31}$$

where $[m/2]$ is the integer part of $m/2$ and α_l is a coefficient equal to 1 for $l = [m/2]$, and equal to 2 otherwise. In summary Eq. (7.6.26) can be rewritten as

$$\begin{aligned} \left(\frac{\partial^2}{\partial t^2} - \frac{\partial^2}{\partial x^2} \right) \langle \eta_2 \rangle &= \sum_{m=1}^{\infty} ik_m A_m \beta_m e^{i\theta_m} + \sum_{m=1}^{\infty} ik_m \frac{dA_m}{dx''} e^{i\theta_m} \\ &\quad - \sum_{m=1}^{\infty} \frac{3}{8}\omega_m^2 e^{i\theta_m} \left[\sum_{l=1}^{\infty} 2A_l^* A_{m+l} + \sum_{l=1}^{[m/2]} \alpha_l A_l A_{m-l} \right] \\ &\quad + \sum_{m=1}^{\infty} \frac{\nu}{6} k_m^4 A_m e^{i\theta_m} + \text{c.c.}\,. \end{aligned} \tag{7.6.32}$$

To ensure solvability of the preceding equation, secular terms proportional to $\exp(i\theta_m)$ must be set to zero. With the help of the dispersion relation $\omega_m = k_m$ we get

$$\frac{dA_m}{dx'} + \beta_m A_m - i\frac{\nu}{6} k_m^3 A_m + \frac{3}{8} i\omega_m \left[\sum_{l=1}^{\infty} 2A_l^* A_{m+l} + \sum_{l=1}^{[m/2]} \alpha_l A_l A_{m-l} \right] = 0\,, \tag{7.6.33}$$

for $m = 1, 2, \ldots \infty$. This result constitutes an infinite number of nonlinearly coupled equations governing the slow spatial evolution of harmonic amplitudes and extends the theory of Bryant (1973) for harmonic generation over a smooth seabed (see also Mei, 1989). The linear terms with complex coefficients β_m represent the effects of multiple scattering by disorder.

With a prescribed $\Gamma(\xi)$, the series above can be truncated at some finite but large n for numerical computations, yielding

$$\boxed{\frac{dA_m}{dx'} + \beta_m A_m - i\frac{\nu}{6} k_m^3 A_m + \frac{3}{8} i\omega_m \left[\sum_{l=1}^{n-m} 2A_l^* A_{m+l} + \sum_{l=1}^{[m/2]} \alpha_l A_l A_{m-l} \right] = 0}\,,$$

$$m = 1, 2, \ldots, n\,. \tag{7.6.34}$$

7.6.3. *Gaussian Disorder*

For illustration we assume a Gaussian correlation function:

$$\Gamma(\xi) = \exp\left(-\frac{\xi^2}{2l^2}\right), \tag{7.6.35}$$

where l is the dimensionless correlation distance normalized by $1/K$. After some algebra, Grataloup and Mei (2003) have shown that

$$\frac{\beta_m l}{\sigma^2} = k_m^2 l^2 \frac{\sqrt{2\pi}}{8}(1 + e^{-2k_m^2 l^2}) \\ - \frac{i}{2} k_m l \left(1 - \frac{k_m l}{\sqrt{2}} e^{-2k_m^2 l^2} \int_0^{\sqrt{2}k_m l} e^{u^2}\, du\right). \tag{7.6.36}$$

Figure 7.6 shows the dependence of the real and imaginary parts of $\beta_m l/\sigma^2$ on $k_m l$. They also proved the following general relation on the first-order wave energy,

$$\frac{d}{dx'} \sum_{m=1}^{n} |A_m|^2 = -2 \sum_{m=1}^{n} \mathcal{R}e\,(\beta_m)|A_m|^2 , \tag{7.6.37}$$

where n is any integer representing the highest harmonic in the truncated differential system. Physically, due to multiple scattering by disorder, the

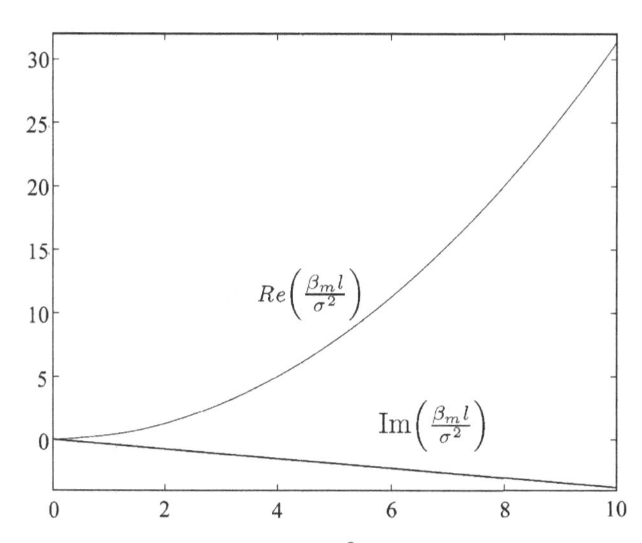

Fig. 7.6. Real and imaginary parts of $\beta_m l/\sigma^2$. From Grataloup and Mei (2003), *Phys. Rev. E.*

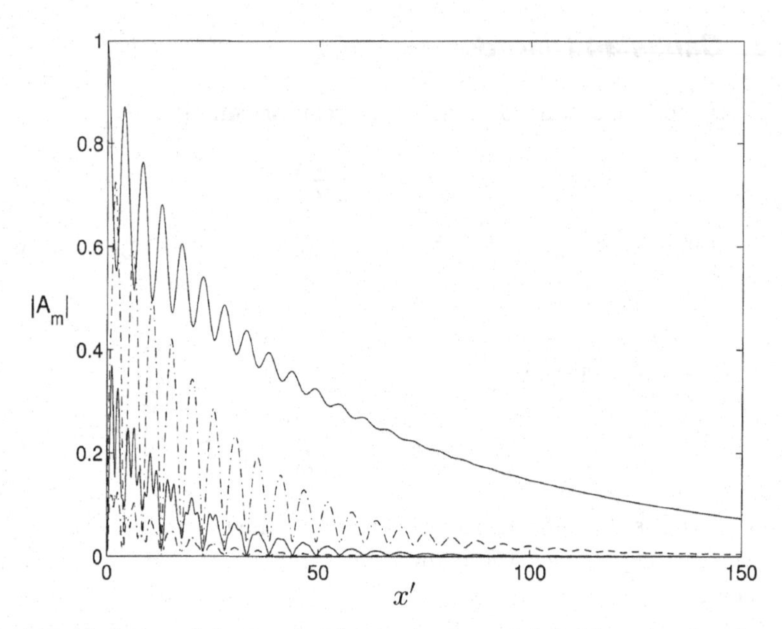

Fig. 7.7. Evolution of the first four harmonics ($m = 1, 2, 3, 4$ from top-down) over a semi-infinite region of disorder. $\sigma = 0.2$, $\nu = 1$, and $l = k_1 = 1$. From Grataloup and Mei (2003), *Phys. Rev. E.*

total wave energy of all leading-order harmonics decreases with propagation distance.

Based on finite-difference solution of Eq. (7.6.34), numerical results of the first four harmonics are displayed in Fig. 7.7 for a semi-infinite region of disorder with $\sigma = 0$, $x' < 0$; $\sigma = 0.2$, $x' > 0$. Incident waves with only the first harmonic arrive from $x' < 0$. The input parameters are: $k_1 = 1$, $l = 1$, and $\nu = \mu^2/\epsilon = 1$. At the start the second and higher harmonics are grown at the expense of the first. But they all attenuate with distance, with the higher harmonics dying out sooner. Once the second and higher harmonics are sufficiently diminished, energy exchange with the first harmonic becomes insignificant; the latter then attenuates monotonically with distance.

An earlier study on nonlinear long waves i over a randomly rough seabed was made by Rosales and Papanicolaou (1983). Homogenization technique has been applied to other aspects of shallow water waves over a randomly rough seabed by Mei and Li (2004), Alam and Mei (2007, 2008).

References

Ablowitz, M. J. and H. Segur (1981). *Solitons and the Inverse Scattering Transform*, SIAM Phila.

Alam, M.-R. and C. C. Mei (2007). Localization of long interfacial waves over a randomly rough seabed. *J. Fluid Mech.* **587**: 73–96.

Alam, M.-R. and C. C. Mei (2008). Ships advancing near the critical speed in a shallow channel with a randomly uneven bed. *J. Fluid Mech.* **616**: 397–417.

Anderson, P. W. (1958). Absence of diffusion in certain random lattices. *Phys. Rev.* **109**: 1492–1505.

Ashcroft, M. W. and N. D. Mermin, (1976). *Solid State Physics*, Brooks/Cole, Thomas Learning, New York.

Bryant, P. J. (1973). Periodic waves in shallow water. *J. Fluid Mech.* **59**: 625–644.

Calflisch, R. E., M. J. Miksis, G. C. Papanicolaou and L. Ting (1985a). Effective equations for wave propagation in bubbly liquids. *J. Fluid Mech.* **153**: 259–273.

Calflisch, R. E., M. J. Miksis, G. C. Papanicolaou and L. Ting (1985b). Wave propagation in bubbly liquids at finite volume fraction. *J. Fluid Mech.* **160**: 1–14.

Chernov, L. A. (1960). *Wave Propagation in a Random Medium*, Dover, 168 pp.

Franken, P. A., A. E. Hill, C. W. Peters and G. Weinreich (1961). Generation of optical harmonics. *Phys. Rev. Lett.* **7**: 118–119.

Frisch, U. (1968). Wave propagation in random media, *Probabilistic Methods in Applied Mathematics*, Vol. 1, Academic.

Grataloup, G. and C. C. Mei (2003). Localization of harmonics generated in nonlinear shallow water waves. *Phys. Rev. E* **68**: 028314, 1–9.

Hu, X. and C. T. Chan (2005). Refraction of water waves by periodic cylinder arrays. *Phys. Rev. Lett.* **95**: 154501, 1–4.

Ishimaru, A. (1997). *Wave Propagation and Scattering in Random Media*, Oxford and IEEE Press, 574 pp.

Karal, F. C. and J. B. Keller (1964). Elastic, electromagnetic and other waves in a random medium. *J. Math. Phys.* **5**(4): 537–547.

Keller, J. B. (1964). Stochastic equation and wave propagation in random media. *Proc. 16th Symp. Appl. Math.*, 145–170. *Amer. Math. Soc.*, Rhode Island.

Levy, T. and E. Sanchez-Palencia (1983). Suspension of solid particles in a Newtonian fluid. *J. Non-Newtonian Fluid Mech.* **13**: 63–78.

Li, Y. and C. C. Mei (2004). Bragg scattering by a line array of small cylinders in a wave guide. Part I. Linear aspects, *J. Fluid Mech.* **583**: 161, 187.

Li, Y. and C. C. Mei (2007). Multiple resonant scattering of water waves by two-dimensional array of vertical cylinders. *Phys. Rev. E* **76**: 016302, 1–23.

Maloney, J. V. and A. C. Newell (2004). *Nonlinear Optics*, Westview Press, 440 pp.

Mei, C. C. (1989). *Applied Dynamics of Ocean Surface Waves*, World Scientific, Singapore, 700 pp.

Mei, C. C. and M. J. Hancock (2003). One-dimensional localization of surface water waves over a random seabed. *J. Fluid Mech.* **475**: 247–268.

Mei, C. C. and Y. Li (2004). Localization of solitons over a randomly rough seabed. *Phys. Rev. E* **70**: 016302, 1–11.

Mei, C. C. and J. H. Pihl (2002). Localization of nonlinear dispersive waves in weakly random media. *Proc. R. Soc. Lond. A* **458**: 119–134.

Mei, C. C., M. Stiassnie and D. K.-P. Yue (2005). *Theory and Applications of Ocean Surface Waves*, Vol. I, *Linear Aspects*, and II, *Nonlinear Aspects*, World Scientific, Singapore.

Nachbin, A. (1997). The localization length of randomly scattered water waves. *J. Fluid Mech.* **296**: 353–372.

Nachbin, A. and G. C. Papanicolaou (1992). Water waves in shallow channels of rapidly varying depth. *J. Fluid Mech.* **241**: 311–332.

Papanicolaou, G. C. (1995). Diffusion in random media, *Surveys in Applied Mathematics*, eds. J. B. Keller, D. W. McLaughlin and G. C. Papanicolaou, Plenum Press, N.Y.

Papanicolaou, G. C. and S. R. S. Varadhan (1995). Boundary-value problems with rapidly oscillating random coefficients, *Random Fields*, eds. J. Fritz, J. Lebowitz and D. Szaxa, Janos Bolyai Series, North-Holland, pp. 835–873.

Pihl, J. H., C. C. Mei and M. J. Hancock (2002). Surface gravity waves over a two-dimensional random seabed. *Phys. Rev. E* **66**: 016611-1–016611-11.

Rosales, R. R. and G. C. Papanicolaou (1983). Gravity waves in a channel with a rough bottom. *Stud. Appl. Math.* **68**: 89–102.

Sheng, P. (1995). *Introduction of Wave Scattering, Localization, and Mesoscopic Phenomena*, Academic, 339 pp.

Soong, T. T. (1973). *Random Differential Equations in Science and Engineering*, Academic, 327 pp.

van Wijngaarden, L. (1968). On equations of motions for mitures of liquid and gas bubbles. *J. Fluid Mech.* **13**: 465–476.

van Wijngaarden, L. (1972). One-dimensional flow of liquids containing small gas bubbles. *Ann. Rev. Fluid Mech.* **4**: 369–394.

Whitham, G. B. (1987). *Linear and Nonlinear Waves*, Wiley-Interscience.

Additional References on Homogenization Theory

For readers interested in more mathematical aspects of homgenization theory and multiscale mechanics, the following books may be consulted.

Allaire, G. (2001). *Shape Optimization by the Homogenization Method*, Springer.

Attouch, H. (1984). *Variational Convergence for Functions and Operators*, Pitman.

Balkhalov, N. and G. Panasenko (1989). *Homogenization: Averaging Processes in Periodic Media*, Kluver, Dordrecht.

Bensoussan, A., J. L. Lions and G. Papanicolaou (1978). *Asymptotic Analysis for Periodic Structures*, Studies in mathematics and its applications, North-Holland, Amsterdam.

Braides A. and A. Defranceschi (1999). *Homogenization of Multiple Integrals*, Oxford Lecture Series in Mathematics and Its Applications.

Cherkaev, A. (2000). *Variational Methods for Structural Optimization*, Springer Verlag, New York.

Cherkaev, A. and R. Kohn (eds.) (1997). *Topics in the Mathematical Modeling of Composite Materials*, Birkhäuser Verlag.

Cioranescu, D. and P. Donato (2000). *An Introduction to Homogenization*, Oxford Lecture Series in Mathematics and Its Applications, Vol. 17.

Cioranescu, D. and J. Saint Jean Paulin (1999). *Homogenization of Reticulated Structures*, Springer-Verlag, New York.

Dal Maso, G. (1993). *Introduction to Γ-Convergence*, Birkhäuser.

Dormieux, L., D. Kondo and F. J. Ulm (2006). *Microporomechanics*, Wiley.

Ene, H. and G. Pasa (1986). *Homogenization Method. Applications to the Composite Materials Theory*, Academy Publishing House, Bucharest.

Ene, H. and D. Polisevski (1987). *Thermal Flow in Porous Media*, Kluwer.

Gilbert, R. P. and K. Hackl (eds.) (1995). *Asymptotic Theories for Plates and Shells*, CRC Press.

Hornung, U. (ed.) (1996). *Homogenization and Porous Media*, Springer.

Lewinsky, T. and J. J. Telega (2000). *Plates, Laminates and Shells. Asymptotic Analysis and Homogenization*, World Scientific, River Edge, New York.

Marchenko, V. A. and Y. A. Khruslov (2005). *Homogenization of Partial Differential Equations*, Birkhäuser, Boston.

Milton, G. W. (2002). *The Theory of Composites*, Cambridge University Press, Cambridge.

Nemat-Nasser, S. and M. Hori (1999). *Micromechanics: Overall Properties of Heterogeneous Materials*, Elsevier, Amsterdam.

Oleinik, O. A., A. S. Shamaev and G. A. Yosifian (1992). *Mathematical Problems in Elasticity and Homogenization*, North Holland, Amsterdam.

Pankov, A. A. (1997). *G-Convergence and Homogenization of Nonlinear Partial Differential Operators*, Springer.

Sanchez-Palencia, E. (1980). *Non-Homogeneous Media and Vibration Theory*, Lecture Notes in Physics, Springer, Berlin.

Torquato, S. (2002). *Random Heterogeneous Materials*, Springer-Verlag, New York.

Subject Index